VARIORUM COLLECTED STUDIES SERIES

Science in the Medieval Hebrew and Arabic Traditions

Gad Freudenthal

Science in the Medieval Hebrew and Arabic Traditions

Published in the Variorum Collected Studies Series by

Ashgate Publishing Limited
Gower House, Croft Road,
Aldershot, Hampshire
GU11 3HR
Great Britain

Ashgate Publishing Company
Suite 420
101 Cherry Street
Burlington, VT 05401–4405
USA

Ashgate website: http://www.ashgate.com

ISBN 0–86078–952–7

British Library Cataloguing in Publication Data
Freudenthal, Gad
  Science in the medieval Hebrew and Arabic traditions. –
  (Variorum collected studies series)
  1. Jewish scientists – History 2. Muslim scientists – History
  3. Science, Medieval 4. Civilization – Greek influences
  I. Title
  296.3'75

Library of Congress Cataloging-in-Publication Data
Freudenthal, Gad.
  Science in the medieval Hebrew and Arabic traditions / Gad Freudenthal.
  p. cm. – (Variorum collected studies series ; 803)
  Text in English and French.
  Includes bibliographical references.
  ISBN 0–86078–952–7 (alk. paper)
  1. Science, Medieval – Philosophy. 2. Judaism and science. 3. Jewish Science.
  4. Judaism – History – Medieval and early modern period, 425-1789. 5. Science –
  Arab countries – History. I. Title. II Collected studies ; CS803.

  Q124.97.F74 2004
  509'.02–dc22   2004053659

The paper used in this publication meets the minimum requirements of the American National Standard for Information Sciences – Permanence of Paper for Printed Library Materials, ANSI Z39.48–1984. ∞ ™

Printed and bound in Great Britain by TJ International Ltd, Padstow, Cornwall

VARIORUM COLLECTED STUDIES SERIES CS803

Contents

This volume contains xx + 350 pages

Publisher's Note

The articles have been reprinted without change, except for article VI, in which a number of typos have been corrected, and article XVI, to which a new note has been added. Because of constraints imposed at their first printing, articles VI and X carry no notes. Articles VIII and IX have been translated into English for this volume.

Preface

For a historian of ideas, collecting between two covers studies written over a period of nearly twenty years is an occasion, indeed an invitation, to look back and trace the evolution of his or her research. For me it is also a welcome opportunity to acknowledge some intellectual debts not evident in mere bibliographical references. This is my purpose in what follows.

During my studies in mathematics, physics, and the history and philosophy of science at the Hebrew University of Jerusalem, I was especially marked by my contact with three individuals. Yehoshua Bar-Hillel, the analytical philosopher of language, instilled in me the quest for precise, "clear and distinct" thinking and an almost obsessive respect for precision in one's choice of words. Yehuda Elkana passed on to me his high regard for intellectual history and enthusiasm for non-positivistic, non-Whiggish approaches to the history of science. Joseph Ben-David, a leading authority on the sociology of science, made me aware of the crucial importance of the social structures within which scientific knowledge is produced—structures of which the historian of science must always be cognizant. Ben-David's faith in the university as the best available institutional locus for solving conflicts, through rational discussion rather than violence, impressed me as well. I expressed my indebtedness to him, as well as my personal affection, by publishing a collection of his most significant papers (1991a; includes a brief intellectual biography) and two studies devoted to his thought (1987c, 1988j).[1] I have explicitly applied his theoretical outlook in some historical studies, but it underlies others as well. My later studies at the Freie Universität Berlin brought me into contact with so-called Continental philosophy, especially the hermeneutics of Hans-Georg Gadamer and the Critical Theory of Jürgen Habermas.

My first published work, an introduction to the philosophy of science (1976a), tried to meld all of these approaches to the study of science. At a later stage, they also informed my appreciation of Hélène Metzger's methodological essays on the historiography of science and moved me to

---

[1] Bibliographical references refer to the List of Selected Publications, to be found after this Preface. The Roman numerals refer to the chapters of this book.

rescue them from oblivion (1987a). I also offered a construal of Metzger's and Gadamer's hermeneutical theories as complementary (1988b, condensed English version in 1988f). There is no doubt, however, that my devotion to Metzger's memory (1988a–f; 2003h, 2004/5g) was enhanced by the fact that she perished in the Shoah, as did my uncle (both made the same journey in sealed wagons from Drancy to Auschwitz).

My doctoral dissertation dealt with the history of electrical theories in the seventeenth and eighteenth centuries. The initial *problématique* was sociological, but my attention was soon drawn to purely intellectual aspects. Specifically, I realized that an awkward problem beset theories about electricity in the age of mechanical philosophy. The fundamental empirical fact that the early "electricians" (as they were called) sought to explain was the observation that when certain bodies, like amber or glass, are rubbed, they attract light bodies such as chaff or bits of paper (what we now know as "static electricity"). In the context of seventeenth-century mechanistic theories, the problem was how particles presumably issuing *from* a body (after it is rubbed) produce a motion *toward* it? (This was a special case of the general problem of how corpuscular theories of matter can account for what appear to be attractive forces.) One answer given to this "puzzle" was that the particles issuing from an "electrified" body are oily or unctuous elastic threads that stick to the light body and, when they cool down and contract, draw it to their source. I eventually discovered that the idea of such an oily or unctuous substance was no metaphor, but rather derived from an entrenched tradition of thinking about the causes of cohesion of matter that can be traced back to ancient Greece. Subsequent research on the history of this tradition proved long, but I think also fruitful.

In between, in the early 1970s, the violence in Israel/Palestine induced me to seek a more peaceful place under the sun and I found in France a true *terre d'accueil*. Here the encounter with Roshdi Rashed was decisive. Rashed awakened me from my euro-centric slumber and urged me to study the history of science among medieval Jews, especially in southern France, a field to which I could bring my knowledge of Hebrew, on top of my training as a historian of science. Following Rashed's advice, and later joining the Paris research group he headed, proved exceedingly beneficial: almost none of the papers in this volume would have been written otherwise. Having made Hebrew into one of my research tools, it was inevitable that I meet Bernard R. Goldstein. From him I learned, in addition to innumerable facts of detail, to harbour an unremitting methodological circumspection about "established" truths and to approach a text with questions about its sources.

From that time, my work followed two research programs that often cross-fertilized one another: a thematically defined inquiry into the history of theories of matter and a culturally and chronologically defined study of the history of science in Hebrew. The quest for the genealogy of theories of cohesion (of material substances or of the cosmos) led me to Anaximander, the Presocratic philosopher who asked why the cosmos does not fall apart, even though it consists of opposite and mutually annihilating forces (1986b [= XI]). Inasmuch as Aristotle integrated Anaximander's premises into his own theory of matter, the theoretical challenge posed by the persistence of composite substances remained on the agenda until the demise of Aristotelianism. This is why the paper on Anaximander could conclude with a quotation from Gersonides, the Provençal Jewish philosopher-scientist who lived almost two millennia later. The probe into "Anaximander's problématique" ultimately yielded my book on Aristotle's theory of matter (1995a), which in turn underlies much of my subsequent work on the medieval Arabic and Hebrew traditions. To this category belong 1991b, 1991c [= XII], 1993c, 1994a, 1996b [= XIII], 1998c [= XIV], 1998d, 2000b [= XVI], 2002a [= XV], 2004/5b, and 2004/5c. In two early studies on the history of electricity, which derive from my doctoral dissertation (1981a, 1983a), the perspective is still limited to the seventeenth century; later articles on the same topic (1986a, 1991b, 1993c), however, include input from my studies of earlier theories of matter. The fact that I focused on topics related to the cohesion of matter naturally led to a quest for the traces of Stoic physics in medieval Arabic and Hebrew thought (1993c, 1994a, 1996b, 1998c).

In parallel, my research on the medieval Hebrew scientific-philosophical tradition followed a number of lines. Some papers seek to offer a sociologically informed picture of scientific activity among medieval Jews writing in Hebrew (notably 1993a, 1995b [= I], 2001c [= II]). Other studies are devoted to various medieval Jewish scholars who either wrote in Hebrew or whose works have exerted major influence in their medieval Hebrew versions. To this category belong notably studies on Maimonides (1993b [= III], 2000a, 2003c, 2003g, 2004/5b, 2004/5e), Gersonides (1989c, 1990a [= VI], 1992a, 1992b [= V], 1992c, 1995c, 1996a [= IV], 2003f, 2004/5f), and lesser luminaries (1989a [= VII], 1989b [= VIII], 1991d [= IX], 1998b). Still others concern texts whose Hebrew versions (translations from the Arabic) are valuable for the study of the original versions (1988g, condensed English version in 1990b [= X], 1988h, 1988i, 2004/5d) or examine the history of Hebrew texts translated from Arabic (2003d, 2003e). More recently I turned to the study of the role played by Jewish medieval science and philosophy in the Haskalah (Jewish Enlightenment) (2003i, 2004/5a).

Serendipity, or at least occasional fits of interest for new topics, yielded works falling outside the framework defined by the theoretical problématiques. This is notably the case of my foray into the legal and philosophical Jewish responses to the AIDS pandemic (1998a). In a very different vein, I devoted some rewarding effort to the republication of long-forgotten articles, published in French between 1948 and 1986, on the roles played by Jewish doctors in the history of medicine and, especially, to researching the biographies of their authors (2003a). Last but not least, editing festschrifts in honor of esteemed scholars is a particularly agreeable part of academic life and I willingly lent my hand to two such enterprises (2001a–b, 2003b–c).

My last debt—but certainly not the smallest—is to the Centre National de la Recherche Scientifique (CNRS), the French government-funded agency that provides scientists and scholars of all disciplines with tenured positions and total freedom to pursue research as they see fit, their sole obligation being to expand human knowledge. I realize today, far more than I did 21 years ago, when I was fortunate to enter the CNRS, how privileged I have been. At the same time I also realize that this privilege has a price: the acute awareness that the distance between achievements and aspirations is a result only of one's own shortcomings. The following studies, however, are (re-)offered to readers in the hope that they can help diminish our ignorance on some points of the intellectual history of Hebrew and Arabic medieval civilization.

This book is dedicated, with loving gratitude, to the ever-dearer memory of my parents.

GAD FREUDENTHAL

Châtenay-Malabry, near Paris
December 2003

LIST OF SELECTED PUBLICATIONS*

1976

a. *Introduction to the Philosophy of Science* (Hebrew) (Tel Aviv: The Israeli Open University, 1976).

1979

a. "Littérature et sciences de la nature en France au début du XVIIIe siècle: Pierre Polinière, l'introduction de l'enseignement de la physique expérimentale à l'Université de Paris et l'*Arrêt burlesque* de Boileau," *Revue de synthèse* 99–100 (1979), 267–95.

1981

a. "Early Electricity Between Chemistry and Physics: The Simultaneous Itineraries of Francis Hauksbee, Samuel Wall, and Pierre Polinière," *Historical Studies in the Physical Sciences* 11 (1981), 203–29.

1983

a. "Theory of Matter and Cosmology in William Gilbert's *De magnete*," *Isis* 74 (1983), 22–37.

1986

a. "Die elektrische Anziehung im 17. Jahrhundert zwischen korpuskularer und alchemischer Deutung," in: Christoph Meinel (ed.), *Die Alchemie in der europäischen Kultur- und Wissenschaftsgeschichte (Wolfenbütteler Forschungen*, Band 32) (Wiesbaden: Otto Harrassowitz, 1986), 315–26

b.* "The Theory of the Opposites and an Ordered Universe: Physics and Metaphysics in Anaximander," *Phronesis* 31 (1986), 197–228.

c. "Cosmogonie et physique chez Gersonide," *Revue des études juives* 145 (1986), 295–314.

1987

a. (Ed.) Hélène Metzger, *La Méthode philosophique en histoire des sciences. Textes 1914–1939* (Paris: Fayard [Corpus des œuvres de philosophie en langue française], 1987). Translated into Italian as: *Il metodo filosofico nella storia delle scienze. Testi 1914–1939*, raccolti da Gad Freudenthal, introduzione di Mario Castellana, postfazione di Arcangelo Rossi (Manduria: Barbieri Editore, 2002).

---

* An asterisk next to an article's number indicates that the article in question is reprinted in this volume.

b. "Épistémologie, astronomie et astrologie chez Gersonide," *Revue des études juives* 146 (1987), 357–65.

c. "Joseph Ben-David's Sociology of Scientific Knowledge," *Minerva* 25 (1987), 135–49.

1988

a. (Ed.) *Études sur/Studies on Hélène Metzger* (= *Corpus* [Paris] no. 8/9 [1988]; reprinted: Leiden: Brill [Collection de travaux de l'Académie internationale d'histoire des sciences, no. 32], 1990).

b. "Épistémologie des sciences de la nature et herméneutique de l'histoire des sciences selon Hélène Metzger," in: *Études sur/Studies on Hélène Metzger*, 161–88.

c. "Hélène Metzger: Éléments de biographie," in: *Études sur/Studies on Hélène Metzger*, 197–208.

d. (Ed.) "Hélène Metzger, Extraits de lettres," in: *Études sur/Studies on Hélène Metzger*, 247–69.

e. "Bibliographie complète d'Hélène Metzger," in: *Études sur/Studies on Hélène Metzger*, 270–74.

f. "The Hermeneutical Status of the History of Science: The Views of Hélène Metzger," in: Edna Ullmann-Margalit (ed.), *Science in Reflection. [The Israel Colloquium: Studies in History, Philosophy and Sociology of Science*, vol. 3. *Boston Studies in the Philosophy of Science*, vol. 110] (Dordrecht: Kluwer Academic Publishers, 1988), 123–44.

g. "La Philosophie de la géométrie d'al-Fârâbî: Son commentaire sur le début du 1er livre et le début du Ve livre des *Éléments* d'Euclide," *Jerusalem Studies in Arabic and Islam* 11 (1988), 104–219.

h. "Maimonides' *Guide of the Perplexed* and the Transmission of the Mathematical Tract 'On Two Asymptotic Lines' in the Arabic, Latin, and Hebrew Medieval Traditions," *Vivarium* 26 (1988), 113–40. (Reprinted in: Robert S. Cohen and Hillel Levine (eds.), *Maimonides and the Sciences* [= *Boston Studies in the Philosophy of Science*, vol. 211] [Dordrecht/ Boston/London: Kluwer Academic Publishers, 2000], 35–56.)

i. "Pour le dossier de la traduction latine médiévale du *Guide des égarés*," *Revue des études juives* 117 (1988), 167–72.

j. (With Ilana Löwy) "Ludwik Fleck's Roles in Society: A Case Study Using Joseph Ben-David's Paradigm for a Sociology of Knowledge," *Social Studies of Science* 18 (1988), 625–51.

1989

a.* "Sur la partie astronomique du *Liwyat Ḥen* de Lévi ben Abraham ben Ḥayyim," *Revue des études juives* 148 (1989), 103–12.

b.* "Distinguishing Two R. Joseph b. Joseph Naḥmias: The Commentator and the Astrologer" (Hebrew), *Qiryat sefer* 62 (1989), 917–19.

c. "Human Felicity and Astronomy: Gersonides' Revolt Against Ptolemy" (Hebrew), *Daat* (Bar-Ilan University, Israel) 22 (1989), 55–72.

1990

a.* "Levi ben Gershom as a Scientist: Physics, Astrology and Eschatology,"*Proceedings of the Tenth World Congress of Jewish Studies*, Division C, vol. 1: *Jewish Thought and Literature* (Jerusalem: World Union of Jewish Studies, 1990), 65–72.

b.* "Al-Fârâbî on the Foundations of Geometry," *Knowledge and the Sciences in Medieval Philosophy. Proceedings of the Eighth International Congress of Medieval Philosophy (S.I.E.P.M.)*, Vol. III (=*Acta Philosophica Fennica 48*), edited by Monika Asztalos, John E. Murdoch and Ilkka Niiniluoto. (Helsinki, 1990), 52–61.

1991

a. (Ed.) Joseph Ben-David, *Scientific Growth: Collected Essays on the Social Organization and Ethos of Science* (Berkeley/Los Angeles/Oxford: University of California Press, 1991). Translated into French as: *Éléments d'une sociologie historique des sciences*. Textes réunis et présentés par Gad Freudenthal, traduction de Michelle de Launay, revue par Jean-Pierre Rothschild (Paris, Presses universitaires de France [Collection "Sociologies"], 1997).

b. "The Problem of Cohesion Between Alchemy and Natural Philosophy: From Unctuous Moisture to Phlogiston," in: Z.R.W.M. von Martels (ed.), *Alchemy Revisited. Proceedings of the International Conference on the History of Alchemy at the University of Groningen, 17–19 April 1989* (Collection de travaux de l'Académie internationale d'histoire des sciences, vol. 33) (Leiden: Brill, 1991), 107–16.

c.* "(Al-)Chemical Foundations for Cosmological Ideas: Ibn Sînâ on the Geology of an Eternal World," in: Sabetai Unguru (ed.), *Physics, Cosmology and Astronomy, 1300–1700: Tension and Accommodation* (= *Boston Studies in the Philosophy of Science*, vol. 126) (Dordrecht/Boston/London: Kluwer Academic Publishers, 1991), 47–73.

d.* "Two Notes on *Sefer Meyasher 'aqob* by Alfonso, alias Abner of Burgos" (Hebrew), *Qiryat sefer* 63 (1991), 984–6.

1992

a. (Ed.) *Studies on Gersonides—A Fourteenth-Century Jewish Philosopher-Scientist* (Leiden: Brill [Collection de travaux de l'Académie internationale d'histoire des sciences, vol. 36], 1992).

b.* "Sauver son âme ou sauver les phénomènes: Sotériologie, épistémologie et astronomie chez Gersonide," in: *Studies on Gersonides*, 317–52.

c. "Rabbi Lewi ben Gerschom (Gersonides) und die Bedingungen wissenschaftlichen Fortschritts im Mittelalter: Astronomie, Physik, erkenntnistheoretischer Realismus, und Heilslehre," *Archiv für Geschichte der Philosophie* 74 (1992), 158–79.

1993

a. "Les sciences dans les communautés juives médiévales de Provence: Leur appropriation, leur rôle," *Revue des études juives* 152 (1993), 29–136.

b.* "Maimonides' Stance on Astrology in Context: Cosmology, Physics, Medicine, and Providence," in: Fred Rosner and Samuel S. Kottek (eds.), *Moses Maimonides: Physician, Scientist, and Philosopher* (Northvale, NJ London: Jason Aronson Inc., 1993), 77–90.

c. "Clandestine Stoic Concepts in Mechanical Philosophy: The Problem of Electrical Attraction," in: J.V. Field and Frank A.J.L. James (eds.), *Renaissance and Revolution: Humanists, Scholars, Craftsmen and Natural Philosophers in Early Modern Europe* (Cambridge: Cambridge University Press, 1993), 161–72.

1994

a. "'The Air Blessed Be He and Blessed Be His Name' in *Sefer ha-Maskil* by R. Shlomo Simḥa of Troyes: Some Characteristics of a Stoically-Inspired Midrashic-Scientific Cosmology of the Thirteenth Century," Part One: *Daat* (Bar-Ilan University, Israel) 32–33 (1994), 187–234 (Hebrew; English abstract, pp. LXVII–LXVIII); Part Two: ibid., 34 (1995), 87–129.

1995

a. *Aristotle's Theory of Material Substance. Form and Soul, Heat and Pneuma* (Oxford: Clarendon Press, 1995).

b.* "Science in the Medieval Jewish Culture of Southern France," *History of Science* 33 (1995), 23–58.

c. (With Henri Hugonnard-Roche), "Gersonide logicien," *Revue philosophique* 4 (1995), 485–91.

1996

a.* "Levi ben Gershom (Gersonides), 1288–1344," in: S.H. Nasr and O. Leaman (eds.), *The Routledge History of Islamic Philosophy* (London/New York: Routledge, 1996), 739–54.

b.* "Stoic Physics in the Writings of R. Saadia Ga'on al-Fayyumi and its Aftermath in

Medieval Jewish Mysticism," *Arabic Sciences and Philosophy* 6 (1996), 113–36.

1998

a. (Ed.), *AIDS in Jewish Thought and Law* (Hoboken, NJ: Ktav Publishing Co., 1998).

b. "The Study of Mathematics as a 'Great Religious Secret': The Commentary on the Beginning of Euclid's *Elements* by Abraham ben Shlomo of Lunel. A Commented Critical Edition" (Hebrew), in: Aviezer Ravitzky (ed.), *Joseph Barukh Sermoneta Memorial Volume* (Jerusalem: The Hebrew University of Jerusalem, 1998) (= *Jerusalem Studies in Jewish Thought*, vol. 14), 129–58.

c.* "L'héritage de la physique stoïcienne dans la pensée juive médiévale (Saadia Gaon, les Dévots rhénans, *Sefer ha-Maskil),*" *Revue de métaphysique et de morale* (1994), 453–77.

d. "Öl," in: Claus Priesner and Karin Figala (eds.), *Alchemie. Lexikon einer hermetischen Wissenschaft* (München: Verlag C.H. Beck, 1998), 259–60.

2000

a. "Jérusalem Ville sainte? Le dilemme maimonidien," *Revue d'histoire des religions* 217 (2000), 689–705.

b.* "Providence, Astrology, and Celestial Influences on the Sublunar World in Shem-Tov ibn Falaquera's *De'ot ha-filosofim,*" in: Steven Harvey (ed.), *The Medieval Hebrew Encyclopedias of Science and Philosophy* (Amsterdam: Kluwer Academic Publishers, 2000), 335–70.

2001

a. (Ed. with Jean-Pierre Rothschild and Gilbert Dahan), *Torah et Science: Perspectives historiques et théoriques. Études offertes à Charles Touati* (Louvain: Peeters, 2001).

b. "Révélation et Raison, Torah et Madda dans quelques écrits récents," in: *Torah et Sciencs*, 239–67.

c.* "Holiness and Defilement: The Ambivalent Perception of Philosophy by its Opponents in the Early Fourteenth Century," in: *Gli Ebrei e le Scienze. The Jews and the Sciences* (=*Micrologus* IX) (Florence: Sismel/Edizioni del Galluzo, 2001), 169–93.

2002

a.* "The Medieval Astrologization of Aristotle's Biology: Averroes on the Role of the Celestial Bodies in the Generation of Animate Bodies," *Arabic Sciences and Philosophy* 12 (2002), 111–37.

2003

a. (Ed. with S. Kottek), *Mélanges d'histoire de la médecine hébraïque. Études choisies de la* Revue de l'histoire de la médecine hébraïque, *1948–1985* (Leiden: Brill, 2003).

b. (Ed. with Peter Barker, Alan C. Bowen, José Chabás, and Y. Tzvi Langermann), *Astronomy and Astrology from the Babylonians to Kepler: Essays Presented to Bernard R. Goldstein on the Occasion of his 65th Birthday* (= *Centaurus* 45 [1–4] [2003]; 46 [1] [2004]).

c. "'Instrumentalism' and 'Realism' as Categories in the History of Astronomy: Duhem vs. Popper, Maimonides vs. Gersonides," in: *Astronomy and Astrology from the Babylonians to Kepler*, 227–48.

d. *"Ketav ha-da'at* or *Sefer ha-Sekhel we-ha-muskalot*: The Medieval Hebrew Translations of al-Fârâbî's *Risâlah fi'l-'aql*. A Study in Text History and in the Evolution of Medieval Hebrew Philosophical Terminology," *Jewish Quarterly Review* 93 (2003), 29–115.

e. *"La Quiddité de l'âme,* traité populaire néoplatonisant faussement attribué à al-Fârâbî: Traduction annotée et commentée," *Arabic Sciences and Philosophy* 13 (2003), 173–237.

f. "Gersonide, génie solitaire," in: Colette Sirat, Sara Klein-Braslavy and Olga Weijers (eds.), *Les méthodes de travail de Gersonide et le maniement du savoir chez les scolastiques* (Paris: Librairie philosophique J. Vrin, 2003), 291–317.

g. "Four Implicit Quotations of Philosophical Sources in Maimonides' *Guide of the Perplexed,*" *Zutot: Perspectives on Jewish Culture,* vol. 2 (2002) (Dordrecht: Kluwer Academic Publishers, 2003), 114–25.

h. (With Cristina Chimisso), "A Mind of Her Own. Hélène Metzger to Émile Meyerson, 1933," *Isis* 94 (2003), 477–91.

i. "Israel ben Moshe Halewi Zamosc," in: Andreas B. Kilcher and Otfried Fraisse (eds.), *Metzler Lexikon jüdischer Philosophen und Theologen* (Stuttgart: Verlag J. B. Metzler, 2003), 174–6.

2004–2005 (Scheduled for Publication)

a. "Sephardi Medieval Science on Polish Soil: Toward an Intellectual Biography of Rabbi Israel ben Moses Halevy of Zamosc (c. 1700–1772)," in: R. Fontaine, A. Schatz, I.E. Zwiep (eds.), *Sepharad in Ashkenaz. Medieval Knowledge and Eighteenth-Century Enlightened Jewish Discourse* (Amsterdam: Edita, in print).

b. "Maïmonide: La détermination biologique et climatologique (partielle) de la félicité humaine," in: Tony Lévy et Roshdi Rashed (eds.), *Maïmonide, philosophe et savant* (Louvain: Peeters, 2004), 79–129.

c.   "Averroes' Changing Mind on the Role of the Active Intellect in the Generation of Animate Beings," in: Ahmed Hasnaoui and Roshdi Rashed (eds.*), La pensée philosophique et scientifique d'Averroès dans son temps* (Louvain : Peeters, in print).

d.   (With Tony Lévy), "De Gérase à Bagdad: Ibn Bahrîz, al-Kindî, et leur recension arabe de l'*Introduction arithmétique* de Nicomaque, d'après la version hébraïque de Qalonymos b. Qalonymos d'Arles," in: Régis Morelon and Ahmed Hasnaoui (eds.) *De Zénon d'Élée à Poincaré. Recueil d'études en hommage à Roshdi Rashed* (Louvain-Paris: Éditions Peeters, 2004), 479–544.

e.   "Maimonides' Philosophy of Science," in: Kenneth Seeskin (ed.), *The Cambridge Companion to Maimonides* (Cambridge: Cambridge University Press, in print).

f.   (With José Luis Mancha), "Levi ben Gershom's Criticism of Ptolemy's Astronomy. Critical Editions of The Hebrew And Latin Versions and an Annotated English Translation of Chapter Forty-Three of the *Astronomy (Wars of The Lord*, V.1.43)," *Aleph: Historical Studies in Science and Judaism* (forthcoming).

g.   "Hélène Metzger," in: Paula Hyman and Dalia Ofer (eds.) *Jewish Women: A Comprehensive Historical Encyclopedia* (Jerusalem: Shalvi Publications [on CD ROM], forthcoming).

Acknowledgements

The author and publisher gratefully acknowledge the permissions to reprint material in this volume given by the editors or publishers of periodicals and collective volumes. In particular: Essay IV © 1996 Routledge Ltd. Reprinted with kind permission of Routledge Ltd; Essays V and XI © 1992 and 1986 by Brill Academic Publishers. Reprinted with kind permission of Brill Academic Publishers; Essays XII and XVI © 1991 and 2000 by Kluwer Academic Publishers. Reprinted with kind permission of Kluwer Academic Publishers; Essays XIII and XV © 1996 and 2002 by Cambridge University Press. Reprinted with kind permission of Cambridge University Press.

I

SCIENCE IN THE MEDIEVAL JEWISH CULTURE OF SOUTHERN FRANCE

1. INTRODUCTION: THE SOCIOLOGICAL STUDY OF SCIENCE AND RELIGION AND THE NOTION OF THE AUTONOMY OF SCIENCE

"The belief in the value of scientific truth is not derived from nature but is a product of definite cultures": this quotation from Max Weber, with which Robert K. Merton begins his seminal essay on "Science and the social order",[1] epitomizes the basic postulate of the sociology of science in the Weberian-Mertonian tradition. In a searching and illuminating analysis of a paradigmatic building-block of this sociology — namely of Merton's celebrated thesis on the "Protestant spur to science" — the late Joseph Ben-David has shown that the Merton thesis in fact consists of two distinct theses:[2] one affirms that seventeenth-century Protestantism was conducive to the *legitimation* within society of the autonomous pursuit of scientific knowledge, indeed to putting empirical science on top of the intellectual hierarchy, and thus to institutionalizing science within society; the other states that Protestantism gave a number of first-rate individuals a powerful *motivation*, on the level of their individual psychology, to engage in science. The two theses are certainly related, but distinguishing them clarifies the issue: in discussing the 'spur' (or otherwise) given to science by a religious view, or by any other cultural system, one should set apart (i) its incidence on the legitimation within society of scientific pursuit as a social practice and institution; and (ii) the motivation it gave to individuals to invest themselves in scientific research. This double-faceted thesis provides the theoretical matrix for the following paper.

Here a word of methodological caution should be added. In studying the incidence of a religion on the fortunes of science we must not consider only or even primarily that religion's explicit stands on science: Weber and Merton have shown that the 'official' doctrinal positions of a religion may be very different from, indeed opposite to, the incidence of that religion on *actual* social behaviour. The preached moral and the real ethos may differ greatly. Issues such as whether, say, a religion has a positive or a negative influence on the development of science, or whether it accords science autonomy or construes it as a 'handmaiden' to theology, are ones whose study must go beyond assembling the relevant pronouncements from the pens of that religion's spokesmen.[3] We should rather try to determine the incidence of the normative statements on real behaviour. In analogy to the heuristic recommended by Albert Einstein — "if you wish to learn from the theoretical physicists anything about the methods which they use, then I suggest that you follow the principle: do not listen to their words, rather examine what they do"[4] — we should aim to determine the incidence of a given belief or theology on real human behaviour.

Let me illustrate the above with an example that will be of great importance to our subsequent analysis. I have in mind the emergence of the autonomy of science within the Protestant tradition as analysed by Michael Heyd.[5] The founders of Protestantism were little interested in the study of nature and of course never preached the autonomy of scientific inquiry. Yet following Melanchthon's insistence that God can and should be known through His works, science within Protestantism gained the status of a legitimate, independent and autonomous sphere of knowledge. During the seventeenth century, Heyd argues, the Protestant tradition ascribed to science a positive religious value, namely as a means to acquire knowledge of God or His works. Science was construed as a 'soteriological bridge', allowing man to create a link with the transcendent. As a result of being invested with this positive religious value, science came to be recognized as a valuable independent sphere of knowledge. Thus, and this is the general lesson to be gained from Heyd's analysis, the fact that science was ascribed a religious role — a role that both legitimized it and supplied Protestant men of science with motivation — is perfectly consistent with the intellectual and institutional autonomy of science. The notorious question whether in a given tradition science was or was not the 'handmaiden' of theology is thus a *question mal posée*, for these are not necessarily mutually exclusive alternatives: science can be legitimized through religion and yet be pursued for its own sake and have intellectual autonomy.[6] It is of course a factual question whether, and to what extent, science in a given context was autonomous or not.

From our perspective, we should note that had we taken into consideration only doctrinal pronouncements (by scientists, theologians, etc.) on the religious significance of scientific knowledge, we would necessarily have arrived at the erroneous conclusion that Protestant science was a 'handmaiden' to theology. We would have missed the point that the scientific practice of the Protestant scientists, although religiously legitimated and motivated, was autonomous; that once science was accorded a religious role and dignity, it could go its own ways, independently of the religious belief that set it going. To obtain a sociological view of the influence of medieval Judaism on the attitudes towards science will be my goal in what follows, even where — by necessity or by habit — I will be discussing ideas held or propounded by various thinkers.

2. THE PROBLEM: JUDAISM AND 'GREEK WISDOM' IN THE MIDDLE AGES

Intellectual activity of Jews has traditionally gravitated around the study of the Jewish traditional texts.[7] Historically, indeed, intellectual quests having their origin in other cultures — notably science and philosophy — were perceived and referred to as 'alien [sometimes: Greek] wisdom': the most prevalent attitude of Judaism towards knowledge other than that which was sanctified and legitimized through its own tradition has been one of circumspection, indeed more often than not even of hostility.[8] The following quotation from the pen of the erudite Talmudist and very prestigious R. Asher ben Yeḥiel (Rosh), who first encountered science and philosophy when, early in the fourteenth century, he emigrated to Spain (fleeing persecutions in Germany), reflects a long-standing central attitude within Judaism:

> The science of philosophy and the science of the Torah are not the same. The Torah was donated to Moses at Mount Sinai.... Philosophy being a natural science, it was inevitable that the philosophers deny the Torah, for the Torah is not a natural science, but rather [a science] received and transmitted [qabbalah].... [The Torah and philosophy] are indeed contraries, two rival wives who cannot be at one and the same place.[9]

Indeed, a line of transmission and reception (qabbalah), linking the present to the Revelation received by the Prophet Moses, was invoked in order to legitimize traditional knowledge and, concomitantly, fend off alternative bodies of belief. Those historical episodes in which Judaism, or rather sections of it, displayed a receptive attitude towards philosophy and science are relatively few and are in need of explanation.[10]

Here we will be concerned with one such episode, in fact with the major one among them (at least prior to modernity). Between the early tenth and the fifteenth centuries, central parts of the Jewish communities integrated within their worldviews significant elements borrowed from the heritage of Greek-Arabic philosophy and science. The process has two components which somewhat overlap temporally. It began in the Islamic lands, where Jews were integrated into their cultural environment to a degree that was never reached again before the Enlightenment.[11] Contrary to what was the case in Latin Europe, Jews spoke, read and wrote in the language of the majority (more precisely: in Judeo-Arabic rather than in Arabic) and thus had direct access to the ambient culture, with no linguistic barrier hampering the process of cultural transmission. Saadia Gaon (882–942) at Baghdad was the first to compose (in Arabic) two philosophical works with the aim of showing that there is no incompatibility between Moses's revealed Law and the results of rational philosophical investigation. Many other authors, in both East and West (notably Spain), followed suit. The process reached its peak with R. Moses ben Maimon, Maimonides (1135–1204), whose *Guide of the perplexed* created a lasting and most influential platform for a synthesis between Judaism and Greek-Arabic philosophy.

The second component of the process has its starting point in the first half of the twelfth century in southern France. For some two-and-a-half centuries, Jewish scholars knowledgeable in Arabic — encyclopedists, translators and philosophers

— put a considerable portion of the corpus of Greek-Arabic philosophy and science at the reach of their brethren unfamiliar with Arabic or Latin. Scores of works by Greek and Arab authors were translated from Arabic into Hebrew, in what was an unprecedented large-scale appropriation of 'alien wisdom' within Jewish communities which beforehand had been entirely devoted to traditional learning.

This paper will be concerned only with this last process, that is, with the appropriation of science and philosophy within the Jewish communities that resided in southern France and northern Spain (a region I will for brevity also term, if inaccurately: Provence) and whose cultural language was Hebrew.[12] I thus leave out of consideration both the Jews of Islam and the Jews of northern France and Germany: the attitudes towards the sciences within each of these communities followed distinct cultural dynamics, whose study can and should be separated from those to be examined here.

In what follows, I wish first to *delineate* and then to *explain* the contours of the reception of science and philosophy within the Jewish communities of Provence. My questions will be the following: To what extent was science appropriated and naturalized within these communities? Where did that appropriation have its *limits*: specifically, which scientific disciplines *were*, and which *were not*, transmitted and appropriated? Furthermore, did the Jewish scholars seek to make their own contributions to the sciences they had appropriated, or were they content to absorb the sciences as they received them? Last but not least, I will try to offer a sociological account of the phenomena observed.

Medicine will unfortunately be left out of the account, mainly because its history among medieval Jews has not yet been studied sufficiently.[13] I believe, however, that its inclusion would require a refinement, but not a profound revision, of the following analysis.

3. THE APPROPRIATION OF SCIENCE AND PHILOSOPHY BY THE JEWS OF PROVENCE: SOME HISTORICAL FACTS

I begin with a very brief *factual* historical *aperçu* which makes no claims to either completeness or originality. Its modest aim is merely to sketch the evolution of the landscape with which we will be concerned.[14]

The reception of Arabic science and philosophy within the Jewish community of Provence can usefully be divided into three phases: (i) the start of the process of transmission around the beginning of the twelfth century, when Spanish scholars translated scientific works from Arabic into Hebrew for the Jews living north of the Pyrenees; (ii) the period in the second half of the century when the process gathered momentum, as Andalusian Jewish scholars immersed in Arabic culture fled the Almohad persecutions to Provence, accelerating considerably the translation into Hebrew of philosophical works; and (iii) the first decades of the thirteenth century, when the process received a new and decisive impetus as Maimonides's writings, notably the Hebrew translation of the *Guide of the perplexed* (1204), acquired influence in southern France. This period of intensive translation lasted until about the middle of the fourteenth century, when the philosophical-scientific activity among the Jews of southern France and Spain began to decline.

(i) During the first half of the twelfth century, Abraham bar Ḥiyya (d. c. 1145) of Barcelona, a political leader and scholar knowledgeable in the sciences of his day, wrote a number of works summarizing these sciences in Hebrew. He apparently wrote these works at the request of Jewish notables of Provence, who felt a need for instruction in various subjects. Bar Ḥiyya thus offered basic courses in such immediately useful disciplines as practical geometry (teaching the division of lands), and astronomy (calculation of the calendar), but also composed an encyclopedia of general culture, with no immediate practical utility.

A younger contemporary of Bar Ḥiyya, who also contributed much to the spread of science and philosophy among Jews, is Abraham Ibn Ezra (1089–1164), one of the greatest Hebrew poets of all time. Ibn Ezra's characteristic mystic bent of mind gave a special flavour to his arithmetical or mathematical writings and helped to introduce them even into circles (including in northern Europe) that were generally hostile towards all 'alien wisdom'. In addition to his mathematical writings, Ibn Ezra composed and translated astronomical and astrological works. His numerous Biblical commentaries often invoked scientific or philosophical notions. Owing to the great popularity of these commentaries throughout Europe, they contributed much to spreading and legitimizing the view that 'Greek learning' not only was not incompatible with the teachings of the Torah, but was in fact indispensable to a correct understanding of the latter.

(ii) Fleeing the Almohad persecutions in Spain, a number of erudite Jewish families settled in Provence during the late 1140s. They brought with them a culture that was altogether different from that of their co-religionists: whereas the latter were still fully absorbed in traditional, Talmudic, learning,[15] the immigrants were at home in Arabic poetry, literature, grammar, philosophy, and science. The encounter between the two Jewish cultures was followed by a massive translation movement, which was to last for some two centuries and during which the newcomers and their descendants were to render from Arabic into Hebrew a rich body of literature. The translated works were at first essentially in Jewish religious philosophy, but gradually works of general philosophy and science by pagan or Muslim writers were translated as well.[16]

The first part of this massive process of transmission took place during the second half of the twelfth century, that is, prior to the spread and decisive impact of Maimonides's philosophy. This stage is of capital importance for our subject. It in fact initiated the gradual acceptance of the Greek-Arabic philosophy and science by certain circles within the Jewish communities of Provence, an acceptance that was to be consolidated during the thirteenth century. It must regretfully be said, however, that, the efforts of some distinguished historians notwithstanding,[17] this process is not well understood: specifically, it is not clear why the relatively few newcomers succeeded in 'converting' to their own *Weltanschauung* significant segments of the host society, rather than being 'assimilated' by it. The half-century between the arrival in Provence of the refugees from Andalusia and the impact of Maimonides's *Guide* witnessed a highly important process of cultural transformation that still calls for further research.

(iii) A large portion of the translations of works in philosophy and science was done by scholars belonging to the Tibbonid family. To this illustrious family belong notably: Yehudah ibn Tibbon (*c.* 1120–90), the so-called "father of the translators", his son Shmuel (1150–1230), and the latter's son Moshe (*fl. c.* 1240–85). Shmuel Ibn Tibbon translated notably, in 1204, Maimonides's *Guide of the perplexed*, whose immense influence was to reinforce the philosophical movement and the translation activity that it nourished. Also associated with this family are Yaqob Anaṭoli (1194–1256) and Yaqob ben Makhir (*c.* 1236–1304). A number of other translators of scientific and philosophical texts were active too, most notably Qalonimos ben Qalonimos (1286–*p.* 1328).

Thanks to the labours of these professional translators, a Hebrew-reading scholar living at the beginning of the fourteenth century could have access to an impressive body of scientific writings, comprising a large share of the theories known to the Arabs. To name but a few of the more important authors available in often excellent Hebrew translations: in logic al-Fârâbî, Ibn Sînâ, and Ibn Rushd, in mathematics Euclid and Archimedes, in astronomy Ptolemy, Jâbir ibn Aflaḥ, and al-Biṭrûjî, and in natural science, physics and metaphysics almost all of Ibn Rushd's commentaries on Aristotle. This process of transmission was remarkably continuous, testifying to a steady *demand* for scientific and philosophical writings. It must however be kept in mind that the interest in science and philosophy was limited to certain segments of the Jewish communities and that, moreover, this 'pro-philosophy camp' was under continued criticism and attacks from the traditionalists quarters. We will come back to this below.

Two observations confirm the widespread interest in science among Jews of Provence in the thirteenth and fourteenth centuries. The first is this. As is well known, the years 1303–6 witnessed a renewed outbreak of a virulent controversy over the study of philosophy: it was triggered off by the attempt of the opponents of philosophy to have the study of science and philosophy forbidden, at least before the age of twenty-five.[18] The spiritual leaders who took the initiative for the ban would hardly have attempted to proscribe something no-one was doing or intended to do, and we may conclude that the study of science and philosophy was widespread. A second important indication that the study of science was indeed prevalent is given by the sheer number of Hebrew manuscripts that have survived. Let me take only some of Ibn Rushd's writings as examples:[19] the Hebrew translation of his Epitome of Aristotle's *Physics* has survived in some twenty manuscripts, the Epitome of the *De caelo* in eighteen, the Middle Commentary of the *De caelo* in no fewer than thirty-six, the Epitome of the *Parva naturalia* in some twenty-five, while even the Hebrew version of the pseudo-Aristotelian *De plantis* is extant in some nine manuscripts:[20] these numbers compare well with the some one hundred and seventy extant manuscripts, from all periods and places, of the Hebrew translation of Maimonides's *Guide of the perplexed*, the most studied Jewish philosophical work.[21] It should also be mentioned that almost all of Ibn Rushd's commentaries have themselves been the subject of commentaries, testifying to an intense and continuous activity of study and teaching.

4. THE LIMITS OF THE APPROPRIATION OF SCIENCE AND PHILOSOPHY BY THE JEWS
    OF PROVENCE: MORE FACTS

Against the background of the constant interest in science and philosophy among the Jews of Provence, the following statements, which I will seek to substantiate in what follows, appear surprising. First, some of the most innovative and advanced parts of medieval Arab (and Latin) science were not transmitted in Hebrew and remained unknown to the Hebrew-reading Jewish scholars. Second, except in astronomy, Jews contributed little to the advancement of medieval science: there are no Jewish counterparts to such scientific geniuses as, say, al-Birûnî, Ibn al-Haytham, and Thâbit Ibn Qurra in the Arabic culture, or Robert Grosseteste, Roger Bacon and Nicole Oresme in the Latin. Let us consider the limits of the appropriation of scientific knowledge in the various disciplines in some detail.

A. *Logic*. The numerous logical treatises in Hebrew testify to a continued interest in this discipline among medieval Jewish scholars.[22] A close examination reveals, however, that these works are essentially translations, commentaries or epitomes of fairly elementary Arabic works, notably by al-Fârâbî, Ibn Sînâ (transmitted through al-Ghazâlî), and Ibn Rushd. The only Hebrew work in logic that is truly original is the *Sefer ha-heqqesh ha-yashar* ("Book of the correct syllogism"), by R. Levi ben Gershom (Gersonides), to whom we will return.

B. *Astronomy* is certainly the discipline which more than any other engaged the efforts of medieval Jews and to which Jews contributed most, quantitatively and qualitatively. Most of the fundamental astronomical texts were available in Hebrew, and their readers in turn composed new treatises, calculated astronomical tables, and invented new astronomical instruments. The interest in astronomy among Jews is remarkably constant: Bernard R. Goldstein has shown that astronomical activity can be detected at all periods in most Jewish communities, including those that had no interest in philosophy and science.[23]

C. *Mathematics*. The interest in mathematics is far less than that in astronomy.[24] To be sure, a number of basic treatises were translated into Hebrew, the most important among them being Euclid's *Elements*. Yet central domains of medieval mathematics, most notably algebra and number theory, remained altogether unknown to the Jewish scholars. Whereas algebra was a major preoccupation of the best Arab mathematicians and in fact constitutes one of the foremost innovations of medieval science, there are no traces of it in the Hebrew medieval literature. The first translations of works on algebra into Hebrew are due to Mordekhai Finzi as late as the middle of the fifteenth century. Nor was Apollonius's *Conics*, which played an important role in Arabic mathematics, translated into Hebrew. Consistent with this absence of translations of advanced works is the failure to write original ones. As the only exception to the rule we find again Gersonides, and perhaps also Immanuel ben Yaqob Bonfils of Tarascon.

Similarly, there is a virtual absence of interest in music and optics, the two other subjects of the quadrivium which were of great importance in the Arabic and Latin traditions.[25]

D. *Physical sciences*. Physics proper, that is the study of the subjects covered by Aristotle's *Physics*, was of course the object of constant attention by Jewish philosophers in the context of discussions of God's existence and incorporeality and of the eternity of the world. But there apparently is no mention in the Hebrew philosophical literature of the more recent and innovative developments of Arab (or Latin) physics — as, most notably, the impetus theory.[26] Similarly, Jewish philosophers writing in Hebrew scarcely went beyond what they had received through translations: they did not venture to make contributions of their own. They were content to interpret what they learnt from others without seeking to change it. To this there are two exceptions: Gersonides (yet again) and R. Ḥasdai Crescas, both of whom to some extent confirm the rule. I will come back to them later.

Let us now consider another physical study, namely alchemy, one of the most flourishing medieval disciplines among both the Arabs and the Latins, and one whose impact on Renaissance and early modern science was momentous. Moritz Steinschneider and Gershom Scholem have already pointed out that there are virtually no Hebrew medieval writings on alchemy: almost nothing alchemical was translated into Hebrew and the Jewish medieval philosophers do not know even the names of such illustrious alchemists as Jâbir Ibn Ḥayyân, al-Râzî, or the Latin Geber.[27] Nor was a single alchemical work written in Hebrew during the Middle Ages. Jewish philosophers occasionally drew on the notion of transmutation as a *metaphor*, but none of them, it seems, sought to come to grips with it.

No less noteworthy is the fact that the Hebrew scientific literature almost nowhere invokes the famous sulphur-mercury theory of matter. As is well known, this theory — which is independent of the postulate of transmutation — played a pivotal role in medieval and Renaissance natural philosophy. Yet the Jewish authors are almost entirely unaware of it.

This all too brief survey of the limits of the appropriation of science within the Jewish communities of Provence during the Middle Ages thus leads to the following, somewhat paradoxical, conclusion: on the one hand, science was very widely studied; on the other hand, some sciences or scientific disciplines were not at all appropriated, and to those sciences that were appropriated, astronomy excepted, the Hebrew-writing scholars made few original contributions. This picture can certainly be nuanced, yet I believe that *grosso modo* it reflects the historical reality.

This state of affairs is surprising. Jewish society was presumably literate to a considerable extent,[28] and a non-negligible fraction of the Jewish communities in southern France supported the study of philosophy. It follows that, unlike what happened in the other medieval societies, the study of the sciences by medieval Jews was not limited to an intellectual élite of philosophers, but was quite widespread (we will come back to the reasons for this below). One would therefore have expected a considerable spin-off of creative scientific activity. Why, then, did the study of science by many individuals lead so few of them to pursue investigations of their own? And why did the lively interest for science stop at the threshold of certain disciplines? In short, how can the fact be explained that *the dialectic of research and translation*, characteristic of the development of Arab science,[29] did not evolve within the Hebrew medieval culture as well?

5. TOWARDS AN EXPLANATION OF THE FACTS: TWO SOURCES OF INTEREST IN, AND LEGITIMATION OF, SCIENCE

5.1 *A Desideratum: The Translators as a Strategic Research Site*

A fine-grained answer to the above questions would require that we begin by considering the genesis of the body of translations into Hebrew. Indeed, the *corpus* of philosophical and scientific writings was largely the work of a relatively small group of often highly proficient and productive translators, who more often than not were scholars in their own right who composed original works. From the vantage point of our subject, these translator-scholars, and their occasional patrons, are to be viewed as the persons whose presumably individual, sporadic and uncoordinated decisions concerning the identity of the works that would or would not be translated determined the *corpus* of works that was to become available in Hebrew. The translators are, so to speak, the filter that determined which contents would, and which would not, be allowed to enter into contact with the Jewish culture. The question of the criteria of choice followed by the translators, or by their patrons, is therefore of immediate relevance to our subject. Yet only very little systematic research on this subject has been done so far and we know little of the considerations underlying the selection of works for translation. Nor do we know much of the economical aspects of this activity, although it would seem that the translators were not paid for their labours.[30]

A noteworthy exception is a recent illuminating study by Aviezer Ravitzky.[31] In answer to the question, why as early as 1210 did Shmuel Ibn Tibbon choose to translate into Hebrew Aristotle's *Meteorology*, a rather obscure text (in its Arabic version) and one whose utility for whatsoever purpose is not immediately apparent, Ravitzky most convincingly argues that this choice was motivated by the translator's highly rationalist version of Maimonideanism, and specifically by his wish to give a naturalistic account of creation. Unfortunately this study remains an almost isolated instance of what should become an entire research program.[32] The role of the translators in shaping the Hebrew philosophical and scientific tradition is thus a subject in need of further research, and I must content myself with a view taken from a higher altitude.

5.2 *The Theoretical and the Practical Knowledge-Interests*

I thus offer the following general hypothesis. Within the medieval Hebrew-writing Jewish communities, interest in science derived from two distinct sources, with science practised by two different, although often overlapping, groups. (i) One knowledge-interest was religious-theoretical: those guided by it were animated by a desire to acquire a theoretical understanding of the world, albeit principally those aspects of it that have a bearing on metaphysics or on religious philosophy. This theoretical knowledge-interest underlay the quests of scholars who adhered to the Maimonidean interpretation of Judaism and who studied science as a part of the philosophical curriculum. (ii) The second knowledge-interest was practical: it guided practitioners of disciplines affording useful scientific know-how.

Let us consider these two sources of motivation in turn. I will first enunciate both parts of my hypothesis in general terms, and then consider how it accounts for the shape taken by the different sciences considered above.

(i) The explanation for the observed constant and relatively broad theoretically-motivated interest for science and philosophy among the Jews of Provence is quite obvious. Following the translation into Hebrew of Maimonides's *Guide of the perplexed* about 1204, important segments of the Jewish communities adopted the Maimonidean interpretation of Judaism which implied a positive attitude towards Greco-Arabic philosophy and science. Maimonides's philosophy made the study of science and philosophy into a religious obligation[33] — a stance that from the vantage point of traditional Judaism was nothing short of a revolution. The sciences now became part and parcel of the intellectual equipment of whoever wanted to attain perfection in the Maimonidean spirit. "[I]t is certainly necessary for whoever wishes to achieve human perfection to train himself first in the art of logic, then in the mathematical sciences according to the proper order, then in the natural sciences, and after that in the divine science", Maimonides wrote.[34] As is well known, Maimonides repeatedly affirmed (and in the present context it is immaterial whether or not he himself believed it) that the perfection of the human soul and hence the afterlife depended upon the acquisition of metaphysical knowledge — the knowledge of the separate existents — and the study of the sciences was a necessary step towards that end.[35] Maimonides therefore stressed that only truth counted, and that it was of no significance who discovered or taught it: "Listen to the truth from whoever says it",[36] he urged. For demonstrable, scientific truth, only reason, not tradition, was to determine which knowledge-claims were to be accepted within Judaism and which not. Like many other authors, including Moslems and Christians, Maimonides believed that Greek science and philosophy originated with the Jews, who subsequently lost them during the tribulations of the exile.[37]

Maimonides's view was adopted by his followers. A theoretical knowledge-interest thus underlies the pursuit of the sciences by the Jewish medieval students of philosophy, for whom Maimonides's philosophy provided the basic platform. From the beginning of the thirteenth century onwards, then, there is a considerable theoretically-motivated *appropriation* of science within the Jewry of Provence, which is a direct consequence of the acceptance by large circles within it of the Maimonidean philosophy.

Let us now try to account also for the *limits* of that theoretically-motivated appropriation. To this end, I suggest, we should realize that the Maimonidean framework severely curtailed the scientific subjects that should and could be studied; in addition, it did not allow for the pursuit of science as an end in itself.

The very introduction into Judaism of the philosophical notion that the perfection of the soul hinges on scientific and metaphysical study (and not on the study of the traditional texts) was in continuous need of legitimation. This need was all the more compelling as the study of all 'alien sciences' was continually criticized and condemned by the proponents of a traditionalist interpretation of Judaism (we will come back to this below). Medieval Jewish students of science and philosophy therefore constantly endeavoured to elaborate a *religious* philosophy, that is, to show how Moses's Revelation can be interpreted philosophically: Jewish medieval philosophy is, to take the term of its great historian Julius Guttmann, a *philosophy of Judaism*.[38] This entailed, in Guttmann's words, that "[w]hereas the Moslem Neoplatonists and Aristotelians treat the entire scope of philosophy, most Jewish researchers draw for the problems of general philosophy on the works of their Moslem predecessors, limiting themselves to the investigation of some philosophical-religious questions".[39] The Jewish scholar's primary purpose was to illuminate the revealed truth of the Scriptures or to discuss subjects of theological import such as creation, providence, God's justice, or the reasons of the commandments, drawing to this end on established philosophy. *The Jewish philosophers' interest for science was limited to such subjects as had a theological bearing.*[40]

But even when there was no need for legitimation, that is, when one took for granted the compatibility of the revealed truths and the results of scientific and philosophic inquiry, Maimonides's philosophy discouraged its followers from devoting themselves to science. Not mathematics or physics, but metaphysics was to be the ultimate goal of their investigations. In his famous parable on the degrees of human perfection, Maimonides makes clear his view that those who have studied the mathematical sciences alone, as well as those who have studied the physical sciences too, have reached at most the antechambers of the 'ruler's place'; that is, they have not reached the final end, the *telos*, of man's existence, the proximity to God.[41] The "true human perfection", Maimonides declares, "consists in the acquisition of rational virtues (al-faḍâ'il al-nuṭqîya) — I refer to the conception of intelligibles, which teach true opinions concerning divine things. This is in true reality the ultimate end [of man]; this is what gives the individual true perfection, a perfection belonging to him alone; and it gives him", Maimonides adds, "permanent perdurance [al-baqâ' al-da'im, i.e. afterlife]".[42] Not the apprehension of natural entities, composed of matter and form, then, but only the intellection of separate — divine — entities, is man's true perfection and results in the survival of the soul.[43]

For Maimonides, then, all science was legitimate and desirable only as a propaedeutic to the veritable science, the divine science of metaphysics. The sciences *are* indispensable for one's intellectual *cum* religious perfection, but necessary and legitimate are only those sciences having a *metaphysical* relevance. This, I believe, is the essential reason why the theoretical knowledge-interest generated by the Maimonidean philosophy resulted in the observed selective attention to the various scientific disciplines on the part of Jewish scholars. I will later raise the question why Maimonidean philosophy was only very rarely reinterpreted so as to legitimate the autonomous pursuit of the particular sciences.

(ii) A second source of motivation to study the 'alien sciences' was of course the practical utility of some of them. This practical knowledge-interest needs no lengthy explanation. It applies as we will see notably to logic and astronomy (as well as to medicine) and accounts, I will try to show, for the observed patterns of involvement of the medieval Jewish scholars in these disciplines.

The distinction between the theoretical and the practical knowledge interests, let me point out, is not the historian's construct; these Weberian notions are not imposed on the historical material from without. Rather, these are categories that were on the minds of the historical actors themselves. We find them lucidly described by Baḥya ibn Paqûda in his *Duties of the hearts* (second half of the eleventh century):

> All departments of science, according to their respective subjects, are gates which the Creator has opened to rational beings, so that they may attain to a comprehension of revealed religion (ha-Tora) and of the world (ha-ᶜOlam). But while some sciences satisfy primarily the needs of religion, others are more requisite for the benefit of the world.
>
> The sciences specially required for the affairs of the world are the lowest division — namely the science that deals with the natures and accidental properties of physical substances — and the intermediate division — namely the science of mathematics. These two branches of knowledge afford instruction concerning the secrets of the [physical] world and the uses and benefits to be derived from it, as well as concerning arts and artifices needed for physical and material well-being.
>
> But the science that is needed primarily for revealed religion (ha-Tora) is

the highest science, namely the divine science, which we are under obligation to study in order to understand our revealed religion (*toratenu*) and to
reach up to it. To study it, however, for the sake of worldly advantages is
forbidden to us.[44]

5.3 *How the Two Knowledge-Interests Structured the Appropriation of the Sciences*

Let us now consider how each of these two knowledge-interests, the theoretical
and the practical, shaped and delimited the appropriation of the different sciences
within the medieval Jewish communities. We will examine the main disciplines,
trying to discover what reasons have determined the Jewish scholars' attitudes
towards them.

A. *Logic* was studied for more than one reason. As a propaedeutic discipline it
was an integral part of the scientific-philosophical curriculum and thus was studied both by students of a practical science or art, as for example medicine, and by
students of philosophy whose purpose was theological. Yaqob Anaṭoli, for instance, the translator of Averroes's Middle Commentary on Porphyry's *Isagoge*
(1232), considers logic to be nothing less than "a crucible for refining the silver of
understanding and a furnace for refining the gold of belief". "Everyone who truly
desires to seek God", he claims, "stands in need of the science of logic".

Anaṭoli ascribed to logic also an immediate practical utility. He contended, and
this statement was to be repeated often, that logic was essential in disputations
with Christian scholars: "Logic is indispensable if one is to stand up to opponents
who employ sophistic arguments.... It is evident that without studying the science
of logic none of us will be able to hold the ground against the representatives of
other nations who oppose us."[45]

Logic thus had a double utility, practical and theoretical. It responded to practical needs, experienced by students of the quadrivium in the practice of their art
(astronomy, medicine) and by those who had to engage in theological disputations. It also satisfied a theoretical need, on the part of those who intended to
study philosophy in order to perfect their souls. This multipurpose character of
logic presumably explains the strong interest in this discipline among medieval
Jews, as reflected in the great number of translated treatises and surviving manuscripts. At the same time, the subservient role of logic explains why the scope of
the intensive interest for it remained limited, most of the translations being merely
elementary compendia. Logic did not become an object of investigation in its own
right, as it did, exceptionally, for Gersonides.

B. *Astronomy* poses a particular problem.[46] The causes for the remarkably consistent interest in this discipline are difficult to determine with certainty.
Astronomers themselves typically invoke two motives: the practical importance
of astronomy for the correct determination of the Hebrew calendar (an important
religious need), and the relevance to metaphysics of the study of the noble celestial

bodies. Yet the astronomical material at our disposal (notably tables) apparently affords no independent evidence confirming the astronomers' actual involvement in the determination of the calendar, and one may therefore wonder whether their allegations reflect reality or are merely a rhetoric aiming to legitimate the subject. Similarly, although the study of metaphysics certainly directs attention to the heavens, the astronomy required for that purpose is elementary and certainly does not call for any *practice* of astronomy. Moreover, it is well known that the philosophic world-picture was in fact incompatible with the astronomical one and it is therefore doubtful whether astronomy was studied as a means towards attaining one's eternal happiness.

The issue certainly calls for and deserves further investigation. For the moment let me content myself with noting that during the recurring debates within the Jewish communities over the legitimacy of the study of 'alien sciences', even the opponents of the study of philosophy regularly excluded astronomy (along with medicine) from their projected bans on those studies: this I take to be a telling indication that it was generally perceived both as practically useful and as devoid of potentially 'dangerous' implications. In this context it should also be mentioned that according to Maimonides, the members of the Sanhedrin (that is, of a legislating body) must be experts in astronomy.[47] We may thus perhaps conclude tentatively that although we do not understand exactly the social role of the Jewish astronomers and thus are unsure of the reasons for the massive investment on the part of Jewish scholars in the practice of astronomy, it yet seems that it is somehow connected to practical needs, probably related to the establishment of the calendar. The existence of controversies between Jewish and Christian experts in astronomy over the correct establishment of the calendar seems to confirm this surmise.[48] At any rate, it is indisputable that the — supposed or real — utility of astronomy for religious purposes supplied it with *social legitimation* that was never challenged. Only the supposed existence of a continuous social need for astronomy, one that is independent of the vicissitudes of cultural values (as for example the acceptance of the philosophical interpretation of Judaism), can, I think, explain the ubiquity of interest in astronomy within Jewish communities of all periods and of very divergent cultures.

C. *Mathematics* offers a particularly good opportunity to observe the workings of the two different knowledge-interests. Medieval Jews studied mathematics either because they regarded mathematics as a propaedeutic to metaphysics, preparing the intellect to apprehend abstract truths, or because they needed it since it was a prerequisite for the study of mathematical astronomy. Now both of these considerations converged in giving ample legitimation to the study of Euclid's *Elements*, which indeed was widely and continuously studied. But these justifications implied that the study of a field such as algebra was pointless, indeed harmful. Medieval algebra was construed as a mere 'device', or technique, allowing one to solve equations, and as such it had no philosophical value;[49] nor was it apparently of practical use. From the vantage point of the medieval Jewish scholar, therefore, algebra was simply irrelevant. The well-known twelfth-century Aristotelian philosopher Abraham ibn Daud of Toledo brings the point out eloquently. As one knowledgeable in Arab science and philosophy, he was certainly aware of what algebra was, but considered it valueless, indeed injurious. Among those who spend their time on vanities, thereby depriving their soul of afterlife, he writes, is he who

> consumes his time with number and with strange stories like the following: A man wanted to boil fifteen quarters of new wine so that it be reduced to a third. He boiled it until a quarter thereof departed, whereupon two quarters of the remaining [wine] were spilled; he again boiled it until a quarter vanished in the fire, whereupon two quarters of the rest were spilled. What is the proportion between the [quantity] obtained and the [quantity] sought?[50]

Precisely the same stance is expressed in more general terms by Maimonides. In his highly influential *Treatise of the eight chapters*, Maimonides explicitly cites "questions of arithmetics, the books on conics and on devices [i.e. algebra], and multiplying the questions on geometry and on the [science of] weights" as instances of inquiries that must not be pursued as ends in themselves. They are commendable only inasmuch as they "follow the aim of sharpening the intellect and getting the intellect used to demonstrations ... which is the way through which man achieves the knowledge of the true existence of God".[51] In the *Guide of the perplexed* Maimonides himself provides a fine example for the good use of mathematics: he draws on the demonstrated existence of asymptotes to show that imaginability is not a criterion of existence and he exclaims: "Hear what the mathematical sciences have taught us and how capital are the premises we have obtained from them!"[52]

Both the theoretical and the practical knowledge-interests made Euclid's *Elements* into the mathematical text *par excellence*. Its study, however, *qua* propaedeutic for the study of either metaphysics or astronomy, was not prone to lead students to engage in mathematical investigations, quite the contrary. For once the student had completed his mathematical studies and trained his intellect to apprehend separate entities, he could, at long last, come to the study of physics and then of metaphysics. In the medieval division of sciences the latter disciplines of course occupied a higher rank, and had a higher social prestige, than mathematics. Similar remarks apply to the study of the *Elements* by future astronomers. Therefore, the study of mathematics necessarily remained an intermediary step on the way to the higher realms of metaphysical knowledge or of astronomy.

In short, therefore, the pursuit of mathematics as useful in itself was not an accepted social value: this accounts for the fact that medieval Jewish scholars were not interested in mathematical fields such as algebra, and it also explains why they made almost no contributions of their own even to the mathematical domains they did study; those few and short Hebrew mathematical writings that may be original are for the most part offshoots of work in astronomy. An exception to the rule confirms the above analysis: in the wake of Maimonides's remark on asymptotes referred to above, a number of Jewish scholars tried their hands at limited investigations of this mathematical topic: the religious value conferred upon *this* specific mathematical property made it a legitimate subject of study.[53]

Similar considerations explain the virtual absence of interest among Jewish scholars in optics and music, two disciplines devoid of either practical or theological interest,[54] Gersonides being again a notable exception to the rule.

D. Clearly, the study of metaphysics presupposed that of *physics*. Nonetheless the study of physics motivated by a philosophical-religious interest remained limited to the physics of Averroes. There was no dynamic of research likely to encourage the investigation of new problems. For the purpose at hand, the intellectual means offered by Maimonides and by Averroes seemed sufficient.

In fact, with respect to physics Maimonides advocated an "image of science" — that is, a descriptive and normative view of what science is and should be[55] — which presupposed that contemporary physical science had attained the limits of what could become known to man, thereby presumably contributing to hampering the development of research in this domain. In fact, Maimonides suggests two different, indeed opposite, images of science, according to whether they relate to the physics of (i) the sublunar or (ii) the supralunar realms.

(i) Maimonides believed that "everything that Aristotle has said about all that exists from beneath the sphere of the Moon to the centre of the Earth is indubitably correct, and no one will deviate from it".[56] Indeed, he held that Aristotle reached the "upper limit of knowledge attainable by man".[57] The description of the sublunar world given by Aristotle was sufficient and there was no reason to try to improve upon it. Thus in the introduction to Book Two of the *Guide of the perplexed* he enumerated the twenty-five premises needed for establishing the existence and incorporeality of the deity and affirmed that they are all "demonstrated without there being a doubt as to any point concerning them".[58]

(ii) Concerning the human possibilities of attaining knowledge of the celestial realm, Maimonides is, by contrast, profoundly sceptical: "Regarding all that is in the heavens, man grasps nothing but a small measure of what is mathematical.... [T]he deity alone fully knows the true reality, the nature, the substance, the form, the motions, and the cause of the heavens." The heavens, Maimonides holds, are "too far away from us and too high in place and in rank" for us to know them, so that "to fatigue the minds with notions that cannot be grasped by them and for the grasp of which they have no instrument, is a defect in one's inborn disposition or some sort of temptation". Moses alone, Maimonides stated, attained the knowledge of the celestial realm, namely through divine overflow.[59]

The message carried by the *Guide of the perplexed* was thus unequivocal: whatever could be known by man was known already. Maimonides's philosophy made no allowance for the idea of progress of human knowledge of nature, sub- or supralunar.[60] Maimonides supplied legitimation for the very study of physical science, but it was to be studied only as an already *completed and closed description of the world.* For the perfection of one's soul it was both necessary and sufficient to apply one's intellect to the already available description of the universe. Maimonides makes no allowance for investigations seeking new information on the physical world.[61]

E. The fortunes of *alchemy* in Hebrew pose a distinct problem, in fact something of a puzzle. The disregard of Jewish scholars for alchemy is surprising in view of the already mentioned early and considerable interest they had for questions concerning the sublunar world,[62] the fact that alchemy claimed to have had Jewish 'founding fathers',[63] and the fact that many alchemical writings were attributed to Aristotle. Let me emphasize that the reason for this lack of interest is certainly not a shared denial of transmutation, which some Jewish philosophers upheld and others denied. How, then, is one to account for this phenomenon? One possible explanation is that the Jewish communities, many of whose influential members were involved in economic activities, were apprehensive of the presence of alchemists, who were regarded as counterfeiters of money.[64] Further, alchemy presumably had no contribution to make to the religious philosophy in its Maimonidean version, and spiritual alchemy may even have appeared as a threat to Maimonides's kind of religiosity. Surely, too, alchemical manuscripts were particularly difficult to acquire. Yet to me this striking absence of alchemy from the fields of knowledge cultivated by Jews is still something of an enigma.

We may sum up the thesis so far developed as follows. Within the medieval Jewish communities whose cultural language was Hebrew, Greek-Arabic science attracted interest for two distinct reasons and had two independent sources of legitimation, with the development of the corresponding disciplines following two different, if interfering, dynamics. On the one hand there were physics and kindred areas, which, being closely related to metaphysics, the Maimonideans held to have eudaemonic value: their study was accepted as legitimate only by the followers of Maimonides within the Jewish community, that is, by those who adhered to the notion that eternal bliss in the afterlife is conditional upon metaphysical knowledge. Yet this knowledge-interest nurtured only such subjects as were of relevance to metaphysics, and the students motivated by it excluded from their purview all other disciplines. That thirteenth-century informed critic of philosophy, R. Moses ben Naḥman (Naḥmanides), observed that "[the philosophers] themselves acknowledge that the great utility of [the sciences] comes to the fore when in their investigations they [the philosophers] reach the science which they call 'metaphysics', which is the science of divinity".[65] On the other hand, there were practical disciplines, notably astronomy and logic (and, we may add here, medicine), which did not claim to contribute to the happiness of the soul in the afterlife, but which, by virtue of their real or alleged practical utility, thrived even where traditionalist circles suppressed other 'alien studies'. Some disciplines, notably logic and mathematics, were nurtured by both the theoretical and the practical knowledge-interests.

6. TWO APPARENT EXCEPTIONS TO THE RULE: R. LEVI BEN GERSHOM AND R. ḤASDAI CRESCAS

My thesis can be illustrated by a brief consideration of two apparent exceptions to the rule — those of Gersonides and R. Ḥasdai Crescas. I will begin with the latter, but devote more attention to the former.

Crescas's main work, *Or ha-Shem* ("The light of the Lord"), completed in 1410, contains a penetrating and influential criticism of Aristotle's physics, which has been analysed in full detail by H. A. Wolfson.[66] Yet Crescas by no means belies the thesis that Maimonidean philosophy curtailed the development of science. Crescas was motivated by a desire to undermine Maimonides's interpretation of Judaism and to this effect he sought to *refute* Aristotle. Thus, ironically, the most important contribution to physics by a medieval Jew does not come from a philosopher endeavouring to improve upon received theory, but rather from someone who aimed totally to uproot Greek philosophy and science from within Judaism. This underscores my contention: the adherence to the values of Maimonidean religious philosophy by the philosophers made it difficult for them to transcend it, whence the absence of contributions to physics from those who studied it as part of their Maimonidean program of studies.[67]

Levi ben Gershom, Gersonides (1288–1344), I believe, is the only medieval Jewish thinker who was a follower of Maimonides and yet invested himself in theoretically-motivated scientific research. In fact, as we will see, his own personal version of Maimonidean philosophy gave rise to a commitment to the idea of the dignity and legitimacy of autonomous scientific inquiry into nature. Specifically, I will suggest that Gersonides's attitude to science differed from that of his co-religionists because of his views on the route to the soul's perfection and after-life, and that his case in fact highlights and confirms my general thesis.

Gersonides was certainly one of the greatest medieval scientists, at least in Europe.[68] He devised an original astronomical system and is one of the very few medieval men of science to have made astronomical observations; he invented astronomical instruments to carry out these observations, of which the Jacob's Staff is the best known; he composed mathematical treatises which are in part original; he wrote commentaries upon most of Averroes's scientific works; and it seems he conducted botanical experiments and, at least on paper, invented a sort of proto-microscope designed to magnify the parts of insects too small to be observed with the naked eye.

We must now ask what allowed Gersonides to make science his main focus of interest, something which, according to my account, the Maimonidean philosophy barred the other Jewish scholars from doing.[69] The answer, I believe, lies in two consequential twists Gersonides gave to Maimonides's view of the way to achieve perfection of the soul. First, while accepting the conventional notion that the soul's afterlife depends upon the knowledge of separate entities, Gersonides had it depend specifically on the knowledge of the active intellect. Now Gersonides construes the active intellect (and occasionally even the Godhead) as the world's *nomos*, the plan of the entire natural order. The consequence is that, to a certain extent, knowledge of God is within the reach of man, and that the way to it passes through the acquisition of factual knowledge about the material world: by investigating God's works empirically one acquires knowledge, albeit partial, of the active intellect and of God, thereby securing the afterlife of one's soul. Contrary to Maimonides, then, Gersonides holds that the route to the knowledge of God passes through empirical research rather than through metaphysics.

This brings us to a second point on which Gersonides departed from Maimonides. Gersonides holds that, since a person's acquired intellect is the set of intelligibles he has acquired, that acquired intellect is *individual* and it perseveres in its individuality after death. Most Aristotelian philosophers, it will be remembered, followed Ibn Bâjja's and Averroes's view, which apparently was also shared by Maimonides, that after death the individual's acquired intellect attains conjunction with the active intellect whereby it loses its individuality.[70]

There are thus two innovations in Gersonides's view of how the soul's afterlife is to be acquired. First, the soul's *eudaemonia* is held to depend upon the empirical study of nature, and, second, its afterlife is taken to be *personal*. Let us consider their social implications in turn.

The significance of the first idea is clear: whereas Maimonides's philosophy directed man's desire for knowledge to the transcendent, ascribing to the knowledge of the material world a subordinate value only, Gersonides's philosophy warrants a decidedly inner-worldly attitude. In addition, whereas Maimonides was thoroughly sceptical with respect to the possibility of acquiring any new knowledge germane to man's afterlife, Gersonides is confident in the progress of the sciences of both the sub- and supralunar realms and in the power of the rational soul to attain eternal felicity. Gersonides thus embraces an 'image of science' that is radically different from that advocated by Maimonides. As a result, for Gersonides science "assumed a soteriological role as a bridge towards the knowledge of transcendental reality", to use Professor Heyd's fitting formulation,[71] precisely as was to be the case, a little over three centuries later, with the Protestant thinkers involved in the New Science. And indeed Gersonides adumbrates what was to be on a large scale the attitude towards science in the Protestant tradition: the acquisition of scientific knowledge about the world becomes for him a legitimate end in itself. Science has a religious value, but this value legitimizes it *qua* an independent inquiry: because all knowledge of God's works has a religious significance, Gersonides upholds the legitimacy of the autonomous pursuit of science. For Gersonides, just as for the Protestant men of science, the religious function of science and its autonomy are dialectically related.

The second innovative component of Gersonides's theory of the soul, the view that in the afterlife one's soul survives individually, certainly gave Gersonides the strongest possible motivation to immerse himself in scientific research, on a par with his philosophical reflections which had originally allowed him to establish the way to eternal bliss. On this count Gersonides seems to resemble a number of his illustrious Muslim predecessors, a point to which I will return. Gersonides's great astronomical treatise, significantly an integral part of his philosophical *opus*, bears witness to endless nights passed in observations and calculations. (Gersonides, it should be added, held that astronomy was studied for its cognitive value as a description of reality, and not for a possible practical value. His view that God exercises His providence through influences emanating from the heavenly bodies made the celestial motions into the most noble object of study for anyone wishing to apprehend the *nomos* of God's created world.)

As an apparent exception to the rule, Gersonides's case in fact underscores my general thesis: that Gersonides took to science to an extent unparalleled by any other Jewish scholar is due to his distinctive theory of soul. The contrast between Gersonides and most other Jewish philosophers highlights the role of Maimonidean philosophy in thwarting science and it underscores the role of Gersonides's own theory of soul in fostering interest in scientific knowledge. Gersonides's philosophy gave him a powerful personal motivation to devote his life to science, but he remained a lone rider, in fact an outsider; scientific research *per se* did not acquire social legitimacy even within the 'pro-philosophy camp' of the Jewish community.

At this point, the following question may be raised: Gersonides certainly considered himself a follower of Maimonides,[72] so what was it that initiated his deviation from mainstream Maimonideanism? To this question, I must concede, I have no satisfactory answer. Indeed, the following section, which tries to offer some explanation for the commitment of most Jewish philosophers to the Maimonidean program, makes the problem more acute.

7. THE ABSENCE OF INTEREST IN SCIENCE: A SOCIOLOGICAL PERSPECTIVE

We must now enlarge our perspective. Ascribing the limits of the interest in science among medieval Jews to their adherence to the Maimonidean theology is not a sufficient explanation. For we must ask what made the commitment to the framework of Maimonidean religious philosophy so strong that not even a few scattered individuals freed themselves from it. Moreover, why was Maimonides's philosophy usually accepted in its canonical interpretation with no-one (as far as I know) following Gersonides's reinterpretation of it? What made the consensus over the aims of Maimonides's religious philosophy so compelling? The answer cannot but be sociological, and in what follows I will attempt nothing more than a few remarks indicating where, I think, the answer should be sought.

To begin with, it must be emphasized that Maimonides was not simply one, be it a most distinguished, Jewish philosopher among others, nor even the first among equals. Rather, his *Guide of the perplexed* was invested with the incomparable prestige, indeed the authoritative charisma, of its author, who for medieval Judaism was first and foremost the great codifier of the Jewish Law. This means that while the legal authority of Maimonides in the first place gave Jewish scholars the *social legitimation* to introduce philosophy into Judaism and to study 'alien wisdom' with no practical utility, it at the same time kept that legitimation within the narrow confines of a religious philosophy. Throughout the period with which we are concerned, *the normative world-view of the* Guide *continued to define the limits of what was socially commendable and acceptable within the 'pro-philosophy camp' inside the Jewish community.* Concomitantly, of course, Maimonides's philosophy denied its followers any personal *motivation* to devote themselves to a kind of study that was not conducive to afterlife, specifically to the study of the particular scientific disciplines. In sum, then, even for the followers of Maimonides, there was neither social legitimation nor personal motivation for the pursuit of scientific study other than for a propaedeutic purpose. Within Maimonideanism, science remained the 'handmaiden' of religion in the fullest sense.

Maimonides's philosophy in its canonical interpretation thus exerted a sort of centripetal attraction, hindering the emergence of radical, more naturalistic (Averroist) interpretations of it, as these are, exceptionally, exemplified in the thought of Samuel ibn Tibbon in the thirteenth century and in that of Gersonides in the fourteenth. In addition to this centripetal force, which acted so to speak from within the Maimonidean philosophy itself and was sustained by Maimonides's authority, there were also powerful external social factors that tipped the balance in favour of a conservative interpretation of Maimonideanism.

Medieval Jewish society was a traditional society in the sense that it was, as Professor Jacob Katz has put it, "a society which regards its existence as based upon a common body of knowledge and values handed down from the past".[73] In the absence of territorial concentration and of State authority, in the absence also of a religious hierarchy disposing of means to suppress deviations centrally, the various systems of symbols accepted as meaningful by members of any Jewish community tended to exert centrifugal tendencies. Philosophy, to be sure, was one of these systems of symbols threatening social cohesion, all the more so since it was often perceived as threatening the observance of *halakha* (the Law) itself. This indeed is why the controversies over the legitimacy of philosophy (to which we will presently come) were so virulent: at stake was nothing less than the identity, and therefore the social unity, of the Jewish community. This state of affairs is noted with great perspicacity at the beginning of the fourteenth century by Qalonimos b. Qalonimos, the great translator of scientific works: "Each district", he writes in his *Even Bohan*, "upholds its own persuasion, ... each condemning the other saying: 'I am afraid there is some heresy [in its opinions]. My God is not its God'"; in a word: "*our Gods are as numerous as our towns*".[74]

In this traditional society, the persistent centrifugal tendencies were counterbalanced essentially by concentration on the study of the authoritative texts of Judaism (Bible and Talmud, notably); for the traditionalists, indeed, these were the only legitimate objects of study. The Jewish philosopher, too, had to assume this social role and constantly had to appeal for legitimation to the authoritative texts: there are almost no Jewish philosophers who did not write commentaries on the Bible or other sacred texts. For the philosopher this was a social necessity if he wished to continue to be a part of the Jewish community; at the same time, this was an absolute necessity if social cohesion was to be maintained. In brief: the Jewish scholar was tied to the traditional texts by social bonds, making it scarcely possible for him to dissociate himself from them without leaving his social framework at the same time. In other words: Maimonideanism in its canonical form, that is, as a religious philosophy in which science and philosophy are the 'handmaiden' of religious revelation, marks the limit of what the Jewish community in its entirety (that is, including the traditionalist camp) could tolerate. This, I believe, is a fundamental reason why medieval Jewish philosophy remained essentially a philosophy of Judaism.

A brief look at the controversies over the study of philosophy that lingered on within the Jewish communities of Provence during the entire period with which we are concerned substantiates the above analysis. By highlighting the continued and unceasing pressure exerted on the 'rationalist' camp, that of the followers of Maimonides, by the 'anti-rationalist', traditionalist, circles,[75] it will also help to account for the virtual absence of a more 'radical', scientifically-oriented Maimonideanism, of the kind exemplified by Gersonides.

Less than three decades after the appearance of its Hebrew translation, Maimonides's *Guide of the perplexed* was harshly criticized and condemned by a few proponents of the traditional interpretation of Judaism.[76] In this first "Maimonidean controversy" (as it is known) the rationalists had the upper hand, but this was largely due to the maladroit manoeuvring of their adversaries, as well as, to be sure, to the great prestige of Maimonides as the codifier of the Jewish Law. Indeed, although the traditionalists henceforth had grudgingly to accept the existence of a Maimonidean faction within the Jewish communities, they held the rationalists under constant suspicion and pressure. Thus, even after the first controversy had apparently been brought to an end, the antagonism between the two camps remained endemic.

Over the decades the lines of demarcation shifted subtly, however: when the controversy broke again into the open in 1304–5, the anti-rationalists had assimilated certain Maimonidean positions (notably on the need to interpret allegorically some passages of the Scripture) and they no longer sought to outlaw the *Guide* itself (which, one yet feels, they were still far from cherishing). Instead, they now opposed the study of 'alien' science and of philosophy, that is, of the numerous works that had been translated from Arabic into Hebrew during the preceding decades in the wake of the *Guide*. After long debates and negotiations, a *ḥerem* (ban) was issued on 26 July 1305 against "any member of the community who, being under the age of 25 years, shall study the works of the Greeks on natural science or metaphysics".[77] Excepted from the ban were only works in the practically useful sciences, namely astronomy and medicine.

The ban was apparently not followed by any significant consequences and no one is known to have been persecuted as a result of its proclamation (a fact that in all likelihood has much to do with the expulsion of the Jews from the Kingdom of France by Philippe the Fair in 1306). Nonetheless the point that matters to us here is the very fact that the opposition between rationalists and anti-rationalists, between supporters of the study of 'Greek wisdom' and its opponents, persisted throughout the period with which we are concerned (and indeed to this very day). This continued struggle over the limits of the use of reason in the interpretation of the world and of the Scriptures, and hence over the true interpretation of Judaism, compelled the rationalists to remain constantly on their guard: an imprudent move, an excess of allegorization, was liable to lead to considerable social pressure.[78] The Jewish rationalist scholar thus had to exercise his reason within strict boundaries and under constant supervision. Although theoretically Maimonidean philosophy could be developed in different directions — including one of extreme rationalism — the concrete social ecology within which it actually evolved in southern France between the thirteenth and the fifteenth century favoured the unfolding of a conservative Maimonideanism, a Maimonideanism, that is, that allocated the use of reason a strictly delimited space and ruled out its autonomous exercise.

That indeed the study of science and philosophy was curtailed by the pressure of the traditionalist camp was perceptively noted in 1394 by a 'participant observer', Leon Joseph of Carcassonne, a Jewish physician whose discontent with the state of science — particularly of medicine — among the Jews of his day was enough to lead him to study at the University of Montpellier. In the introduction to one of his translations of medical books from Latin into Hebrew he wrote:

> [The sciences (ha-ḥokhmot)] contain rational matters. These are as far removed from what our [scil. Jewish] masses are familiar with as east is from west. Even still farther are [the sciences] removed from the foundations of the Torah and of religious faith. Therefore some of our scholars avoided studying and knowing them and understanding their contents.
>
> Even those few whom God has granted the grace of studying [the sciences] ... had to do this in secret and in hiding.... This is due to the fact that they were apprehensive of what the mass of the ignoramuses would say. For they are very few and the others are many.
>
> To this [apprehension of the masses] is added the fear of some of the Talmudists [toraniyim], who proscribed the superfluous from our souls and who have the upper hand over those who are engaged in rational study: not because of their [intellectual] power or because of the breadth of their knowledge, but only because of the power of their arms and their numerous subterfuges. For the above-mentioned masses obey them, *believing that the sciences and those who study them exclude themselves from the community of those who adhere to the Torah.*[79]

The limits of the consensus within the Jewish community were largely drawn and socially enforced by the opponents of the introduction of philosophy into Judaism.

Other social factors contributed to maintaining the hegemony of a domesticated Maimonideanism. One crucial factor was the absence of organized teaching of philosophy within the Jewish communities. Whereas the traditional Jewish subjects were taught in southern France in flourishing *yeshivoth* at least from the beginning of the tenth century,[80] philosophy and science seem to have been taught only privately.[81] To see how important were the consequences of this absence of organized study of science, a glance at the contemporary European universities is sufficient. For one thing, among the Jewish scholars there are no lines of descent of masters and students, as the universities, and also the *yeshivoth*, naturally produced; consequently, there was little or no continuity in the transmission of knowledge and in the pursuit of inquiries. "Perchance once in my life-time ... the Lord may create for me a worthy disciple to whom I could bequeath my weak and poor secrets", Joseph Ibn Kaspi cries out in the fourteenth century.[82] More important, the medieval university institutionalized certain modes of discussion among scholars and students which, through their very structure, were conducive towards the emergence of a new type of knowledge: discussions of the *potentia Dei absoluta et ordinata* or of what is and is not possible *secundum imaginationem*, paved the way to a kind of inquiry emancipated from the bonds of theological or philosophical postulates.[83] It is precisely this emancipation that is wanting in the reflections of the Jewish philosophers and which kept them within the confines of the received doctrines of their religious philosophy. In other words: by its very structure, the medieval university gave rise to a new type of inquiry and to a new social role — that of the 'professional' scholar devoted to the autonomous search of truth;[84] owing to the absence of analogous structures in the Jewish community, a similar process did not take place there.

That the absence of institutionalized teaching of the sciences was one of the factors that hampered the scientific activity of Jewish scholars is again perceptively noted by the physician and translator Leon Joseph of Carcassonne. The state of scientific knowledge among the Jews is inferior to that among the Christians of his time, Leon says, a fact that is *inter alia* due to the absence of public teaching of the sciences, as this is practised in the universities: of 'the rationalists', that is, proponents of the study of philosophy, he says that they

> were not allowed to teach science in the market-places and in the streets, nor to discuss [the sciences] and to show their reasonableness, nor to form a public *yeshiva*. Thus they could not have the truth emerge perfectly, for *truth cannot be known but through its opposites.*[85]

As against this,

> [the Christians'] exchanges on these sciences is unceasing, and they miss nothing of what is worth investigating. Indeed, they leave out nothing when

it is a question of debating the truth and even the falsehood [of a proposition]. Through their rigorous scrutinizing questions and answers by way of disputation (*wikuah*) and by explaining everything through two contrary [opinions], they have the truth emerge from the centre [of the contradiction] as a lily among the thorns.[86]

Indeed, Leon adds, this method of (scholastic) disputation is precisely the one exercised by the ancient Jewish sages of the Talmud, but unfortunately the afflictions suffered by the Jews during the exile caused them to forget it.[87]

But how are we to account for this absence of institutionalized teaching of philosophy? After all, organized, systematic teaching was a traditional institution within Judaism, and we should ask why it was not extended to the new type of knowledge.

One important reason for the absence of organized tuition of the 'alien sciences' is again to be sought within Maimonidean philosophy itself.[88] Maimonides's image of science and philosophy is that of an *esoteric knowledge*, that is, of a knowledge to which access is selective and restricted. Maimonides famously held that science and philosophy are reserved to a tiny élite, and are not for the masses. He identifies (Aristotelian) physics with what in the Mishnah is called *maaseh bereshit* (the work of the beginning, that is, of creation), and metaphysics with *maaseh merkavah* (the work of the Divine Chariot). Now these traditional terms denoted two mystical and esoteric domains of learning, whose study was perilous and whose public teaching was forbidden.[89] By identifying physics and metaphysics with the *maaseh bereshit* and *maaseh merkavah* of the ancient tradition, Maimonides hit two birds with a single stone: while conferring legitimacy upon the two 'alien' disciplines, he at the same time severely restricted access to them. Thus, in his *Mishneh Torah*, that is, in a binding code of law, Maimonides explicitly says that *maaseh merkavah* must not be taught even to a single individual (one may at most communicate some rudiments of it to a wise man), whereas the *maaseh bereshit* may be taught, but to individuals only.[90] Thus, although access to Aristotelian physics is construed as less restricted than to the nobler science of metaphysics, still there was a formal interdiction against teaching it publicly. Maimonidean philosophy at the same time legitimized the study of the 'alien sciences' and imposed a taboo on their being taught publicly. Presumably this formal interdiction goes a long way to explain why the study of science and philosophy by Jews was done only on an individual basis.

A second, complementary, factor limiting the development of a public teaching of the sciences is, of course, the pressure exerted by the traditionalist camp. The Jewish community tolerated the activity of philosophers as individuals, but frustrated the institutionalization of philosophical education: the public teaching of philosophy would probably have gone beyond the limits of what could be tolerated by those who were apprehensive of the social consequences of the Maimonidean interpretation of Judaism. The above quotation from the pen of Leon Joseph confirms that indeed the rationalists "were *not allowed* to teach science in the market-places and in the streets".[91] Since Maimonides had taught that philosophy should not be taught publicly, there can be little wonder that 'alien wisdom' was at best the subject of private teaching only.

We may conclude that the social structure of the Jewish communities of Provence, specifically the ongoing opposition between two camps of which one was Maimonidean while the other remained opposed to the study of 'Greek wisdom', largely explains why there developed little autonomous scientific research in disciplines other than those that were considered useful (astronomy and medicine). Maimonidean philosophy itself tended to curtail interest in science for its own sake; since, in addition, the study of science was under the continuous scrutiny and suspicion of the traditionalists, any tendencies to develop an extreme rationalist brand of Maimonideanism were presumably nipped in the bud. Within the medieval Jewish communities of Provence there was no place for scholars pursuing an autonomous scientific research guided by a theoretical knowledge-interest.[92]

The well-known fourteenth-century convert Alfonso de Valladolid, formerly Abner of Burgos, seems to illustrate this last point. "Since my youth and until my old age", he writes in his philosophical-mathematical work *Meyashsher ᶜAqov* ("*Straightening the curved*"), "I begged of God ... one single thing, namely to know whether it is possible to find a rectilinear surface equal to the surface of the circle, according to the truth and not approximately".[93] This is precisely the kind of motive for which one looks in vain in the writings of medieval Jewish scholars. Whatever other reasons may have led Abner to convert, it seems to me that it is not mere chance that this scholar, nurtured by a strong desire to acquire purely theoretical knowledge, followed an intellectual itinerary that ended outside the Jewish community: medieval Jewish scholars as a rule valued theological discussions or talmudic questions higher than the demonstration of a mathematical theorem.[94]

8. A CONCLUDING REMARK: THEORIES OF AFTERLIFE AND THE PROGRESS OF SCIENCE — A COMPARATIVE VIEW (ISLAM, JUDAISM, CHRISTIANITY)

The circumstances that structured the appropriation of science within the medieval Hebrew-writing Jewish communities can usefully be compared with those that have conditioned, first, the rise and decline of Arab science and, second, the scientific revolution of the seventeenth century. I limit myself to a very brief remark.

According to a suggestive analysis by Professor A. I. Sabra, the rise of Arab science between the eighth and the twelfth centuries is due to scholars such as al-Fârâbî, Ibn Sînâ, Ibn al-Haytham, al-Bîrûnî, and Ibn Rushd, who embraced a philosophy holding that the route to the eternal felicity of one's soul passes *via* the autonomous search for truth.[95] In the terms introduced above they are those whose quests were guided by the theoretical knowledge-interest; they constitute a sociological type which Sabra calls "philosopher-scientist". Arab science declined when their philosophy gradually became replaced by a view, propagated notably by al-Ghazâlî, according to which felicity is attained through theological research only, science being legitimate only if it is immediately useful. The social role of the man of science changed: the 'philosopher-scientist' was replaced by the 'jurist-scientist', whose main concern was jurisprudence and not philosophical-scientific truth about the world and whose research was guided by the practical knowledge-interest.

From a sociological point of view, the conditions that led to the decline of Arab science bear strong resemblance to those that have *permanently* prevailed within the Jewish communities: there, indeed, the social role of the scholar was continually centred around the study and teaching of the Bible; science was studied for its utility (practical or propaedeutic), rather than as an end in itself. The only true scientist (professional astronomers excepted) among the Hebrew-writing philosophers is in fact Gersonides and it is significant that he and his Muslim counterparts not only share the notion of autonomous scientific research, but also ground it similarly in the view that felicity can be attained through scientific and philosophical search for truth only.

This parallelism acquires further historical significance in light of the 'Merton thesis' on the social conditions underlying the scientific revolution of the seventeenth century.[96] As mentioned at the beginning of this paper, Merton has argued that the institutionalization of the idea of autonomous scientific research (that is, its acceptance as a legitimate, indeed worthy social practice), which is one of the factors that have greatly contributed to the emergence of the New Science, was favoured by Protestant theologies teaching that salvation is to be achieved through the acquisition of knowledge of the physical world. The empirical study of nature was taken to reveal God's power and providence and as such it assumed a theological significance. Moreover, as Heyd has argued, that very theological significance of the research into nature gave science a (relatively) autonomous status as an independent sphere of knowledge. There thus is a striking analogy between the factors which in our three cases led to the rise of scientific activity: the great Muslim scientists, Gersonides, and the Protestant scientists of the seventeenth century were all committed to autonomous scientific quest as the indispensable 'soteriological bridge'. Their respective theologies both legitimized natural science as an end in itself and supplied a forceful motivation to engage in it. This is precisely the kind of legitimation and motivation that was proscribed by the Maimonidean theology dominant in the philosophical circles within the Jewish medieval communities.

9. A LAST CAVEAT

Before concluding, let me add a caveat. First, my suggestion here was that medieval Jews were interested in science only inasmuch as it was either propaedeutically or practically useful and that this was so because those who were at all interested in science did so as followers of Maimonides, who valued theology higher than the individual sciences. Now in a scientistic age like ours, in which science is rightly ascribed a high moral value, this statement may all too easily be taken to carry an implicit anti-Maimonidean or even 'anti-religious' value-judgement. This would be a gross and regrettable misunderstanding. Gersonides in no sense less 'religious' than any other Jewish philosopher, and he certainly considered himself a disciple of Maimonides — indeed he copied for himself the entire *Guide of the perplexed*![97] Yet through his personal reinterpretation of Maimonides, scientific research became for him invested with religious significance (a 'soteriological bridge', in Heyd's terminology), so that his religious commitment — he wrote commentaries on the greater part of the Bible — and the commitment to scientific search for truth could reinforce, rather than contradict, one another. Much the same, as just mentioned, holds of many of the seventeenth-century Protestant scientists, of whom Robert Boyle is an excellent example. Thus, my thesis by no means suggests any inherent irreconcilable opposition of science on the one hand and the Jewish religion, or even Judaism in its Maimonidean interpretation, on the other. Rather, my suggestion is that between the thirteenth and the fifteenth century Maimonidean philosophy (and the social structures sustaining it) set a low value on science *per se*, channelling talent and originality elsewhere. Indeed, the greatest manifestations of creativity by medieval Jews are to be found, not in the writings of the philosophers, but rather in those of the kabbalists and the talmudists.

ACKNOWLEDGEMENTS

For helpful comments on drafts of this paper I am much indebted to the following colleagues and friends: Amos Funkenstein, Ruth Glasner, Menachem Kellner, Y. Tzvi Langermann, Tony Lévy, David C. Lindberg, Jean-Pierre Rothschild, David Ruderman, and Josef Stern. My special thanks go to Bernard R. Goldstein of the University of Pittsburgh who patiently and unflaggingly read successive drafts of this paper and never tired of critically reacting to them and of making constructive suggestions.

An abridged version of this paper was presented at the conference on "Science and theology in medieval Islam, Judaism, and Christendom" (University of Madison, Wisconsin, 15–17 April, 1993): I am grateful to the organizers, David C. Lindberg and A. I. Sabra, for the invitation to participate at this meeting. The ideas contained in the paper had been presented orally on a number of previous occasions, some of which were followed by preliminary short expositions: "The place of science in medieval Jewish communities" (in Hebrew), *Zemanim* (Tel Aviv), no. 42 (Summer 1992), 40–51; "The role of science in the medieval Jewish culture of Provence: A sociological view", in *La ciencia en las España medieval*, ed. by Lola Ferre *et al.* (Granada, 1992), 127–44; "The place of science in medieval Hebrew-writing Jewish communities: A sociological perspective", in *Rashi, 1040–1990: Hommage à Ephraïm E. Urbach*, ed. by Gabrielle Sed-Rajna (Paris, 1993), 599–613. My full-length study in French: "Les sciences dans les communautés juives médiévales de Provence: Leur appropriation, leur rôle", *Revue des études juives*, clii (1993), 29–136, contains much more factual information than I could incorporate in the present paper, in which, by contrast, the analysis is carried farther than in the French one, which was written in 1989/90.

The research underlying this paper was generously supported by an individual grant from the Memorial Foundation for Jewish Culture during the years 1991/92 and 1992/93, for which I am most grateful. I also express my gratitude to the Sidney M. Edelstein Center for the History and Philosophy of Science, Technology and Medicine at the Hebrew University, Jerusalem, for the material facilities it regularly puts at my disposal during my stays in Jerusalem.

REFERENCES

1. Robert K. Merton, "Science and the social order" (1938), reprinted in his *The sociology of science: Theoretical and empirical perspectives*, ed. and with an introduction by Norman W. Storer (Chicago, 1973), 254–66, p. 254.

2. Joseph Ben-David, "Puritanism and modern science: A study in the continuity and coherence of sociological research", in *Comparative social dynamics: Essays in honor of S. N. Eisenstadt*, ed. by E. Cohen, M. Lissak and U. Almagor (Boulder, Col. and London, 1985), 207–23, reprinted in his *Scientific growth: Selected essays on the social organization and ethos of science*, ed. by Gad Freudenthal (Berkeley, 1991), 343–60.

3. A particularly extreme example for this approach is provided by Felix Klein-Franke, *Vorlesungen über die Medizin im Islam* (*Sudhoffs Archiv*, Beiheft 23; Wiesbaden, 1982). *Cf.* my review in *History and philosophy of the life sciences*, ix (1987), 119–22.

4. "Höret nicht auf ihre [the physicists'] Worte, sondern haltet Euch an ihre Taten", Albert Einstein, "Zur Methode der theoretischen Physik", in his *Mein Weltbild* (Amsterdam, 1934), 176–87, p. 176; English translation: *On the method of theoretical physics* (The Herbert Spencer Lecture delivered at Oxford 10 June 1933; Oxford, 1933), 5 (translation modified).

5. Michael Heyd, "The emergence of modern science as an autonomous world of knowledge in the Protestant tradition of the seventeenth century", in *Cultural traditions and worlds of knowledge: Explorations in the sociology of knowledge*, ed. by S. N. Eisenstadt and I. Friedrich Silber (*Knowledge and society: Studies in the sociology of culture past and present*, vii; Greenwich, Con. and London, 1988), 165–79.

6. Heyd refers to a situation where science has positive relevance to central cultural and religious concerns as one in which science possesses 'positive autonomy'; 'negative autonomy', by contrast, means simply that science is not subjected to doctrinal control. *Cf. ibid.*, 166, 171, 173.

7. Isadore Twersky holds that Talmudism is the Jewish intellectual activity *par excellence* and states that the "hallmark of Judaism is halakhocentrism". See his "Religion and law", in *Religion in a religious age*, ed. by S. D. Goitein (Cambridge, Mass., 1974), 69–82; reprinted in Twersky, *Studies in Jewish law and philosophy* (New York, 1982), 203–16. *Halakhah* is the Jewish Law, and the term 'halakhocentrism' is intended to capture a double-faceted state of affairs: (i) the behaviour of Jews until the onset of the modern period was regulated by the *halakhah*; and (ii) in some communities, the intellectual activity of scholars consisted mainly in the interpretation and elaboration of the Talmud and the Law. Twersky's

characterization certainly holds of medieval Judaism in e.g. northern Europe (*ashkenaz*), but it is more problematic with respect to Spain and southern France, which is the region at the centre of our interest in this paper.

8. Much source material can be found in Dov Rafal, *Shevaᶜ ha-ḥokhmot: Ha-wikuaḥ ᶜal limudey ḥol ba-yahadut* (Jerusalem, 1990).

9. *She'elot u-teshuvot le-ha-rav Rabenu Asher z.l.* (Jerusalem, 1981), §55, p. 53ᵛᵃ.

10. My concern here is with those instances and periods in which science was accepted within Judaism and the above remarks must by no means be understood as an attempt to establish a typology of attitudes towards science within Judaism. Between the extremes — a full acceptance and an outright rejection — there are many intermediary positions. One of them, that of the outstanding thirteenth-century thinker and kabbalist Naḥmanides, has been sensitively analysed in: Y. Tzvi Langermann, "Acceptance and devaluation: Naḥmanides's attitude toward science", *Jewish thought and philosophy*, i (1992), 223–45.

11. *Cf.* Abraham S. Halkin, "The Judeo-Islamic age", in *Great ages and ideas of the Jewish people*, ed. by L. W. Schwarz (New York, 1956), 213–63; *idem*, "Judeo-Arabic literature", *Encyclopedia Judaica* (16 vols, Jerusalem, 1972), x, 410–23.

12. For the notion of appropriation *cf.* A. I. Sabra, "The appropriation and subsequent naturalization of Greek science in medieval Islam: A preliminary statement", *History of science*, xxv (1987), 223–43; Roshdi Rashed, "Problems of the transmission of Greek scientific thought into Arabic: Examples from mathematics and optics", *History of science*, xxvii (1989), 199–209.

13. *Cf.* however the material gathered in Harry Friedenwald, *The Jews and medicine: Essays* (Baltimore, 1944). One looks forward to the publication of Joseph Shatzmiller's *Jewish doctors in the Middle Ages* (Berkeley, forthcoming).

14. For reasons of space, the following factual account is reduced to a minimum, as are also the references to primary sources and to the secondary literature. Much of the information missing here can be found in my paper, "Les sciences dans les communautés juives médiévales de Provence: Leur appropriation, leur rôle", *Revue des études juives*, clii (1993), 29–136. The present paper largely draws on it for the facts, but goes beyond it in their interpretation.

15. *Cf.* B. Z. Benedict, "Caractères originaux de la science rabbinique en Languedoc", in *Juifs et judaïsme de Languedoc*, ed. by M.-H. Vicaire and B. Blumenkranz (Toulouse, 1977), 159–72; *idem, Merkaz ha-torah bi-Provans* (Jerusalem, 1985).

16. The best account is still Moritz Steinschneider's monumental *Die hebraeischen Übersetzungen des Mittelalters und die Juden als Dolmetscher* (Berlin, 1893).

17. Isadore Twersky, "Aspects of the social and cultural history of Provençal Jewry", *Journal of world history*, xi (1968), 185–207, reprinted in Twersky, *Studies* (ref. 7), 180–202; Joseph Shatzmiller, "Rationalisme et orthodoxie religieuse chez les Juifs provençaux au commencement du XIVᵉ siècle", *Provence historique*, xxii (1972), 261–86.

18. For an overview and bibliography *cf.* H. H. Ben-Sasson, "Maimonidean controversy", *Encyclopedia Judaica* (ref. 11), xi, 745–54.

19. *Cf.* Harry A. Wolfson, "Plan for the publication of a *Corpus commentariorum Averrois in Aristotelem*", *Speculum*, xxxvi (1961), 88–104, reprinted in his *Studies in the history of philosophy and religion*, ed. by I. Twersky and G. H. Williams (Cambridge, Mass., 1973), i, 430–44.

20. *Cf.* H. J. Drossaart Lulofs and E. L. J. Poortman (eds and transl.), *Nicolaus Damascenus, De plantis. Five translations* (Amsterdam, 1989), 347–463.

21. Colette Sirat, "Les manuscrits en caractères hébraïques. Réalités d'hier et histoire

d'aujourd'hui", *Scrittura e civiltà*, x (1986), 239–88. Madame Sirat (*ibid.*, 263–72) surmises that the ratio of the number of Hebrew manuscripts that have existed to that of the manuscripts that are extant today is about 5:100.

22. For an exhaustive overview *cf.* S. Rosenberg, "Logic and ontology in Jewish philosophy in the 14th century" (Ph.D. dissertation, Hebrew University of Jerusalem, 1973; in Hebrew).

23. Bernard R. Goldstein, "The medieval Hebrew tradition in astronomy", *Journal of the American Oriental Society*, lxxxv (1965), 145–8; *idem*, "The role of science in the Jewish community in fourteenth-century France", in *Machaut's world: Science and art in the fourteenth century*, ed. by M. Pelner Cosman and B. Chandler (*Annals of the New York Academy of Sciences*, xxxiv; New York, 1978), 39–49; *idem*, "Scientific traditions in late medieval Jewish communities", in *Les juifs au regard de l'histoire: Mélanges en l'honneur de Bernard Blumenkranz*, ed. by G. Dahan (Paris, 1985), 235–47; *idem*, "Descriptions of astronomical instruments in Hebrew", in *From deferent to equant: A volume of studies in the history of science in the ancient and medieval Near East in honor of E. S. Kennedy*, ed. by David A. King and George Saliba (*Annals of the New York Academy of Sciences*, l; New York, 1987), 105–41.

24. For the literature *cf.* Moritz Steinschneider, *Mathematik bei den Juden*, ed. by Adeline Goldberg (Hildesheim, 1964); Gad B. Sarfatti, *Mathematical terminology in Hebrew scientific literature of the Middle Ages* (in Hebrew) (Jerusalem, 1968).

25. All Hebrew texts relating to music have been collected in I. Adler, *Hebrew writings concerning music in manuscripts and printed books from the geonic times up to 1800* (*International inventory of musical sources / Répertoire international des sources musicales*, vol. B IX²; Munich, 1975). *Cf.* also A. Sendrey, *Bibliography of Jewish music* (New York, 1951), 56–62. Gersonides is again an exception to this generalization.

26. *Cf.* however Fritz Zimmermann's conjecture in his "Philoponus' impetus theory in the Arabic tradition", in *Philoponus and the rejection of Aristotelian science*, ed. by Richard Sorabji (London, 1987), 121–9, pp. 128–9.

27. Steinschneider, *Die hebraeischen Übersetzungen* (ref. 16), 273; Gerschom Scholem, "Alchemie und Kabbala", *Eranos Jahrbuch*, xlvi (1977), 1–96 (replaces the article with the same title in *Monatsschrift für Geschichte und Wissenschaft des Judentums*, lxix (1925), 13–30, 95–110, 371–4). *Cf.* also B. Suler, "Alchemy", *Encyclopedia Judaica* (ref. 11), ii, 542–9.

28. *Cf.* S. D. Goitein's pertinent observations in his *A Mediterranean society*, ii: *The community* (Berkeley, 1971), 171–3 and 192–5, which apply to the Jewish communities in Europe too.

29. Rashed, "Problems of the transmission" (ref. 12).

30. *Cf.* Steinschneider, *Die hebraeischen Übersetzungen* (ref. 16), "Allgemeines", p. XVI; J. Shatzmiller, "Livres médicaux et éducation médicale: A propos d'un contrat de Marseille en 1316", *Medieval studies*, xlii (1980), 463–70, pp. 468–9.

31. Aviezer Ravitzky, "Aristotle's *Meteorologica* and the Maimonidean exegesis of creation" (in Hebrew), in *Shlomo Pines jubilee volume on the occasion of his eightieth birthday*, part II (*Jerusalem studies in Jewish thought*, ix; Jerusalem, 1990), 225–50.

32. Another fine study, which however is mostly biographical and contains few clues concerning the motivations of its object, is Lawrence V. Berman, "Greek into Hebrew: Samuel ben Judah of Marseilles, fourteenth-century philosopher and translator", in *Jewish medieval and Renaissance studies*, ed. by A. Altmann (Cambridge, Mass., 1967), 289–320. Much information is contained also in J.-P. Rothschild, "Motivations et méthodes des traductions en hébreu du milieu du XIIᵉ à la fin du XVᵉ siècle", in *Traduction et traducteurs au moyen âge*, ed. by G. Contamine (Colloque international du CNRS, IRHT, 26–28 mai 1986; Paris, 1989), 279–302.

33. Herbert A. Davidson, "The study of philosophy as a religious obligation", in *Religion in a religious age* (ref. 7), 53–68.

34. Moses Maimonides, *Guide of the perplexed*, I.34; quoted after the translation by S. Pines (Chicago, 1963), 75. In what follows, quotations from the *Guide* will be from Pines's translation and will be referred to by part, chapter, and page number in the translation.

35. The late Shlomo Pines has argued that between writing the *Mishneh Torah* and the *Guide of the perplexed* Maimonides read al–Fârâbî's (lost) Commentary on Aristotle's *Nicomachean ethics* and thereupon dramatically changed his views on man's capacity to attain metaphysical knowledge and, hence, afterlife: after this "Copernican revolution", Pines suggested, Maimonides ceased to believe that immortality of the intellect was possible and henceforth considered man's perfection to be political (only). *Cf.* Shlomo Pines, "Le discours théologico-philosophique dans les œuvres halachiques de Maïmonide comparé avec celui du *Guide des égarés*", *Délivrance et fidélité, Maïmonide: Textes du colloque tenu à l'Unesco en décembre 1985 à l'occasion du 850ᵉ anniversaire du philosophe* (Toulouse, 1986), 119–24. In the present context it is also of no bearing whether Maimonides held that intellectual perfection, inasmuch as it is possible, can be attained by anyone or only by those among the Jews who have accomplished legal (Talmudic) studies. For this view *cf.* Menachem Kellner, *Maimonides on human perfection* (Atlanta, 1990).

36. Maimonides, *Eight chapters*, Introduction. This dictum was often rehearsed by followers of Maimonides: *cf.* Twersky, "Aspects" (ref. 17), 190 (with n. 20); Shatzmiller, "Rationalisme" (ref. 17), 277; R. Jospe, *What is Jewish philosophy?* (Tel Aviv, 1988), 16–20. On Maimonides's attitude to 'alien wisdom' *cf.* also Isadore Twersky, *Introduction to the Code of Maimonides* (New Haven and London, 1980), 215–19.

37. N. Roth, "The 'theft of philosophy' by the Greeks from the Jews", *Classical folia*, xxxii (1978), 53–67.

38. This idea was elaborated by Julius Guttmann mainly in his classical book *Philosophie des Judentums* (Munich, 1933); *cf.* notably pp. 9–11 (English translation: *Philosophies of Judaism* (Philadelphia, 1964), 3–5).

39. *Ibid.*, 63 ff. (English translation: 55).

40. This has been also noted by the late Georges Vajda: "en dernière analyse", he wrote, "la connaissance est tournée vers Dieu, la connaissance de la nature étant seulement un degré préparatoire de la connaissance métaphysique dont l'objet propre et suprême est la divinité". Georges Vajda, *Introduction à la pensée juive du moyen âge* (Paris, 1947), 143.

41. Maimonides, *Guide*, III:51:619.

42. Maimonides, *Guide*, III:54:635.

43. *Cf.* Zev Harvey, "R. Ḥasdai Crescas and his criticism of philosophical felicity" (in Hebrew), *Proceedings of the Sixth World Congress of Jewish Studies* (Jerusalem, 1977), iii, 143–9. As is well-known, Maimonides paradoxically claims that the separate entities, whose knowledge alone leads to perfection of the soul, are in fact unknowable. *Cf.* e.g. S. Pines, "The limitations of human knowledge according to Al-Farabi, Ibn Badja, and Maimonides", in *Studies in medieval Jewish history and literature*, ed. by I. Twersky (Cambridge, Mass., 1979), 82–109, and n. 35 above. *Cf.* also Alexander Altmann, "Maimonides' 'four perfections'", reprinted in his *Essays in Jewish intellectual history* (Hanover and London, 1981), 65–76, and Kellner, *Maimonides on human perfection* (ref. 35).

44. *Ḥovot ha-levavot*, Introduction; translation (modified) quoted from: R. Bachya ben Joseph ibn Paquda, *Duties of the heart*, translated from the Arabic into Hebrew by R. Yehuda ibn Tibbon, with an English translation by Moses Hyamson (1925–1947) (Jerusalem and New York, 1970), i, 14–17.

45. Quoted from Herbert A. Davidson (transl.), *Averroes, Middle commentary on Porphyry's*

'Isagoge' (Cambridge, Mass., 1969), 3–4. This idea is repeated by later scholars, as e.g. Profiat Duran (known as "Efodi"); cf. Maase Efod, Einleitung in das Studium und Grammatik der hebräischen Sprache von Profiat Duran, ed. by Jonathan Friedländer and Jakob Kohn (Vienna, 1865), Hebrew part, p. 15. Cf. also Daniel J. Lasker, Jewish philosophical polemics against Christianity in the Middle Ages (New York, 1977).

46. I acknowledge with gratitude numerous enlightening conversations with Bernard R. Goldstein (Pittsburgh) on the subject of the following paragraph. Needless to say, Professor Goldstein does not bear responsibility for the opinions expressed here.

47. Maimonides, Mishneh Tora, Hilkhot Sanhedrin 2:1.

48. Isaac Israeli, Sefer yesod olam, ed. by B. Goldberg (2 vols, Berlin, 1846, 1848), ii, 36ᵃ ff., reports a controversy he had in 1334 over the correct date of the Passover; his opponent was apparently the well-known convert Abner of Burgos (Alfonso of Valladolid).

49. This stance goes back to al-Fârâbî; cf. e.g. Harry A. Wolfson, "The classification of sciences in medieval Jewish philosophy" (1925), reprinted in his Studies (ref. 19), i, 493–545, esp. nn. 157, 160.

50. R. Abraham Ibn Daud, Emunah ramah, II, Introduction; text in: Das Buch Emunah Rama, ed. and transl. by S. Weil (Frankfurt, 1852), 45 (translation: 57 f.); The exalted faith, translated with commentary by Norbert M. Samuelson, [Hebrew] translation edited by Gershon Weiss (Cranbury, N.J. and Rutherford, 1986), 123a9 ff. It is noteworthy that much the same argument is ascribed by Ibn Khaldûn to the ninth-century physician Abû Bakr al-Râzî: "As for mathematics, I confess that I have only studied this subject to the extent that was absolutely indispensable, not wasting my time on tricks and refinements.... I make [my excuse] boldly on the grounds that what I have done is the right course, rather than the way chosen by the so-called 'philosophers' who devote their whole lives to indulging in geometrical superfluities." Quoted (without reference) in J. L. Berggren, "Islamic acquisition of the foreign sciences: A cultural perspective", The American journal of Islamic social science, ix (1992), 310–24, p. 315. (I am grateful to Prof. B. R. Goldstein for having brought this reference to my attention.)

51. Maimonides, Eight chapters, chap. 5.

52. Maimonides, Guide, I:73:210.

53. Cf. Gad Freudenthal, "Maimonides' Guide of the perplexed and the transmission of the mathematical tract 'On two asymptotic lines' in the Arabic, Latin, and Hebrew medieval traditions", Vivarium, xxvi (1988), 113–40; Tony Lévy, "L'étude des sections coniques dans la tradition médiévale hébraïque: Ses relations avec les traditions arabe et latine", Revue d'histoire des sciences, xlii (1989), 193–239.

54. In addition, Maimonides expressed an explicit negative attitude towards music, which presumably had an incidence on Jewish scholars' attitude towards the study of the theory of music. Cf. Henry George Farmer, "Maimonides listening to music", Journal of the Royal Asiatic Society, 1933, 867–84; Boaz Cohen, "The responsum of Maimonides concerning music", Jewish music journal (New York), ii/2 (May-June 1935), 3–7. Maimonides allows however the use of music for therapeutic purposes. Cf. Shlomo Marcus, "Maimonides on music — in particular its medical use" (in Hebrew), Koroth, v (1972), 819–22 (English summary on pp. CXL–CXLI).

55. This concept has been put forward by Yehuda Elkana in his "A programmatic attempt at an anthropology of knowledge", in Science and cultures, ed. by E. Mendelsohn and Y. Elkana (The sociology of the sciences, v; Dordrecht, 1981), 1–76.

56. Maimonides, Guide, II:22:319.

57. Maimonides, "Letter to R. Shmuel Ibn Tibbon", in Letters and essays of Moses Maimonides (in Hebrew), ed. and transl. by I. Shailat (Maaleh Adumim, 5748 [1988]), 553. Ibn Rushd

goes even further than Maimonides in his veneration of Aristotle: *cf.* the passages cited in S. Munk, *Mélanges de philosophie juive et arabe* (Paris, 1859), 316, 440–41.

58. Maimonides, *Guide*, II, Introduction, 235.

59. *Ibid.*, II:24:326–7. On Maimonides's epistemological scepticism in general *cf.* S. Pines, "The limitations of human knowledge" (ref. 43); *idem*, "Maimonides, Rabbi Moses ben Maimon", *Dictionary of scientific biography*, ix, 27–32. For a different view *cf.* Tzvi Y. Langermann, "The 'true perplexity': The *Guide of the perplexed*, part II, chapter 24", in *Perspectives on Maimonides: Philosophical and historical studies*, ed. by Joel L. Kraemer (Oxford, 1991), 159–74.

60. For a different view *cf.* Joel L. Kraemer, "Maimonides on Aristotle and scientific method", in *Moses Maimonides and his time*, ed. by Eric L. Ormsby (Washington, D.C., 1989), 53–88 [Hebrew translation in *Shlomo Pines jubilee volume*, part II (ref. 31), 193–224].

61. *Cf.* Menachem Kellner, "On the status of the astronomy and physics in Maimonides' *Mishneh Torah* and *Guide of the perplexed*: A chapter in the history of science", *The British journal for the history of science*, xxiv (1991), 453–63.

62. *Cf.* Ravitzky, "Aristotle's *Meteorologica*" (ref. 31).

63. R. Patai, "Biblical figures as alchemists", *Hebrew Union College annual*, liv (1983), 195–226. *Cf.* now also: *idem, The Jewish alchemists* (Princeton, 1994). The wealth of material presented in this book confirms my claim concerning the virtual absence of interest in alchemy among medieval Jews.

64. This point has been well taken by M. Guy Beaujouan: "On oublie trop souvent que, par exemple, dans la célèbre décrétale de Jean XXII (généralement datée de 1317), l'hostilité à l'alchimie est d'abord liée au problème de la fausse monnaie. Dans les communautés juives, les courtiers en métaux précieux n'auraient sans doute guère apprécié les scandales imputables à des coreligionnaires alchimistes." *Cf.* G. Beaujouan, "Les oriéntations de la science latine au début du XIVe siècle", in *Studies on Gersonides — A fourteenth-century Jewish philosopher-scientist*, ed. by Gad Freudenthal (Leiden, 1992), 71–80, p. 80.

65. *Kitvey Rabenu Moshe ben Naḥman*, ed. by Ḥaïm Dov Chavel (Jerusalem, 1963), i, 155f.

66. Harry A. Wolfson, *Crescas' critique of Aristotle* (Cambridge, Mass., 1929).

67. Crescas continues a long Andalusian tradition of opposition to science and philosophy on which *cf.* Bernard Septimus, *Hispano-Jewish culture in tradition: The career and controversies of Ramah* (Cambridge, Mass., 1982). It must be added that Crescas was influenced by scholastic philosophy on the one hand and by Kabbala on the other. *Cf.* S. Pines, *Scholasticism after Thomas Aquinas and the teachings of Ḥasdai Crescas and his predecessors* (*Proceedings of the Israel Academy of Sciences and Humanities*, i/10; Jerusalem, 1967) and W. Z. Harvey, "Kabbalistic elements in Crescas' *Light of the Lord*" (in Hebrew), *Jerusalem studies in Jewish thought*, ii/2 (1982/83), 75–109 (summary in English on pp. IX–XI).

68. On Gersonides's scientific work *cf.* the papers gathered in *Studies on Gersonides* (ref. 64). The volume includes an exhaustive bibliography compiled by Menachem Kellner.

69. The following thesis is developed in greater detail in Gad Freudenthal, "Rabbi Lewi ben Gerschom (Gersonides) und die Bedingungen wissenschaftlichen Fortschritts im Mittelalter: Astronomie, Physik, erkenntnistheoretischer Realismus und Heilslehre", *Archiv für Geschichte der Philosophie*, lxxiv (1992), 158–79; *idem*, "Sauver son âme ou sauver les phénomènes: Sotériologie, épistémologie et astronomie chez Gersonide", in *Studies on Gersonides* (ref. 64), 317–52. For a more general overview *cf. idem*, "Levi ben Gershom (Gersonides), 1288–1344", in *The Routledge history of Islamic philosophy*, ed. by S. H. Nasr and O. Leaman (London, forthcoming in 1995).

70. S. Pines, "Translator's introduction", in Maimonides, *op. cit.*, transl. by Pines (ref. 34), I, ciii f.

I

71. Heyd, "The emergence of modern science" (ref. 5), 172.

72. Cf. below at ref. 97.

73. Jacob Katz, *Tradition and crisis: Jewish society at the end of the Middle Ages* (New York, 1971), 3.

74. Qalonimos b. Qalonimos, *Even bohan*, ed. by A. M. Habermann (Tel-Aviv, 1956), 44.

75. The customary characterization of the two sides as 'rationalist' and 'anti-rationalist' is certainly misleading. Rationality can be appreciated only with respect to accepted premises, and on this score the traditionalist students of the Talmud were no less rational than their adversaries. I use these terms here conventionally as designating proponents of a positive or a negative attitude to 'Greek wisdom'.

76. An excellent concise overview of the history of the controversies discussed in the following paragraph is given in Ben-Sasson, "Maimonidean controversy" (ref. 18).

77. Quotation taken from Ben-Sasson, "Maimonidean controversy" (ref. 18), 752.

78. Cf. e.g. A. S. Halkin, "Why was Levi Ben Hayyim hounded?", *Proceedings of the American Academy for Jewish Research*, xxxiv (1966), 65–76.

79. Leon Joseph of Carcassonne, "Introduction" to his Hebrew translation of Gérard de Solo's Commentary on al-Râzî's *Al-Mansourî*, first printed in: Ernest Renan [A. Neubauer], *Les écrivains juifs français du XIV* siècle* (*Histoire littéraire de la France*, xxxi; Paris, 1893), 771–5, p. 772; new text edition accompanied by an English translation in: Luis Garcia-Ballester, Lola Ferre, and Eduard Feliu, "Jewish appreciation of fourteenth-century scholastic medicine", *Osiris*, 2nd ser., vi (1990), 85–117, "Appendix D", pp. 107–17; the passage is at lines 20–31. (This paper places the text and its author in their historical context; cf. pp. 93 ff.) See also Joseph Shatzmiller, "Étudiants juifs à la faculté de médecine de Montpellier, dernier quart du XIVe siècle", *Jewish history*, vi (1992), 243–55, where our passage is discussed on p. 250. The translation is my own, but in a few places I have adopted that of Garcia-Ballester, Ferre and Feliu, whose text I used in parallel with that of Renan-Neubauer.

80. Benedict, *Merqaz ha-torah bi-Provans* (ref. 15).

81. Joseph Shatzmiller has discovered contracts bearing on the tuition of medicine, concluded between a future teacher and a future student. Cf. Shatzmiller, "On becoming a Jewish doctor in the high Middle Ages", *Sefarad*, xliii (1983), 239–49.

82. Israel Abrahams, *Jewish ethical wills* (Philadelphia, 1926; new edn 1976), 131 (translation slightly modified).

83. Cf. the very concise and clarifying analysis by Jacques Verger, "Condition de l'intellectuel aux XIIIe et XIVe siècles", in *Philosophes médiévaux: Anthologie de textes philosophiques (XIIIe–XIVe siècles)*, ed. by Ruedi Imbach and Maryse-Hélène Méléard (Paris, 1986), 11–49. Also very perceptive and well-informed is Alain de Libera, "Le développement de nouveaux instruments conceptuels et leur utilisation dans la philosophie de la nature au XIVe siècle", in *Knowledge and the sciences in medieval philosophy: Proceedings of the Eighth International Congress of Medieval Philosophy (S.I.E.P.M.)*, ed. by Monika Asztalos, John E. Murdoch and Ilkka Niiniluoto (Helsinki, 1990), i, 158–97. The great importance for the emergence of modern science of the discussions on the *potentia Dei absoluta et ordinata* and on the notion of what is possible *secundum imaginationem* is highlighted by Amos Funkenstein, *Theology and the scientific imagination: From the Middle Ages to the seventeenth century* (Princeton, 1986). For the particularly revealing case of the Oxford calculators, cf. E. Sylla, "The Oxford calculators", in *The Cambridge history of later medieval philosophy*, ed. by Norman Kretzman, A. Kenny, and J. Pinborg (Cambridge, 1982), 540–63; *idem*, "The Oxford calculators in context", *Science in context*, i (1987), 257–79.

84. I am referring to the theory of scientific growth elaborated by the late Joseph Ben-David; cf.

notably his "Scientific growth: A sociological view", *Minerva*, iii (1964), 455–76, reprinted in his *Scientific growth* (ref. 2), 299–320, and his *The scientist's role in society: A comparative study* (2nd edn, Chicago, 1984).

85. Leon Joseph of Carcassonne, "Introduction" (ref. 79): ed. by Renan-Neubauer, 772; ed. by Garcia-Ballester, Ferre and Feliu, lines 24–27.

86. *Ibid.*, ed. by Renan-Neubauer, 773; ed. by Garcia-Ballester, Ferre and Feliu, lines 69–72.

87. *Ibid.*, ed. by Renan-Neubauer, 773; ed. by Garcia-Ballester, Ferre and Feliu, lines 72–81.

88. The idea developed in the following paragraph was suggested to me by Dr Josef Stern (University of Chicago). I am very grateful to him for his advice.

89. For an authoritative concise overview *cf.* Joseph Dan, *Ancient Hebrew mysticism* (in Hebrew) (Tel-Aviv, 1990), 20 ff.

90. Maimonides, *Mishneh Torah, Hilkhot yesodey Torah* 4:11; *cf.* also *ibid.* 2:12 and *Guide*, Introduction.

91. Leon Joseph of Carcassonne, "Introduction" (ref. 79): ed. by Renan-Neubauer, 772; ed. by Garcia-Ballester, Ferre and Feliu, lines 25–26.

92. In the Middle Ages the rigid social structures often prevented non-conformist scholars from residing for any long period at one place. Within Judaism, Abraham Ibn Ezra is a particularly remarkable example. For a highly interesting analysis of his case and of the whole social pattern *cf.* Aryeh Graboïs, "Le non-conformisme intellectuel au XIIe siècle: Pierre Abélard et Abraham Ibn Ezra", in *Modernité et non-conformisme en France à travers les âges*, ed. by M. Yardeni (*Studies in the history of Christian thought*, xxviii; Leiden, 1983), 3–13. To what extent similar phenomena can be detected also in 13th- and 14th-century Provence needs investigation.

93. Alfonso, *Meyashsher ʿaqov*, ed. and transl. by G. M. Gluskina (Moscow, 1983), 139 (= fol. 94ᵃ, lines 4–8 of the manuscript [London, British Library, Add. 26984], which is reproduced in the book). On the identity of the author and the history of the research on this book *cf.* my "Two notes on *Sefer meyashsher ʿaqov* by Alfonso, alias Abner of Burgos" (in Hebrew), *Qiryat sefer*, lxiii (1990–91), 984–6.

94. At this point one may wonder why these social constraints did not stop Gersonides from devoting himself to science and why his example and his interpretation of Maimonides's philosophy were not followed. To these queries I have no satisfactory answer. Little is known at present of contacts Gersonides may have had with other Jewish scholars or leaders, but perhaps further research will shed new light on the social and personal context within which he developed.

95. Sabra, "The appropriation and subsequent naturalization" (ref. 12).

96. *Cf.* notably Robert K. Merton, *Science, technology and society in seventeenth-century England* (1938; New York, 1970). Very important to the understanding of the thesis is Ben-David, "Puritanism and modern science" (ref. 2).

97. Ms Prague, University Library, no. VI 65, fol. 236ʳ. (I consulted the work in microfilm no. 46886 of the Institute for Microfilmed Hebrew Manuscripts at the Hebrew University, Jerusalem.) On the last folio of R. David Qimḥi's *Sefer ha-shorashim*, Gersonides (or his scribe) put down an inventory of his personal library; the *Guide of the perplexed*, "on parchment, in my own hand-writing", is the first book in the section of "books of wisdom" (*sifrey ha-ḥokhma*). The entire inventory has recently been published: Gérard E. Weil, *La bibliothèque de Gersonide d'après son catalogue autographe*, édité par F. Chartrain avec la collaboration d'A.-M. Weil-Guény et J. Shatzmiller (Louvain, 1992).

II

HOLINESS AND DEFILEMENT:
THE AMBIVALENT PERCEPTION OF PHILOSOPHY
BY ITS OPPONENTS IN THE EARLY FOURTEENTH CENTURY

1. *Introduction*

When Maimonides' *Guide of the Perplexed* first appeared on the horizon of the Jewish communities whose cultural language was Hebrew – this happened in southern Europe in 1204 or thereabouts – it elicited extreme reactions: on the one hand outright rejection, as by Solomon ben Abraham of Montpellier and his two disciples, who, in 1232, tried to have the work banned by the Jewish communities[1]; on the other hand whole-hearted acceptance, as by the translator of the *Guide*, Samuel Ibn Tibbon, and his clan, who even made the message conveyed by the *Guide* more radical (at least when read as an overt text). *A priori* there were no intrinsic reasons why this *polarization* should not persist – the antagonism between *hasidim* and *mitnagdim* is almost as fierce today as it was in the eighteenth century. In other words, we might well expect that ensuing generations of Jewish scholars in southern Europe would continue to be split into rival camps of supporters and opponents of Maimonides and his interpretation of Judaism. In reality, however, things took a much more complex turn: by the end of the thirteenth century, Maimonides' writings – both halakhic and philosophical – had been recognized as authoritative texts by practically all spiritual currents in the Jewish communities in Spain and Provence (although not in northern France and Germany). In particular, the need for allegorical interpretation of some Scriptural texts became universally admitted. The subsequent intellectual landscape of Judaism can thus be described as replete with *hybrids* that resulted from the crossing of many possible interpretations of the Maimonidean syn-

1. B. Septimus, *Hispano-Jewish Culture in Transition. The Career and Controversies of Ramah*, Cambridge, Mass. 1982, 63-74.

thesis with diverse doses of traditional Jewish lore and learning. Thus, instead of the community divided into two mutually exclusive camps, as in the 1230s, there emerged a continuum of perspectives on the relationship between Judaism, on the one hand, and philosophy and science, on the other. Thus, when a new controversy over the relevance to Judaism of matters philosophical broke out in 1303, its dynamics differed from those of the controversy of three generations earlier in two fundamental aspects: (i) Although a superficial look suggests that there were, once again, two antagonistic sides, neither was monolithic; instead, each camp comprised a coalition of persons with very different outlooks. (ii) The position of the «conservative» side in the debate was not as logically coherent and clear-cut as had been the stance of the «conservatives» of the 1230s, but integrated different and at times conflicting elements, including some distinctively Maimonidean components.

In this paper I shall not be dealing with this social *cum* ideological reality itself, however. Instead, I will be concerned mainly with the contemporary *perceptions* of that reality, specifically with the perception of philosophy and of its dangers by those who were apprehensive of secular studies. My special aim will be to show that the latter were ambivalent about philosophy; I shall suggest that it was precisely because of this ambivalence that they perceived philosophy as a menace not only to their opponents, the proponents of the philosophical studies, but to themselves as well.

2. *The Ambivalent Perception of Philosophy by the Opponents of Secular Studies*

(i) The Historical Facts

Let me begin by very briefly reviewing the main *personae* involved in the controversy and the course of events[1]. Sometime about 1303 the scholar Abba Mari Astruc of Lunel reached the conclusion that philosophy had penetrated too deeply into the world view of his fellow Jews

---

1. For a competent recent summary of the state of research on the controversy and a bibliography, see D. Berger, «Judaism and General Culture in Medieval and Early Modern Times» in J. J. Schacter, ed., *Judaism's Encounter with Other Cultures: Rejection or Integration?*, (Northvale, N. J. and Jerusalem, 1997, 57-141, esp. 100-8.

in southern France, particularly the younger generations, and that traditional Jewish learning was losing ground. He was particularly
alarmed by the fact that extremist allegorical interpretations were
expounded as a matter of course to the large, i.e. lay, public, notably
on the occasion of Saturday services in the synagogue, at weddings,
and other special events. Looking for a way to put an end to what he
regarded as a threat to the theological soundness and indeed the very
existence of Judaism, he decided to urge the greatest Jewish Talmudic
scholar and communal leader .of his day – Rabbi Solomon ben Abraham Adret of Barcelona (known as «Rashba») – to issue a ban on the
study of philosophy, at least before a certain age. This initiative produced a lively correspondence in which Abba Mari, Rashba, and
many other scholars exchanged long and disputatious letters among
themselves and with their opponents, who defended the dignity and
religious value of science and philosophy. Other subjects, notably the
religious legitimacy of the use of talismans and astral magic, were also
raised during the controversy[1]. On 9 Av 5065 (31 July 1305), Rashba in
Barcelona issued a partial ban: it prohibited the study of philosophy
before the age of twenty-five, but excluded the *Guide of the Perplexed*,
as well as the practical disciplines of astronomy and medicine, from the
interdiction. Soon afterwards, Abba Mari collected most of the letters
that had been exchanged during the controversy under the title *Minhat
Qena'ot* (= Offering of Jealousy). This unique collection of letters,
which has fortunately survived within the *responsa* of Rashba, has
recently been excellently (re-)edited by Hayyim Zalman Dimitrowsky[2].

Important light has been thrown on the controversy in several studies by Dov
Schwartz. See notably his «Beyn shamranut li-sekhaltanut (haguto ha-`iyyunit shel
hug ha-Rashba)», *Da'at*, 32-33 (1994), 143-82; «Changing Fronts in the Controversies over Philosophy in Medieval Spain and Provence», *Journal of Jewish Thought
and Philosophy*, 7 (1997), 61-82; «'Mesharet Moshe' le-Rabbi Qalonimos», *Qoves
'Al-Yad*, NS 14 (1998), 299-394. See also C. Touati, «La controverse de 1303-1306
autour des études philosophiques et scientifiques», *Revue des études juives* 128 (1968),
21-37, reprinted in idem, *Prophètes, Talmudistes, Philosophes*, (Paris) 1990, 201-17; G.
Freudenthal, «The Study of Mathematics as a 'Great Religious Secret': The Commentary on the Beginning of Euclid's *Elements* by Abraham ben Shlomo of Lunel.
A Commented Critical Edition» (Hebrew), *Jerusalem Studies in Jewish Thought*, 14
(1998), 129-58.
    1. D. Schwartz, «Ha-wikkuah al ha-magiyya ha-astralit bi-provans ba-me'ah
ha-y"d», *Zion*, 58 (1993), 141-74.
    2. H. Zalman Dimitrowsky, ed., *Teshuvot ha-Rashba*, pt. 1, vols. 1-2, Jerusalem
1990; *Minhat Qena'ot* occupies 223-883.

Although Abba Mari and Rashba were allies in the debate, there are great differences in their attitudes toward philosophy in general and toward Maimonidean philosophy in particular. Abba Mari can be described as a conservative Maimonidean who, for instance, accepts as indispensable the Maimonidean proofs of the existence, unity, and incorporeality of God[1], as well as the philosophic idea of the survival of the rational soul[2]. Rashba's philosophical-theological outlook is more traditional: true, he too greatly venerates the Master – he was instrumental in having Maimonides' Commentary on the *Mishna* translated into Hebrew – and he accepts some key Maimonidean ideas[3]; on the other hand, Rashba clearly upholds a fideistic theology, which maintains that philosophy and the Torah are irreconcilable[4], accords philosophical investigation a subordinate rank[5], and accordingly affirms that the survival of the soul depends solely on the observance of the religious precepts[6]. Abba Mari's and Rashba's ideologies are thus very far from being the same.

Although the opponents of secular studies constituted a disparate group, for want of a better term I will refer to their camp as that of the «traditionalists»; their opponents I will designate by the (equally misleading) name «rationalists». (The quotation marks will be kept throughout, as equivalent to the qualification «so-called».)

(ii) The Thesis

In what follows I will be concerned with the «traditionalist» side of the controversy, more specifically with its perception of philosophy and its perils. My main thesis is that because the «traditionalists»

1. *Minhat Qena'ot*, 235 ff.
2. *Ibid.*, 235.
3. Rashba follows Maimonides in holding that the tenet that God is Necessary Existence, whose Verity is apprehended by Him alone, is a fundamental premise of Judaism. See R. Shlomo ben R. Abraham Adret, *Hiddushey ha-Rashba: Perushey ha-Hagadot*, ed. by A. L. Feldman, Jerusalem 1991, 56. Cf. Maimonides, *Guide* 1:52; 2: Introduction (premise 20); 2:1, and *Hilkhot Yesodey ha-Torah* 2:8-10. D. Schwartz, «Le-heqer ha-Hugim ha-Pilosofiyim bi-Sefarad u-vi-Provans lifney ha-Gerush», *Pe'amim* 49 (1992), 5-23; 10-11, points out that the view of the Divinity as Necessary Existence is a characteristic of Rashba's circle.
4. *Minhat Qena'ot*, 342. Jacob Ben Makhir correctly understands Rashba's allegation: «You said that the [sciences] are to the Torah like a rival wife, and [so] everyone who occupies himself with them is ... guilty of idolatry» (*ibid.*, 507).
5. Rashba, *Perush ha-Haggadot*, 103.
6. *Ibid.*, 118.

accepted the *Guide* as an authoritative text and embraced some distinctly philosophical ideas, while remaining very reticent about accepting certain consequences that the «rationalists» drew from their own reading of the *Guide*, their image of philosophical knowledge was inherently *ambivalent*. In other words, the ambivalence resulted from the fact that the «traditionalists» recognized that many philosophical notions acquired via Maimonides had become indispensable to them, while at the same time they rejected other ideas that some followers of Maimonides held to be equally entailed by the Master's philosophy. Similarly, the «traditionalists» held secular studies to be extraneous to the «true» Jewish curriculum and were very apprehensive of their introduction into the curriculum for the young, who, they thought, should first be socialized through traditional studies. At the same time, however, they also recognized that the sciences had become indispensable, inasmuch as they supplied the premises needed for the Maimonidean theology, and therefore conceded that they could be studied by adults. Contrary to the anti-Maimonideans of the early thirteenth century, then, their intellectual successors three generations later took an inherently ambivalent position on all questions relating to the study of science and philosophy. In what follows, my main concern will be to analyze the wavering attitude that the «traditionalists» evinced toward philosophy as spawning a two-faceted sense of veneration and fear.

(iii) The Value and Perils of Philosophy according to the «Traditionalists»: Philosophy as «A Barrel of Honey about Which A Dragon is Coiled»

Of particular interest to us is the image of Aristotle held by the «traditionalists». It is well known that Maimonides rated Aristotle very highly – indeed, as the acme of nonprophetic human intelligence – and that Aristotelian philosophy – in its Arabic, notably Avicennian, Neoplatonically infused version – constituted the conceptual and doctrinal backbone of Maimonides' thought. Now Abba Mari, Rashba, and their fellow-«traditionalists» had integrated many of Maimonides' ideas into their thought, with distinctly philosophical concepts among them. At the same time, Aristotle was also the emblematic figure of the philosophical mode of inquiry, with which the «traditionalists» associated notably two views that they loathed – the eternity of the universe and the absence of individual providence. Concerning Aristotle, the «traditionalists» were confronted by the problem of how to throw away the bath water but keep the baby – that is, how to take over the proofs for the existence, unity, and incorporeality of God, without at the same time being committed to the two irksome Aristotelian theses. Indeed, their attitude toward Aristotle is highly ambivalent. «Aristotle will be remembered positively because he provided proofs for the existence and unity of God, blessed be He, and for [God's] being incorporeal and not subject to affections and changes», Abba Mari wrote[1]. Indeed, Abba Mari sees Abraham the Patriarch himself as a proto-philosopher – «the first of the true philosophers» – who provided rational proof of the existence and incorporeality of God (and, unlike Aristotle, of creation as well)[2]. Yet the same Abba Mari also sees Aristotle as the source of much trouble: «Aristotle believed», he writes, «that providence stops at the sphere of the moon, and this is the belief of the heretics among the people of our Torah. And he further believed that the world is eternal... and that any change in the nature of a thing is absolutely impossible. But this implies the negation of all the miracles and the denial of our holy Torah!»[3] How can one reconcile these two antithetical aspects of Aristotle? Abba Mari is not slow to come up with an answer: the world as a rule acts in accordance with Nature ('olam ke-minhago noheg)[4]; miracles being therefore very rare, Aristotle did not have the occasion to witness any himself and, as a result, «his eye led him into error»[5]. Abba Mari has no doubts that «had Aristotle and his comrades been in Egypt and on the [Red] Sea [i.e. had they witnessed the miracles associated with the Exodus], they would not have doubted the thesis of the creation of the world, and Aristotle would have taken his books and arguments in favor of the thesis of the eternity of the world and burned them on the spot»[6].

1. *Minhat Qena'ot*, 257 (quoted in Schwartz, «Changing Fronts», [n. 1], 72).

2. *Minhat Qena'ot*, 251-3.

3. *Ibid.*, 654.

4. This expression (whose origin is Talmudic; see *'Avodah Zarah* 54b) is used by Maimonides a number of times; cf. I. Shailat, ed., *Iggerot ha-Rambam* Jerusalem 1987, 1:362, n. 51.

5. *Minhat Qena'ot*, 257 (quoted in Schwartz, «Changing Fronts», [n. 1], 72). Aristotle should not be castigated for this error, because the belief in miracles is not one of the seven Noachide commandments (*Minhat Qena'ot*, 257).

6. *Ibid.*, 256.

By virtue of the proofs concerning the existence, unity, and incorporeality of God, Abba Mari says, the philosophical books contain hidden «honey and balm», of which the Israelites must not dispossess themselves. But these goods are to be likened to «a barrel of honey about which a dragon is coiled»[1]. For Abba Mari, the study of philosophy is comparable to collecting gems from the heads of vipers and to producing medicine from the flesh of a snake: the end result is beneficial, but the way is fraught with dangers[2] – a parable readily accepted by the «rationalists» as well[3]. Rashba too writes that although the philosophical books imply the impossibility of creation and of the miracles that occurred to the Israelites, we should not therefore say that everything in them is false and that one should believe nothing of what is affirmed in them[4]. To characterize the appropriate attitude toward these books, Rashba uses a well-known talmudic image: they are like pomegranates, of which one eats the inside and throws away the peel[5].

Thus the «traditionalist» opponents to philosophy fully recognized that philosophical wisdom contains «gems» that had become indispensable to contemporary Judaism. Fearing, however, the ideological dangers inhering in the philosophical books, they wished to set limits on their study and make them available only to those who would know how to separate the good from the obnoxious in philosophical knowledge. Thus Rashba writes that the problem with the study of the philosophical sciences is not «that these are evil in and by themselves»; rather, they are prone to cause damage only if there is some «weakness in the one who receives them»[6]. Rashba thus accepts the

1. *Ibid.*, 653.
2. *Ibid.*, 653.
3. *Ibid.*, 510.
4. Here Rashba seems to counter the stance of the «traditionalists» who were more radical than himself and who imposed an *en bloc* ban on everything that smacks of philosophy. This extreme position is described, for example, by the rationalist of the former generation, R. Shem-Tov Ibn Falaqera: «These fools have thought that whoever studies philosophy is deviating from the path of our holy Torah; in their mind [lit. thoughts], which is devoid of all truth, they thus thought that anything mentioned in philosophy is void and false». Shem-Tov Ibn Falaqera, *Sefer ha-Ma'alot*, in Ludwig Venetianer, ed., *Das Buch der Grade von Schemtob b. Joseph Ibn Falaquera*, Berlin 1894, 73.
5. *Minhat Qena'ot*, 297; Dimitrowsky gives the origin of the underlying talmudic statement as *B. Hagigah* 15b.
6. *Minhat Qena'ot*, 480.

premises adduced by the «rationalists» to the effect that just as the Torah permits us to drink wine, although at times it causes harm to some individuals, neither should the study of philosophy be altogether forbidden[1]. Rashba's conclusion from this consideration, however, is that since «not all minds are on a par», the study of the sciences must consequently be forbidden to all[2]. Similarly, Abba Mari writes: «Truly and clearly, although [in themselves] the books of the [foreign] nations contain nothing wicked, it is only with great difficulty and under imperative necessity that their study has been permitted and that it has even been allowed to commend their propositions»[3]. This phrase quintessentially encapsulates Abba Mari's ambivalence toward philosophical knowledge.

The «traditionalists» were fully conscious of the ambiguous worth of philosophy. Rashba concedes that the philosophical works «contain some useful things»[4], but still considers their study to be perilous. One who studies them will not be able to avoid their sword: «These [philosophical] matters bind the heart with ropes of love and end by raising dissension and opposition against the True Torah», he claims[5]. Further, echoing a well-known remark by Maimonides, he views philosophy as a cunning lure[6], which advances arguments from nature as a means of seduction – arguments that are all the more powerful because human beings are predominantly of an earthy nature[7]. Consequently, one who studies the sciences necessarily tends to accord them the central place and relegate the study of the Torah to the periphery[8]. «You

1. *Ibid.*, 481, 845-6.
2. *Ibid.*, 391.
3. *Ibid.*, 658.
4. *Ibid.*, 343.
5. *Ibid.*, 341.
6. Maimonides, *Guide*, Introduction to the First Part: «The human intellect having drawn him [the religious man for whom the validity of the Law has become established in his soul and who has studied the sciences of the philosophers] on and led him to dwell within its province...» (Moses Maimonides, *The Guide of the Perplexed*, translated by S. Pines, Chicago 1963, 5).
7. *Minhat Qena'ot*, 341; cf. also 550. The meaning of the last statement seems to be that man's intellectual capacities are limited because his constitution is material. Cf. the following passage, in which Rashba explains that man's and God's thoughts are incommensurable: «We [men] are of an earthy nature and are composite [bodies]. Therefore our thinking is subject to changes and to corruption, on account of our being a composite body... How can we discuss the thoughts and ways of God, Who is not subject to any change?» (Rashba, *Perush ha-Haggadot*, 105).
8. *Minhat Qena'ot*, 341.

and I know», Rashba writes to one of the sages of Lunel, «that the philosophical sciences are built on nature, which attracts the heart; once the heart has been tinged by this [philosophical] color, nothing will wash it out again»[1]. The philosophical sciences employ artifice to distract human beings from the true knowledge of God[2]. Even a cursory study of philosophy is a danger, for «can a man take fire in his bosom and his clothes not be burnt?»[3] «The day you forsake the study of the Torah for that of the sciences», Rashba warns, «your intention will be good, but your soul will then be walking amid dangerous beasts. ... Your intention is desired [by God], but your deeds are not», he warns, in an obvious echo of the *Kuzari*[4].

The «traditionalists» are keenly aware of the impossibility of drawing a neat demarcation between what is «in» Judaism and what is «outside» it, between what is necessary and what is dangerous. They (correctly) perceived themselves as being on a *slippery slope*: after they had recognized Maimonides' teaching as authoritative, it was difficult to keep out the Greco-Arabic sciences and philosophy from which it sprang, which it commended, and on which it relied. In this anxiety-generating situation, the sense of menace felt by the «traditionalists» is reflected in the imagery they used in order to characterize the potential benefit of the philosophical sciences.

(iv) Constructing Fences About the Torah

One image used by the «traditionalists» to suggest the threat of the philosophical sciences is that of a *fence* or a *wall* that must be constructed around the Torah to protect it from what they perceived as an assault from outside. Implicitly referring to the well-known Mishnaic injunction to make a «fence» around the Torah[5], Abba Mari refers to Rashba's ban as a Godly deed of «making fences around the Torah and closing all breaches»[6]. Conversely, he describes the «rationalists» as «vain and light fellows, slaves and bastards», who dare to breach the bulwarks erected around the Torah[7]. Similarly, Rashba calls upon the

1. *Ibid.*, 465; similarly 477.
2. *Ibid.*, 388.6
3. *Ibid.*, 343; Proverbs 6:27 (Jewish Publication Society 1917 translation).
4. *Minhat Qena'ot*, 387.
5. M. *Avot* 1:1.
6. *Minhat Qena'ot*, 581.
7. *Ibid.*, 581.

sages of Lunel to «establish the house and erect the wall» around the Torah: «you yourselves are the fence and what you say is like towers», he writes[1], suggesting the image of the Torah as a besieged fortress in need of defenders. Rashba's own goal, too, is to erect a wall[2]. The ramparts are indispensable to «save Mount Zion that has been deserted, so that foxes be prevented from strolling in it», he adds, alluding to a powerful image from the Book of Lamentations (5:18)[3].

(v) Philosophy as A Foreign and Rival Woman: Sexual Fidelity and Holiness

We have just cited the evocative image of the personified Torah as threatened with being deserted if philosophy succeeds in drawing away its students. One of the rhetorical means used by the «tradition-alists» is indeed to identify the received, traditional (i.e., chiefly talmu-dic) studies with the «Torah» *tout court* and to personify the latter as a defenseless lonely woman. «How can one remain silent», the sages of Lunel write to Rashba, «when the Torah is girded with sackcloth, seated bereaved and lonesome with no one to kiss her lips? She is like a woman deserted by her husband»[4]. Her former sumptuous clothes have been given to her rival – i.e., to philosophy – leaving her almost naked, with her kingdom taken away from her[5]. This anthropomor-phic imagery is pushed still one step further when the study of philos-ophy is repeatedly compared to an illicit sexual liaison. The sages of Lunel liken philosophy to a «foreign wife» identified with «whore-dom» and call upon Rashba to drive her/it away[6]. Similarly, drawing on the entrenched metaphor of Israel's relationship with the Torah to that between husband and wife, Rashba views the philosophical – alien – sciences as «rival wives»[7] and further exclaims (invoking an unambiguous phallic symbol): «Shall we remain silent when the son of another woman ploughs on our back?»[8] The identification of philos-ophy with an alien woman is omnipresent in the pages of *Minhat*

1. *Ibid.*, 631.
2. *Ibid.*, 672; cf. also 628, 629, 641.
3. *Ibid.*, 631.
4. *Ibid.*, 618.
5. *Ibid.*, 619.
6. *Ibid.*, 617.
7. *Ibid.*, 393.
8. *Ibid.*, 385.

Qena'ot, whose very title, *Offering of Jealousy*, is taken from the section in Numbers (5:11–31) that deals with the procedure to be followed when a husband suspects his wife of infidelity, in order to determine whether she indeed is guilty. Thus Rashba and other sages of Barcelona cry out: «Who had intercourse with the son of a foreign god? [Is it possible that] in view of the whole community they have instituted alien women and sent away the daughter of Judah? That within the sight of everyone they fondled the bosom of the alien girl?!»[1] Rashba also says that he who studies the foreign sciences from his youth has married «the daughter of an alien god» and that he is «stirred up» by his wife, whom he calls «Jezebel», in allusion to Ahab's wickedly idolatrous wife[2]. Indeed, the «rationalists» themselves observe that their opponents, the «traditionalists», view them as having been seduced by foreign – especially Greek – women and as having betrayed «Jacob's daughter», divorced her, and sent her away[3].

To understand the use of these emotionally powerful sexual imagery we may recall that in interpreting Proverbs 7:5-21 Maimonides had likened the matter of the human body, the «cause of all hindrances keeping man from his ultimite perfection», to a harlot, thereby establishing a parallel between the attraction exerted by false ideas and sexual temptation[4]. Maimonides also evoked another connection between certain ideas and the fair sex: in his letter to the sages of Lunel, he referred to the study of the Torah as the love of his youth – i.e., his true love – while describing the philosophical sciences as female servants at the service of the Torah – «perfumers, cooks, and bakers»[5]. Explicitly referring to this Maimonidean statement, Rashba warns that not everyone can cook as well as Maimonides; i.e., know how to circumscribe the influence of foreign sciences[6]. He himself goes on to compare the sciences with «Sarah's maid» who seeks to usurp her mistress's place[7] and he warns of the danger that the Torah be left in a corner and Greek science become the governess of the house[8]. The «rationalists», too, accept this hierarchy (at least out-

---

1. *Ibid.*, 410.
2. *Ibid.*, 629; 1 Kings 21.
3. *Minhat Qena'ot*, 439 (after 1 Sam. 8:13).
4. Maimonides, *Guide*, Introduction to the first part, trans. Pines, 13.
5. Shailat, *Iggerot*, II:502 (see also below, p. 20).
6. *Minhat Qena'ot*, 342.
7. *Ibid.*, 276.
8. *Ibid.*, 478.

II

wardly) and assure their adversaries that the Torah will for ever remain
the mistress to whom all attention is due, whereas the alien sciences,
being maidservants, will be visited only occasionally[1].

The proponents of philosophy adduce another interesting argument
in which sexual imagery looms large. As we have noticed, the «traditionalists» compared the sciences to foreign maids, with whom Israelite men are tempted to have sexual relations[2]. The «rationalists» now
invoke the commonplace notion that the sciences had in fact originated in Israel, whence they passed to the nations[3]. This allows them
to liken philosophy and the sciences to a native Jewish maid who had
been captured and stripped of her original garments: «She remained
nude and her tongue became a foreign one»[4]. But the fact that in cap-
tivity the maids – the originally Jewish sciences – were abused by their
captors does not mean that they should now be avoided, the «rationalists» plead: does the fact that «their breasts have been fondled by Egyptians» really mean that they have been altered or defiled, they ask
rhetorically[5].

The choice of sexual images to convey the sense of danger and
betrayal that the «traditionalists» believed to be involved in the study
of philosophy is significant. It harks back to what David Biale has
called «the political theology of intermarriage» in the Bible[6]. Such
marriages were construed as an immediate threat to the identity and
theological purity of the Israelite people. Thus Deuteronomy says:

When the Lord your God shall bring thee into the land whither thou goest
to possess it, and shall cast out many nations before thee, the Hittite, and the
Girgashite, and the Amorite, and the Canaanite, and the Perizzite, and the
Hivite, and the Jebusite, seven nations greater and mightier than thou; and
when the Lord thy God shall deliver them up before thee, and thou shalt
smite them; then thou shalt utterly destroy them. Thou shalt make no co-
venant with them, nor show mercy unto them; neither shalt thou make marriages with them: thy daughter thou shalt not give unto his son, nor his

1. *Ibid.*, 433.
2. *Ibid.*, 410.
3. E.g. *Minhat Qena'ot*, 432-6, 653; see also N. Roth, «The 'Theft of Philosophy'
by the Greeks from the Jews», *Classical Folia*, 32 (1978), 53-67.
4. *Minhat Qena'ot*, 436.
5. *Ibid.*, 436.
6. D. Biale, *Eros and the Jews. From Biblical Israel to Contemporary America*, New
York 1992 (Basic Books, 20). Biale writes «intermarriage», but in fact the kind of
relationships to which he alludes is larger than the one subsumed under the term
«marriage».

daughter shalt thou take unto your son. For he will turn away thy son from following Me, that they may serve other gods; so will the anger of the Lord be kindled against you, and he will destroy thee quickly. But thus shall ye deal with them: ye break down their altars, and dash in pieces their pillars, and hew down their Asherim, and burn their graven images with fire. For thou art a holy people unto the Lord thy God: the Lord thy God hath chosen thee to be His own treasure, out of the peoples that are upon the face of the earth[1].

This well-known passage makes the holiness and chosenness of the Israelites contingent upon their abstaining from involvement with foreign women. The contact with alien women is identified with the practice of idolatry and thus with loss of the status of a holy and chosen people. Indeed, the prophets often use the metaphor of marriage to describe the relationship between God and Israel and correspondingly construe the Israelites' unfaithfulness as marital infidelity[2]. Similarly, when the exiles returned from the Babylonian exile, Ezra ordered them to divorce the foreign women they had married. The text in Ezra refers to the «Canaanites, Hittites, Perizzites, Jebusites, Ammonites, Moabites, Egyptians, and Amorites»[3]. In the time of Ezra – the fifth century BCE – all or most of these peoples had long since disappeared; still, they were named here explicitly, with the clear intention of recalling the biblical interdiction of foreign women and the connection between conjugal purity and holiness and chosenness. Referring to the peoples that had lived in Canaan at the time of the first conquest was, and has remained, part of the cultural resources on which Jewish authors can draw to add an affective element to the sense of danger and treachery they wanted to attach to contact with foreigners. It is precisely this rhetorical device that the «traditionalists» exploited in their denunciation of philosophy: by setting up a parallel between the study of philosophy and union with foreign women, they sought to evoke feelings related to betrayal of the Torah and to idolatry and thereby mobilize emotional energies in favor of their own, more purist, construction of Judaism. At the same time, the identification of the sciences with seductive foreign women suggests that they appeal to the senses and not to the divine Word as communicated by prophecy. For the «traditionalists», the «rationalistic» synthesis –

1. Deut. 7:1-6 (Jewish Publication Society 1917 translation); quoted in part in Biale, *Eros and the Jews*, 21.
2. *Ibid.*, 21.
3. Ezra 9:1.

embracing both the Torah and philosophy – is ruled out: one has «to keep apart the holy and the profane, the impure from the pure, life from death and chaos», Rashba urges[1]. The message was indeed understood correctly: «In your eyes we are idolaters», a «rationalist» rejoins to Rashba[2]. And Jacob ben Makhir, the great scientist, writes to him indignantly: «It is now ten times that you have insulted [the philosophical sciences] and that you have described their practitioners as [idolatrous]»[3].

(vi) Philosophy – An Esoteric Lore?

Another image that was frequently associated with philosophy was that of an esoteric, secret lore that must be kept away from the multitude. During the dispute of 1303–1305 some «traditionalists» used the identity that Maimonides had posited between Aristotelian physics and metaphysics and the traditional mystical *Ma'aseh Bereshit* and *Ma'aseh*

---

1. *Minhat Qena'ot*, 572.
2. *Ibid.*, 673.
3. *Ibid.*, 608, with reference to 2 Kings 17:31. The use of the idea of the *femme fatale* as a symbol for the dangerous and tempting sciences is widely diffused and calls for further research. Let me only mention that St. Jerome had already compared secular (viz. heathen classic) thought with a desirable woman. Contrary to the medieval Jewish «traditionalists», however, he considered it to be a «captive woman», a «handmaid» whom, because she has been deprived of most of her sexually alluring features (her head has been shaven, her nails pared, etc.), he could legitimately espouse. Consider the following passage from Jerome's letter to Magnus, a Roman orator, written about the year 397, on behalf of his willingness to quote secular (classical) literature in the defense of Christianity.

«You ask me at the close of your letter why it is that sometimes in my writings I quote examples from secular literature and thus defile the whiteness of the church with the foulness of heathenism. I will now briefly answer your question. ... And if this were not enough, that leader of the Christian army, that unvanquished pleader for the cause of Christ [Paul], skilfully turns a chance inscription into a proof of the faith. For he had learned from the true David to wrench the sword of the enemy out of his hand and with his own blade to cut off the head of the arrogant Goliath. He had read in Deuteronomy [21:10-13] the command given by the voice of the Lord that when a captive woman had had her head shaved, her eyebrows and all her hair cut off, and her nails pared, she might then be taken to wife. Is it surprising that I too, admiring the fairness of her form and the grace of her eloquence, desire to make that secular wisdom which is my captive and my handmaid, a matron of the true Israel? Or that shaving off and cutting away all in her that is dead whether this be idolatry, leisure, error, or lust, I take her to myself clean and pure and beget by her servants for the Lord of Sabaoth? My efforts promote the advantage of Christ's family, my so-called defilement with an alien increases the number of my fellow-servants. Hosea took a wife of whoredoms, Go-

Merkavah, respectively, arguing that by teaching the Aristotelian sciences openly, the «rationalists» violated the talmudic ban on revealing esoteric knowledge. Thus, whereas some «rationalists» seem to have propounded (or at least practiced) a notion of knowledge as *open*, their «traditionalist» adversaries advocated a view of knowledge as a *closed* system, accessible only to a small elite[1].

Abba Mari explicitly follows Maimonides in identifying the Aristotelian sciences with the mystical *Ma'aseh Bereshit* and *Ma'aseh Merkavah*[2]. The various sciences, he says, «are most useful for the sages who are pious. For by them one can examine the representation of what exists and so come to know and understand some of the wonders of God, may He be exalted. And following these steps one will progress until he reaches the two uppermost sciences, which are *Ma'aseh Bereshit* and *Ma'aseh Merkavah*. Thus will he come to know and recognize the One by Whose word the world came to be»[3]. Now this identification is double-edged. On the one hand it can be used to argue that the study of the sciences should be forbidden until a student has reached the traditionally specified age of forty years[4]; on the other hand, this identification also attests to the truth of the sciences. Abba Mari accepts both conclusions. He argues, again following Maimonides, that the teaching of *Ma'aseh Bereshit* and *Ma'aseh Merkavah* should be strictly curtailed and that these sciences must not be studied

---

mer the daughter of Dibliam, and this harlot bore him a son called Jezreel or the seed of God [Hosea 1:2-4]. Isaiah speaks of a sharp razor which shaves 'the head of sinners, and the hair of their feet' [Is. 7:20?]: and Ezekiel shaves his head as a type of that Jerusalem which has been an harlot [Ezek. 5:1?], in sign that whatever in her is devoid of sense and life must be removed». (Jerome, *Letters*, trans. W. H. Fremantle, in *Library of Nicene and Post-Nicene Fathers*, 2nd series [New York, 1893], vol. 6, 149).

I would like to thank Prof. Michael McVaugh of the University of North Carolina for calling this passage to my attention.

In the first half of the thirteenth century, Judah ha-Kohen Ibn Matqa also established a parallel between an attractive woman and «heretical wisdom which takes on the appearance of truth and seduces one, like some of the words of Aristotle on the eternity of the world» (quoted by Septimus, *Hispano-Jewish Culture in Transition*, 97).

1. Cf. A. Funkenstein and A. Steinsalz, *Ha-Soziologia shel ha-Ba 'arut* (Tel Aviv, 1987).

2. *Sefer ha-Yare'ah*, chaps. 1 ff., in *Minhat Qena'ot*, 648 ff.

3. *Ibid.*, 649.

4. Cf. M. Idel, «*Le-toldot ha-'issur lilmod qabbalah lifney gil 'arba'im*», *AJS Review*, 5 (1980), Hebrew section, 1-20.

before one has intensively studied the Torah and the Talmud[1]. He considers the Torah as «planted in the King's orchard [*Pardes*][2]» – i.e., esoteric knowledge – and thinks that philosophers who preach in public have all but stripped the Torah of its secrets and «have left it nearly nude»[3]. As for the right to acquire esoteric knowledge, Abba Mari remains ambivalent. On one occasion he says that in general the study of the sciences should be forbidden, but that great sages should be allowed to pursue it, because they know how to separate the fine flour from the coarse meal[4]. Elsewhere, addressing the «rationalists», he says: «It was not our wish to condemn philosophy, God forbid, for this idea did not even cross our minds. We merely wish to uphold your suggestion and stipulate an age [for this study], for [philosophy] is not better than [i.e., is not different from] *Ma'aseh Merkavah*»[5]. Thus, if the native Jewish sciences can be studied only by a perfect individual who has reached middle age, the same holds *a fortiori* for the study of foreign sciences[6].

Abba Mari thought that the identification of the philosophical sciences with *Ma'aseh Bereshit* and *Ma'aseh Merkavah* would be instrumental in fostering his objective, namely, restricting the study of philosophy to sages or at least forbidding it under a certain age. Similarly, R. Abraham ben Abraham of Aix-en-Provence, a supporter of Rashba, complains that the philosophers think they are divinely inspired and cry out the meaning of *Ma'aseh Bereshit* and *Ma'aseh Merkavah* in the marketplace[7]. Others in the «traditionalist» camp, such as the more anti-philosophical Rashba, rejected this identification *in toto* (a position that is hardly surprising for a scholar with kabbalist leanings). Rashba ironically asks why the sages of Israel sealed the secret of the *Merkavah* and severely curtailed access to it if indeed it is true that access to it is open to all, seeing that Aristotle and his friends expound it fully among the nations[8]. No, he contends, in reality the philosophers are not privy to any secrets and hence can divulge

---

1. *Minhat Qena'ot*, 651.
2. *Ibid.*, 313.
3. *Ibid.*, 634.
4. *Ibid.*, 659–60.
5. *Ibid.*, 447.
6. *Ibid.*, 652.
7. *Ibid.*, 553.
8. *Ibid.*, 278.

none. «As for those poor people of whom you wrote», Rashba writes
to Abba Mari, «who are impertinent enough to preach about *Ma'aseh
Merkavah* in synagogues in front of the public, Heaven will reveal their
sin, [namely] that they preach nonsense and make public their insanity.
But concerning your allegation, namely, that they reveal what has
been hidden by God, my heart tells me that they revealed nothing of
what has been concealed. They have therefore not committed the sin
of disclosing anything; indeed their stupidity and ignorance save them
from sinning»[1]. As to the true secrets of the Torah, they are not to be
revealed at all, except perhaps to a very few at an old age: «The mysteries belong to our God, but the plain things are for us and our sons»
(Deut. 29:28)[2].

3. *The Ambivalent Attitude toward the Sciences and its Roots in Mai-monides*

Provence and northern Spain of the end of the thirteenth century
were the scene of the coexistence of an almost continuous spectrum of
interpretations of Judaism. At one extreme there were those who,
continuing a tradition that originated with Samuel Ibn Tibbon and
Jacob Anatoli, thought that the *Guide of the Perplexed*, and philosophical wisdom generally, contained the keys to deciphering the Divine
mysteries – *Ma'aseh Bereshit* and *Ma'aseh Merkavah* – and believed that
they had succeeded in laying bare at least some of these secrets. At the
other extreme there were those who, like R. Asher ben Yehiel and, to
a lesser degree, Rashba, remained apprehensive about the introduction
of philosophy into Judaism, even though most of them had internal-
ized philosophical notions, to one degree or another. Essentially fideistic, they thought that the vast majority of Scriptural texts, including
in particular those recounting miracles, had to be interpreted literally.
In between these two extremes were those who, like Abba Mari,
accepted many (although not all) of Maimonides' fundamental theses
but were horrified by the extremism of the first faction.

Having reviewed this multiplicity of attitudes toward philosophy
and the sciences and having noticed in particular that almost all the

---

1. *Ibid.*, 345.
2. *Ibid.*, 859-60; cf. also 481.

participants in the controversy looked up to Maimonides as the uncontested Master, we should briefly reflect upon the conditions that made this peculiar physiognomy of the controversy of 1303–1305 possible. How could there be so vehement a polemic between two camps when both recognized Maimonides as an authority? Why, in fact, did the «traditionalists» recognize Maimonides as such, instead of entirely rejecting his teaching as had their forebears in the 1230s? In other words, what factors shaped the inherently ambivalent position of the «traditionalists» toward science and philosophy?

The observed uncertainty in the attitudes of the Provençal and Spanish scholars toward philosophy and the sciences, I will now suggest, is the result of two complementary factors: (i) The ambiguity of Maimonides' own positions on cardinal issues that are the underpinning of the balance struck between a «rationalist» and a «traditionalist» position. (ii) The fact that Maimonides' ambiguities entirely passed into the positions of the «traditionalists», a consequence of the fact that they had embraced Maimonidean ideas not so much because they appealed to them as because Maimonides' dominating personality (due chiefly to his halakhic authority) left them no choice.

Let us very briefly consider the message conveyed by the *Guide of the Perplexed* as to the authority of tradition as compared with that of philosophy and science. For the purposes of the present brief discussion we shall ignore the fact that the *Guide* was already considered to be an esoteric text by its first translator into Hebrew, Samuel Ibn Tibbon[1], and follow the great majority of its thirteenth-century readers, who saw it as an overt text[2]. Still, as we shall see, Maimonides' statements on the two questions that were at the core of the controversy of 1303–

---

1. A. Ravitzky, «R. Shmuel Ibn Tibbon we-sodo shel *Moreh Nevukhim*», *Da'at*, 10 (1987), 19-46 (for an earlier English version of this paper see *AJS Review* 6 (1981), 87-123); idem, «Sefer ha-me'te'orologiqa le-Aristo we-darkhey ha-parshanut ha-maimonit le-ma`aseh bereshit», in *Shlomo Pines Jubilee Volume on the Occasion of His Eightieth Birthday*, Part II (= *Jerusalem Studies in Jewish Thought*, vol. IX) Jerusalem 1990, 225-50; D. Schwartz, «*Otot ha-shamayim*. Aspaqlaria la-esoteriqa ha-yehudit bi-yemey ha-beynayim», *Da'at*, 38 (1997), 145-48.

2. Therefore we need not go into the debate between Pines and Davidson over the principled «limitations of human knowledge» according to Maimonides. See Sh. Pines, «The Limitations of Human Knowledge According to Al-Farabi, Ibn Badja, and Maimonides», in I. Twersky, ed., *Studies in Medieval Jewish History and Literature*, Cambridge, Mass. 1979, 82-109; H. A. Davidson, «Maimonides on Metaphysical Knowledge», *Maimonidean Studies* 3 (1992-93), 49-103.

1305 are equivocal enough to allow conflicting and indeed opposing interpretations.

(i) Scientific Investigation: Its Necessity and Limits

Concerning the status of the sciences and their place in the curriculum, Maimonides emphasizes at many places in the *Guide*, in line with the Arabic philosophical tradition, that the philosophical sciences are indispensable for attaining knowledge of the Deity, through which the human soul attains eternal felicity. Maimonides repeatedly stresses that the study of the sciences is a religious obligation [1] and holds the key for unraveling the secrets of *Ma'aseh Bereshit* and *Ma'aseh Merkavah*. Moreover, Maimonides' hermeneutic method turns the demonstrated truths of philosophy into the interpretive instrument for the authoritative texts of Judaism. For Maimonides, philosophy – that is, Arabic Peripatetic philosophy – is the touchstone by which to judge whether a particular biblical verse can be interpreted literally: Whatever philosophy *proves* wrong *cannot* be the intention of the Books received by Revelation. This stance comes to the fore with particular acuteness in Maimonides' remark that had the philosophers demonstrated the eternity of the world he would have no qualms about interpreting Scripture accordingly [2]. This tendency in Maimonides' thought gives a predominant role to philosophical knowledge and is reflected in Maimonides' poor judgment of mere «talmudists» in his parable of the Ruler's Palace [3].

But at the same time Maimonides also sets limits to the primacy of philosophy in its relationship to the Scripture. Thus, to take again the example of creation, Maimonides elaborates a skeptical position according to which Aristotle's proofs of the eternity of the world are

---

1. Cf. H. A. Davidson, «The Study of Philosophy as a Religious Obligation», in S. D. Goitein, ed., *Religion in a Religious Age* Cambridge, Mass. 1974, 53-68. See also G. Freudenthal, «Les sciences dans les communautés juives médiévales de Provence: Leur appropriation, leur rôle», *Revue des études juives,* 152 (1993), 29-136; idem, «Science in the Medieval Jewish Culture of Southern France», *History of Science,* 33 (1995), 23-58.

2. *Guide* II, 25 (see also n. 85 below). The point was clearly perceived by Judah b. Joseph Ibn Alfakhar, an anti-Maimonidean of the controversy of the 1230s: according to Maimonides, he writes, «for whatever biblical verse, if its contrary is demonstrated, then the verse is not to be interpreted normally [i.e. literally]». See A. Lichtenberg, ed., *Qoves Teshuvot ha-Rambam we-Iggerotaw* Leipzig 1859, pt. 3, 1bb.

3. *Guide,* III, 51.

not conclusive, so that the biblical account of creation need not (and therefore must not) be interpreted according to the tenets of philosophy[1]. He thus accepts the traditional, literal reading of Genesis, i.e., the thesis of creation *ex nihilo*, although he knows it to be incompatible with Aristotelian science and philosophy, on which he relies so strongly elsewhere. Maimonides' skeptical stance severely curtails the scope within which philosophy is authoritative – his epistemology denies autonomy to philosophy. Maimonides' famous remark to the effect that the philosophical sciences are nothing but «perfumers, cooks and bakers», maids whose only function is to bring forth the beauty of the Torah[2], highlights this tendency within his thought.

Thus the holders of both attitudes toward the study of the philosophical sciences, as they emerged in the controversy of 1303–1305, could justly view themselves as the legitimate heirs of the Maimonidean heritage. On the one hand there were the «rationalists» who took Maimonides at his word and whose curriculum for young Jews accorded a central place to the study of logic, mathematics, astronomy, physics, and metaphysics, at the expense of the traditional study of the Talmud and its commentators. On the other hand, the «traditionalists», who sought to give priority to traditional Jewish learning and in particular wanted to keep philosophy and allegorical interpretation from being applied to certain parts of Scripture, such as the accounts of creation and miracles, could also see themselves as reaffirming the limits Maimonides had set on philosophical inquiry.

(ii) Allegorical Interpretation of the Scripture: Its Necessity and Limits

Maimonides' ambiguity about the status of philosophy as a touchstone of truth naturally reappears in his views on the scope of the hermeneutic method. To Maimonides, questions of hermeneutics are capital, because he purports to cast his entire reinterpretation of Judaism in hermeneutic terms; the purpose of the *Guide* is merely «to explain the meanings of certain terms occurring in books of prophecy»[3]. The terms in question are those whose literal interpretation loads biblical verses with untenable theological positions, such as,

1. Cf. *Guide*, II, 16-17.
2. See above, p. 11.
3. *Guide*, Introduction, 5.

notably, an anthropomorphic notion of the Deity. Since for Maimonides in most cases it is philosophy that determines which Scriptural passages are to be understood literally and which must be interpreted allegorically, it follows that the limits of the hermeneutic enterprise are determined not by the biblical text itself but by one's philosophical beliefs – specifically, by what one takes to be *demonstrated* truth. In other words, the meaning one ascribes to biblical passages depends upon one's extra-biblical commitments, which determine what the revealed text can be assumed to have meant and what not. Because biblical passages are not overtly labeled as to whether they should be interpreted allegorically or literally, the scope of the hermeneutic enterprise depends on philosophy[1]. Maimonides himself notes that neither prophecy nor a trustworthy tradition going back to the prophets told him which passages were to be interpreted allegorically. This is why he had to rely on the method of reconciling the biblical text with what is possible according to the standards of (Aristotelian) philosophy[2].

It follows – and this is the main point in the present context – that once the principles of Maimonideanism are accepted, the limits of the hermeneutic undertaking are to be determined by each and every student of Scripture, who must strike a balance between the truth-value ascribed to philosophy and the possible (literal or allegorical) meanings of each individual verse. Maimonides himself interpreted numerous passages allegorically; there was no inherent reason in his philosophy, and no explicit statement by him, as to why allegorization should stop precisely there. Maimonides could legitimately be understood as having supplied the paradigm and methodological framework for allegorical interpretations generally and as having thereby licensed his followers to interpret allegorically any Scriptural passage they thought contradicted truths they considered to be established through reason.

1. This stance comes to the fore with particular acuteness in Maimonides' remark, already mentioned, that had the philosophers demonstrated the eternity of the world he would have no difficulty in interpreting Scripture accordingly (*Guide* II, 25). The same conclusion follows even more forcefully from an analysis of Maimonides' numerous remarks on the methodology of his interpretation, scatted throughout the *'Iggeret Tehiyat haMetim* (in Shailat, *'Iggerot*, 1:319-338 [text], 339-374 [Hebrew translation]).

2. *'Iggeret Tehiyat haMetim*, 1:330 (text), 1:360-61 (Hebrew translation).

Nevertheless, the *Guide* also conveyed the opposite message; namely, that only those passages interpreted by Maimonides allegorically had to be so explicated – and no others. In fact, although Maimonides' methodological statements implied that any biblical verse is a candidate for allegorical interpretation, in the *Guide* he gives the strong impression that all verses requiring allegorization to keep them from contradicting verities established by philosophy have already been so interpreted by himself. There seems to be no hint in Maimonides' writings that he recognized the desirability or legitimacy of allegorical interpretations other than his own. Maimonides could thus also be taken as having supplied not a *model* for allegorical interpretations, but rather the definitive *corpus* of all such legitimate interpretations.

Concerning the limits of the allegorical method, too, Maimonides' writings thus authorized the two possible attitudes and their various shades. Those who carried allegorization to the extreme could legitimately view themselves as creatively extending Maimonides' work and, in particular, as using his biblical exegeses as models for other possible allegorizations. Conversely, those who pleaded for minimal allegorization could legitimately hold that Maimonides himself had already expounded all necessary allegorizations and no new ones were admissible.

The fact, therefore, that the opposing parties in the dispute of 1303–1305 could both view Maimonides as an authority is perfectly reasonable: such a diversity of attitudes is a direct consequence of the ambiguities in Maimonides' writings. The «rationalist» position, according to which the study of philosophical writings is mandatory because they hold the key to the eternal felicity of the soul, was a legitimate extension of one strain in Maimonides' thought; and so was its corollary that allegorical interpretations of Scripture are legitimate if their goal is to make the contents of Scripture agree with philosophically established truths. The opposite, more fideist position, according to which one must «concede» to philosophy only the strict minimum and interpret Scripture allegorically only where a literal interpretation would lead to anthropomorphism, and its corollary, the position that the study of traditional Jewish lore takes precedence over the study of the philosophical sciences, could also legitimately be regarded as a consistent extension of currents in Maimonides' thought[1]. This is the reason

---

1. The above by no means implies that all the positions defended during the controversy can be described as Maimonidean.

why the controversy of 1303–1305 pitted two opposing camps that shared allegiance not only to the traditional Jewish texts, which all accepted as binding, but also to Maimonides' authority.

Another fact must be emphasized now, which brings us back to an observation made at the beginning of this paper. In the controversy of the 1230s the anti-Maimonidean group was homogeneous and defended a logically coherent position, rejecting the *Guide in toto.* By contrast, the conservative camp of 1303–1305 was a heterogeneous coalition, whose members had integrated Maimonidean ideas to different degrees and thus held less clear-cut positions[1]. The reason for this, we now understand, is that the evolution of the conservative camp after the 1230s was very largely due to Maimonides' influence. Maimonides' incomparable prestige and attraction, due first and foremost to his *halakhic* authority, made it increasingly impossible to oppose his teachings, including the philosophical ones, overtly. By the beginning of the fourteenth century it had become impossible to deny that some Scriptural passages required allegoric interpretation. To put this differently, the «traditionalists» were not drawn to Maimonides' philosophy by its intellectual merits, but were pushed to adopt it by the intellectual and social impossibility of opposing it. The «traditionalists» integrated philosophical elements in their thought only reluctantly, a fact lucidly recognized by Abba Mari, who (as we already saw) writes: «Truly and clearly, although [in themselves] the books of the [foreign] nations contain nothing wicked, it is only with great difficulty and under imperative necessity that their study has been permitted and that it has even been allowed to commend their propositions»[2]. It follows that any philosophical ideas admitted into their world view by the «traditionalists» were accepted only grudgingly. This implies that the ambiguities and vacillations in Maimonides' thought, evident in the *Guide,* were fully reproduced and even reinforced in the attitudes of the «traditionalists». Having approached the *Guide* from its «conservative side», they naturally fostered the fideist tendency within Maimonides' œuvre.

This explains how the debate of 1303–1305 could take place between two camps whose protagonists all looked to Maimonides as

1. The shift in the positions of the «traditionalist» camp between the 1230s and the 1300s has been described in Schwartz, «Changing Fronts».

2. *Minhat Qena'ot,* 658; quoted above, p. 8.

an incontestable authority. Whereas the «traditionalists» accepted the strict minimum of the philosophical ideas and thus followed and even enhanced the conservative facet of Maimonides' thought, the «rationalists» saw Maimonides' *Guide* as a license to go beyond the doctrines professed by the Master himself.

4. Conclusion

My main objective in this paper has been to analyze the perception of the philosophical sciences by the opponents of their study. We saw that Rashba and Abba Mari applied a common stock of emotionally charged images to the philosophical sciences. The sciences were described as gems and honey, which were, however, located in parlous places, such as the heads of vipers or in a barrel around which a dragon is coiled. The sciences were perceived as being so seductive that it was difficult to escape the allure exerted by their study. The most emotionally loaded images are those in which the sciences were identified with seductive foreign women drawing the Israelites toward idolatry. These images reveal that their authors viewed the sciences with considerable apprehension and anxiety – sentiments that the letters of Abba Mari, Rashba, and their associates seek to impart to their readers.

It seems to me that this anxiety is all the greater to the extent that the «traditionalists» have difficulties in drawing the boundaries distinguishing their own interpretation of Judaism from that of the «rationalists». The «traditionalists», who wished to preserve important portions of the received non-philosophical interpretation of Judaism while availing themselves of some key Maimonidean notions, felt themselves to be on a «slippery slope»: having made a first step in the direction of the appropriation of the sciences, it was difficult not to make further steps too. Having already made concessions to Maimonideanism, the «traditionalists» of 1303 could no longer defend *purist* positions like those of their forerunners in the 1230s. This situation corresponds roughly to those studied by Mary Douglas in her influential *Purity and Danger*, where she writes:

I believe that ideas about separating, purifying, demarcating and punishing transgressions have as their main function to impose system on an inherently untidy experience. It is only by exaggerating the difference between within and without, above and below, male and female, with or against, that a semblance of order is created[1].

Our «traditionalists» conform to this pattern: they wished to demarcate what is «within» Judaism from what is «without» it — «to keep apart the holy and the profane, the impure from the pure, life from death and chaos», as Rashba put it explicitly[2]. But a neat demarcation was no longer possible — the received forms of the study of the Torah had already been «contaminated» by philosophical ideas. It is this sense that the purity of the doctrine was in jeopardy that gave rise to the strong sense of danger and various emotionally charged images.

1. M. Douglas, *Purity and Danger. An Analysis of the Concepts of Pollution and Taboo*, London 1966, 4.
2. *Minhat Qena'ot*, 572; quoted above, p. 14.

Maimonides' Stance on Astrology in Context: Cosmology, Physics, Medicine, and Providence

M aimonides was one of the rare medieval Jewish thinkers to oppose astrology radically. But it has not yet sufficiently been appreciated that from the viewpoint of medieval science such a stance was more problematical than might appear to a mind informed by modern science, for the universally accepted premises of contemporary physics, biology, medicine, and climatology to a considerable extent *warranted* the claims of astrology. The status of astrology in Maimonides' thought thus raises more questions than previously realized. In Maimonides' own view, medieval science indeed provided a rational basis for some of the fundamental claims of astrology. Is it conceivably possible that the author of the "Letter on Astrology" was not as opposed to astrology as he said he was? Can his admonitions against it possibly have had halakhic – i.e., social and pragmatic – significance only, without reflecting his true beliefs concerning its truth as a cosmological theory?

THE GROUNDING OF ASTROLOGY IN MEDIEVAL SCIENCE

All medieval thinkers subscribed to the doctrine that the heavenly bodies exert influences on the sublunar world: generation and corruption, indeed the very existence of sublunar substances, were taken to depend upon celestial influences.[1] Drawing on this doctrine, astrologers argued that their art had a basis in natural philosophy.

The causal influences issuing from the supralunar region were held to be of two different kinds. First, the *bodies* of the planets were held to be *efficient* causes, blending the four sublunar elements and determining their respective proportions in the composite substances. Second, the active *intellect,* one of the separate intellects moving the planets, was construed as the *formal* cause informing sublunar matter.

The first part of the theory goes back to Aristotle, who pointed to the manifest influence of the sun upon the natural sublunar processes: he argued that the sun's daily and annual motions account, respectively, for the very fact that the four sublunar elements have not settled immobile in their natural places, and for the regular growth and decay of living beings.[2] Aristotle also made passing mention of the influences of the moon on natural processes.[3] Medieval Aristotelians, as well as astrologers (following in the footsteps of Ptolemy), generalized this claim. The sun clearly reinforces the element fire and thus increases its share in composite substances on which it acts; similarly, the moon's role in bringing about the tides was taken to demonstrate beyond doubt that it reinforces the element water. By induction, it was then concluded that *all* planets similarly influence the sublunar elements, even though the effects of most planets were admittedly less perceptible than those of the sun and the moon.[4] The material composition of each sublunar substance was thus held to depend causally upon the celestial bodies.

Like all his contemporaries, Maimonides too subscribed to this scheme, although he proposed a slightly modified, allegedly original version of it. He grouped together the five planets in one single sphere, so that the number of spheres bearing stars was four – "the sphere of the fixed stars, that of the five planets, that of the sun, and that of the moon"[5] – each acting on one of the four elements:

> Thus the sphere of the moon moves the water, the sphere of the sun the fire, while the sphere of the other planets moves the air. . . . The sphere of the fixed stars moves the earth. Perhaps the earth is so sluggish in moving to receive the action being brought to bear upon it and in undergoing combinations because of the slowness of the fixed stars in their motion.[6]

The second part of the theory concerns the formal cause. Aristotle's matter is amorphous and passive; it cannot itself give rise to structured substances. Moreover, because structured substances are composed of contrary qualities (hot/cold, moist/dry), they naturally tend to decay.[7] To account for the coming-to-be and the perseverance of structured substances, Aristotle's natural philosophy was in need of a complement. This was of momentous importance: the necessity to add to Aristotle's physical theory an account of the origin and perseverance of forms made room for a bridge, or rather a synthesis, between the theory of sublunar matter on the one hand and metaphysics and theology on the other. In fact, to account for forms, Peripatetics argued that sublunar substances having forms, notably plants and animals, come to exist and persevere because suitably mingled matter is *informed* by an external agent possessing those forms. In other words, since forms do not arise in matter spontaneously, they must be imprinted upon it and maintained by a distinct agent possessing them in actuality.[8] This agent, the formal cause of forms of sublunar substances, was identified as the active intellect, the "Giver of Forms," a notion that has its roots in Aristotle's psychology.[9]

The form of each sublunar substance was thus construed as resulting from the interplay of the efficient and the formal causes. By mingling the matter, the former produces in a given lump of matter a mixture characterized by a specific ratio (*logos*) of the elements composing it. Thereupon the Giver of Forms imprints upon the *mixtum* a form *suitable to its material composition*: the better the balance of the elements in a given mixture, the higher the form it will receive. The form of every sublunar substance is thus conditioned by its material constitution. This physical theory established the notion of God's general governance (or providence) over the sublunar world. Since the separate intellects were associated with the planets as their "motors," this governance, the sum total of the efficient and formal causes, was loosely perceived as "heavenly influences."

Astrologers eagerly drew upon a suitably modified version of the theory concerning the role of the Giver of Forms in producing sublunar substances. For them, however, it was not a transcendent intellect, but rather influences emanating from the planets themselves, that inform sublunar matter (in addition to mixing it).[10]

The physicotheological theory we have reviewed is fully endorsed by Maimonides:

> Know that there is a consensus of all philosophers to the effect that the governance of this lower world is perfected by means of the forces overflowing to it from the sphere . . . and that the spheres apprehend and know that which they govern.[11]

More specifically:

> Heaven in virtue of its motion exerts government over the other parts of the world and sends to every generated thing the forces that subsist in the latter. . . . Know that . . . the forces that come from heaven to this world are four: [1] the force that necessitates the mixture and composition – there is no doubt that this force suffices to engender the minerals;[12] [2] the force that gives to every plant a vegetal soul; [3] the force that gives to every animal an animal soul; [4] the force

that gives to every rational being a rational faculty. All this takes place through the intermediary of the illumination and the darkness [on earth] resulting from the light in heaven and from the heaven's motion round the earth.[13]

This heavenly "governance" over the sublunar world is indispensable for its continued existence. "Just as an individual would die and his motions and forces would be abolished if the heart were to come to rest even for an instant, so the death of the world as a whole and the abolition of everything within it would result if the heavens were to come to rest."[14]

Maimonides was keenly aware of the fact that this physical doctrine – according to which, as he formulates it, the spheres and the stars are "the lords [adonim] of every body other than themselves" (although, to be sure, they themselves are subordinated to God) –[15] is closely related to astrology. Thus, he repeats that

It is known and generally recognized in all the books of the philosophers speaking of governance that the governance of this lower world–I mean the world of generation and corruption–is said to be brought about through the forces overflowing from the spheres.

He then does not shun from quoting a saying from *Bereshit Rabbah* (10:6), of which the Jewish proponents of astrology were particularly fond:

[Y]ou will find likewise that the Sages say: "There is not a single herb below that has not a 'mazzal' in the firmament that beats upon it and tells it to grow." . . . Now they also call a star: *mazzal*.[16]

Maimonides comments on this saying in a sentence that from the pen of any other writer would appear as distinctly astrological: "By means of this dictum they have made it clear that *even individuals subject to generation have forces of the stars that are specially assigned to them.*"[17]

Indeed, Maimonides himself explicitly recognized that the universally accepted Aristotelian doctrine may be taken to provide a scientific basis for astrology:

It also is said with regard to the forces of the spheres that they overflow toward that which exists. Thus the overflow of the sphere is spoken of although the actions come from a body. Hence the stars act at some particular distances; I refer to their nearness to or remoteness from the center or their relation to one another. *From here astrology comes in.*[18]

Maimonides thus endorses even the idea that the actions of the stars upon sublunar bodies depend upon their positions with respect to one another, i.e., on what in the technical vocabulary of astrology is called their "aspects" – a distinctively astrological doctrine. And he notes that this physical thesis may appear to provide astrology with a basis in physical science. The significance of Maimonides' statement is perspicaciously perceived by Moshe Narboni, himself a mild proponent of astrology:[19] "Understand how he generalized and verified [immet] all of astrology [mishpat ha-kokhabim] by saying 'the stars act at some particular distances. . . . From here astrology comes in.' For this is the truth [i.e., essence] of astrology."[20]

A universal consensus thus prevailed over the doctrine that the generation and corruption of all sublunar bodies depend upon the actions of the stars and those of the immaterial intellects, lumped together as "celestial influences." These two-tiered influences were held, first, to determine the material composition of substances and, second, to endow them with appropriate forms. Now this doctrine implies that living beings, too, inasmuch as they are *natural* composite bodies, are affected by the efficient and formal causes of celestial origin. The question thus arises: To what extent does this hold of man, too? Are man's desires, aims, choices, and actions also dependent upon the ubiquitous celestial influences (as astrology would have it)? (Recall that according to Aristotelian doctrine, the forms of living beings are their souls – nutritive, sentient, rational – so that every man and woman owes even his or her rational soul to the active intellect.)

Although man is a being endowed with a rational soul and *deliberates* over his actions, the universally accepted theories of contemporary science entail a largely affirmative answer to our question. To see this we have to look briefly into two further medieval scientific theories: the doctrine of psychophysiology and the related doctrine of climatology, both of which were universally accepted, too, and which also allowed astrology to claim for itself a foundation in natural philosophy.

The Hippocratics, followed by Aristotle, held that certain physiological characteristics of the body – e.g., whether it is hot or cold – determine the character of a person.[21] This idea was elaborated in the doctrine of the four humors, whose equilibrium was held to constitute the person's "temperament," that is, to determine the sum of his or her psychical faculties, including his or her moral traits. This theory was elaborated notably by Galen, one of whose treatises, translated into Arabic, bears the title: "That the faculties of the soul depend upon the mixtures [or temperaments] of the body."[22]

This standard view, part and parcel of medieval biology and medicine, was, of course, shared by Maimonides too. In order to explain why men are psychologically so dissimilar – it is this dissimilarity that makes social organization indispensable – Maimonides said:

[T]here are many differences between the individuals belonging to [the human species], so that you can hardly find two individuals who are in any accord with respect to one of the species of moral habits. . . . The cause of this is the difference of the mixtures [or temperaments], owing to which the various kinds of matter differ, and also the accidents consequent to the form in question.[23]

In other words, every individual has his or her specific "mixture" of the elements or humors, his or her particular complexion or temperaments. Since the material compositions of the individuals are different, their forms (souls) are different too, and so are the accidents – particular psychical properties and dispositions, i.e., traits of character[24] – subsequent upon these forms. Individuals are thus different because they differ in their material composition. Psychical differences – character, dispositions, faculties, and so forth – ultimately depend upon differences in the *material* composition of the body.

Consider an example. The faculty of courage, Maimonides says, "varies in strength and weakness, as do other faculties, so that you may find among people some who will advance upon a lion, while others flee from a mouse." The cause of this, Maimonides goes on to explain, is that there "must necessarily exist a temperamental preparation in the original natural disposition" (the innate physical complexion or temperament) of each individual.[25] The same psycho-physiological doctrine also underlies Maimonides' theory of prophecy, according to which only an individual whose (largely innate) temperament is perfectly well balanced has the perfect imaginative faculty necessary for prophecy.[26]

This theory supplied another important building block for astrology. As we saw, the medieval philosophers held the mixture of the elements within any composite body to depend upon the heavenly bodies. Now *qua* body, man is a natural substance, a composite of the four elements. Therefore, the equilibrium of the elements within a person's body – his or her temperament – depends significantly upon the influences of the heavenly bodies (acting as an efficient cause), along with other factors (such as inherited biological traits and the influence of the environment). Since the temperament in turn determines the psychical faculties, it follows that these faculties (e.g., courage), and consequently the individual's disposition to behave in some manner (e.g., bravely), largely depend upon the heavenly bodies. Astrologers could consequently claim that by knowing how celestial influences affect the temperament, one can predict the behavior of any given person with a reasonable probability.[27] This argument was embraced by natural philosophers, too.[28] Thus the logical (although unstated) corollary of Maimonides' views is that the differences among men, and hence the need for social order, *in fine* are a consequence of the celestial influences affecting men *qua* natural substances.[29]

Maimonides was keenly aware that the theory of the dependence of physicopsychical faculties upon celestial influences is within a hairbreadth of astrology. This can be perceived, for instance, from the fact that in his *Shemonah Perakim,* the short exposition of this theory is immediately followed by a strong admonition against the follies of the astrologers and by an attempt to refute this theory by arguing that the innate temperament can be changed by a suitable "corrective therapy."[30]

The doctrine of the dependence of psychical faculties upon celestial influences underlies and comes to the fore in climatology.

In the Middle Ages, it was universally believed that the physiological equilibrium (of the elements or of the humors) inside a human body, and consequently also the individual's psychical faculties, are strongly influenced by the physical conditions of the environment,[31] some of which in turn depend on the "climate," the geographical latitude of the region, which determines its position with respect to the sun.[32] The impact of these environmental conditions is superimposed, so to speak, upon the direct influences of the heavenly bodies. The doctrine goes back at least to the Hippocratic treatise *Airs, Waters, Places,* which is echoed, for instance, in the *Kuzari,* where Yehudah Halevi puts in the mouth of the philosopher the view that in addition to characteristics a man inherits from his parents, his individual traits are determined by the "influence of winds, countries, foods and water, spheres, stars and constellations."[33] Halevi's rabbi shares this idea and draws on it to foster his view that *Eretz Yisrael* is the most suitable place for prophecy.[34] Similarly, Maimonides, in his *Pirké Moshe,* follows Galen's opinion that the differences of climate are the causes of the differences of the individual temperaments, and consequently also of the differences of the faculties of the speech organs and of the intelligence. He also agrees with al-Fârâbî that the inhabitants of the middle climates are "more perfect in their intellects and in their forms generally," their tongues being in greatest harmony with man's logic.[35] Again, in *Hanhagat ha-Beriut* Maimonides comments: "when the air changes, be it a little, the soul's inquiry changes perceptibly. This is why one finds many people in whom one can perceive how their psychic actions diminish when the air becomes corrupt: I mean that although no change can be perceived in the actions of the vital and the natural functions, yet their understanding and their memory suddenly fail."[36]

Climatology thus supplied a further argument for the astrological doctrine. The sun, the foremost heavenly body, determines the climate and through it influences the physiological temperament of the individual, on which in turn depend his or her psychical faculties. As confirmatory evidence, the astrologers invoked the purported fact, accepted as such by the natural philosophers too, that national collective traits are determined by the climate.[37] Thus even an opponent of judicial astrology as al-Fârâbî could describe the causal chain by which the celestial influences determine psychical qualities in the following general terms:

> The first natural cause for the differences among nations . . . is the difference between the celestial regions. . . . These bring about differences of the parts of the earth [and] differences of the exhalations. . . . Subsequent upon the differ-

ences between the exhalations are differences in the air and differences of the waters. . . . These in turn entail differences of the vegetation [and] of the species of irrational animals, [so that] there are differences in the nutrition of [different] nations. The differences in nutrition entail differences in the matter and the sperm out of which men are constituted . . . causing differences of their forms and natural faculties.[38]

Medieval philosophers of nature thus considered as indisputable the idea that by shaping the natural processes in the sublunar world the heavenly bodies to a large extent determine the "psychology" of every living being. One's temperament – the mixture of the elements (or humors) in one's body – was held to depend both directly and indirectly on the celestial bodies. First, by mixing the elements, the heavenly bodies largely determine the initial composition of every natural substance, including the innate temperament of living beings. Second, by determining the climates, the sun additionally influences the environmental conditions (air, nutrition, etc.), which (along with other factors, such as age and conduct) act upon the initial constitution and eventually modify it. From the theory according to which the temperament conditions the psychical qualities, it further followed that one's disposition to behave in one way or another to a great extent ultimately depends on celestial influences. On this basis it could reasonably be argued that the behavior of individuals is largely influenced by the celestial bodies (even if one supposes that it is not *determined* by them, inasmuch as men are supposed to remain free to behave in ways other than those dictated by their biological makeup). In a further move, it could then be claimed that on the basis of a knowledge of astral influences, the future behavior of individuals can be predicted – not with certainty, but with a degree of probability. All this, I repeat, is based on Aristotelian and medical theory accepted by all medieval philosophers of nature, including, as we have seen, Maimonides.

MAIMONIDES' REJECTION OF ASTROLOGY

In his letter to the sages of Montpellier, Maimonides unequivocally repudiates astrology:

Know, my masters, that every one of those things concerning judicial astrology that [its adherents] maintain – namely, that something will happen one way and not another, and that the constellation under which one is born will draw him on so that he will be such and such a kind and so something will happen to him one way and not another – all those assertions are far from being scientific; they are stupidity.[39]

He rejects astrology with equal clarity and firmness in his *Epistle to Yemen*. "I note," he writes to R. Ya'aqob ben Netan'el al-Fayyûmi, "that you are inclined to believe in astrology and in the influence of past and future conjunctions of the planets upon human affairs. You should dismiss such notions from your thoughts. Cleanse your mind as one cleanses dirty clothes. Accomplished scholars, whether they are religious or not, refuse to believe in the truth of this science. Its postulates can be refuted by real proofs on rational grounds."[40] He had made similar pronouncement in his *Mishneh Torah*[41] and, as already mentioned, in *Shemonah Perakim*. There is thus no ambiguity concerning the firmness of Maimonides' opposition to astrology.

The foregoing analysis of the basis for astrology afforded by contemporary science should make us pause, however. We should ask how and on what grounds Maimonides reconciled his acceptance of a natural philosophy from which, as he himself had noted, "astrology comes in," with his rejection of judicial astrology. Maimonides' attitude toward astrology has often been taken to be self-explanatory: it has been perceived as singularly perspicacious, rational, and modern, deriving from his ingenuity and forward-looking mind. (This approach was particularly tempting when Maimonides' stance was viewed against the background of medieval Jewish philosophy, most of whose proponents to one degree or another accepted astrology.) This, however, is a wrong way to look at the matter. Maimonides' position must be evaluated against the background of the science he himself held to be true and not by reference to later standards of argument. The fact that Maimonides' position is supported by modern science in no way explains why he propounded it some eight centuries ago. Therefore, since judicial astrology justly claimed to have roots in the tenets of natural philosophy – that is, in what is called "natural astrology" – the question we must raise is: what were Maimonides' reasons for his negative stance on judicial astrology? Or could it be that Maimonides' repeated public denial of judicial astrology is addressed to the masses and covers up a secret, "esoteric" position more favorable to this art? This question seems warranted on its own merits, by virtue of the *problematique* described above, independent of any general stance on the significance one should accord to Maimonides' statements in the *Guide* on the one hand, and in his halakhic writings on the other. It obviously gains in urgency if we adopt the Straussian interpretive paradigm. According to Leo Strauss, as is well known, "the *Guide* is 'my speech' revealing 'my opinion,' as distinguished from 'our opinion,' expressed in 'our compilation,' the *Mishneh Torah*, where generally speaking Maimonides appears as the mouthpiece of the Jewish community or of the Jewish tradition," so that "the *Guide* was not addressed to the vulgar, nor the *Mishneh Torah* to the perplexed."[42] Four considerations may *prima facie* seem to militate in favor of such a suggestion.

1. Maimonides discusses astrology explicitly only in writings intended for the masses – halakhic writings and letters to the leaders of communities. In the *Guide,* by contrast, astrology is mentioned only in one or two asides,[43] but is not made into a topic. Did Maimonides deliberately avoid the subject in his most theoretical opus in order to conceal his true views of it?

2. Maimonides nowhere attacks the postulates of judicial astrology head-on, trying to show that on the basis of Aristotelian science its claims are *false*. In his "Letter on Astrology" he insists on the dangers of astrology to the Jewish faith, i.e., on its possible religious and moral harmful consequences, but does not endeavor to *refute* that doctrine. He contents himself with the statement that "there are lucid, faultless proofs refuting all the roots of those assertions" and that "never did one of those genuinely wise men of the nations busy himself with this matter or write on it,"[44] but he does not intimate what these proofs are. In his *Epistle to Yemen*, Maimonides adduces arguments refuting the astrological theory, going back to Mâshâ'allâh (754–813) and propounded in Hebrew, among others, by Abraham Bar Hiyya (d. ca. 1136) and Abraham Ibn Ezra (1089–1164), construing the history of the world as depending upon small, middle, and great conjunctions of the planets.[45] Its statements, Maimonides shows, conflict with the historical evidence. But here, too, Maimonides does not attempt a theoretical refutation of astrology; he contents himself to show that, from the Jewish vantage point, astrology is *harmful*, but does not try to demonstrate that it is a *false* theory. Could it be that the reason for this silence is that while Maimonides regarded astrology as a dangerous doctrine, he was not all that sure of its falsity?

3. We know from Maimonides himself that he had devoted much reading to astrology, and to idolatry in general. Indeed, says Maimonides, "it seems to me [that] there does not remain in the world a composition on this subject, having been translated into Arabic from other languages, but that I have read it and have understood its subject matter and have plumbed the depth of its thought."[46] To justify this interest which runs against an explicit interdiction, Maimonides invokes his desire to understand the reasons for the precepts, whose primary intention, as he believed, is to eradicate idolatry.[47] One can nonetheless ponder the possibility that astrology, the first of these subjects he studied, also presented an intellectual challenge. This seems perfectly plausible if we remember that Maimonides shared theoretical premises from which, as he recognized, "astrology comes in." Whence again our question: could it be that Maimonides did not have *theoretical* arguments against astrology?

4. Last but not least, if we accept the view advanced by two of Maimonides' foremost interpreters, the late Shlomo Pines and Alexander Altmann, that Maimonides upheld determinism,[48] then we would have two further serious reasons to suspect that his true views of astrology were less negative than he professed. First, the causal chain which then *ex hypothesi* determines man's fate is described by the theories we have reviewed above and it cannot but have its beginning in the heavenly bodies, the "lords" (*adonim*) of every sublunar substance, which (or rather, whom) God entrusted with the governance of the lower world. But is it not then plausible to expect that this causal chain can become known to man—that even if at present astrology is deficient, it can in the future become a true science? Second, if Maimonides

III

believed in determinism, then one important reason he invokes in justification
of his hostility toward astrology – namely the fact that it fosters fatalism –
would fall to the ground; in this case, his statements about the perils of
fatalism which astrology engenders must be intended only for the masses.
Therefore, the one who assumes that Maimonides believed in determinism
must take very seriously the possibility that he had a secret positive attitude
toward judicial astrology.

For these reasons I suggest that we should earnestly examine the possibility
that Maimonides harbored an "esoteric" position on judicial astrology that
was more favorable to it than his public admonitions against it suggest. Even
if, after reflection, we discard this possibility, we will, I believe, have acquired
a better insight into the considerations that motivated Maimonides' position.

The above-mentioned doubts notwithstanding, Maimonides' stance on
astrology was as negative as he said it was. The reasons for his attitude,
including his silences, will become clear when we take into consideration the
intellectual and social contexts in which he lived, thought, and acted.

For Muslim philosophers in the East, astrology was not at all an issue.
Although natural philosophy and astrology shared quite a number of theoretical tenets, most philosophers – including al-Fârâbî and Ibn Sînâ – rejected
astrology unambiguously.[49] Maimonides' position, therefore, is simply that
of his cultural milieu, Peripatetic Arabic philosophy. His claim that "there are
lucid, faultless proofs refuting all the roots of those assertions" was indeed
founded on a consensual body of knowledge which he shared.[50] This is
certainly one of the reasons why Maimonides did not deem it necessary to
take up a theoretical discussion of astrology in the *Guide* or elsewhere.[51]

Maimonides' silence on the theory of astrology also has to do with the
circumstances in which he goes into the matter. As we noticed, Maimonides
takes up astrology not in the context of a philosophical discussion – in the
Guide astrology is not at all discussed – but rather in halakhic works and in
replies to inquiries from faraway communities seeking his guidance. This
means that it is not so much Maimonides the philosopher as Maimonides the
community leader who discusses astrology. His correspondents and readers,
Maimonides understood, sought spiritual direction, not instruction in natural
philosophy; they looked up to him as an authority on the *halakhah*, not as an
expert on science of pagan and Muslim origin.

That Maimonides did not seek to refute astrology theoretically is thus
primarily a consequence of the character and the objective of the writings in
which he addressed the subject. He was interested in the possible religious *cum*
social consequences of astrology, not in astrology as a cosmological theory.
According to Maimonides, astrology presents two major dangers: it unwittingly favors star worship and (as already mentioned) it gives rise to fatalism,
both of which he considered as extremely dangerous and sinful.[52] (This
statement obviously remains true even if we assume that Maimonides sub-

scribed to a secret determinist position. Maimonides could insist on the social dangers of true doctrines; in the Straussian perspective, he does so time and again.)

Nor does the assumption that Maimonides accepted determinism imply that he had a positive attitude toward astrology. For even if one supposes man's fate to be predetermined by a causal chain having its beginning in the heavenly bodies, it does not follow that man can obtain knowledge concerning it. Maimonides' epistemological skepticism, in particular his belief that the laws governing the heavenly bodies – their motions and their physical influences on the sublunar world – will never become known to man,[53] thoroughly undermines astrology. Therefore, even a determinist Maimonides would reject astrology. Indeed, according to Maimonides, whatever knowledge of the future is given to man is obtained *via* prophecy and veridical dreams, both of which derive from a divine overflow, not from a human science of the stars.[54]

Last but not least, in point of fact, as it seems to me, Maimonides did *not* uphold determinism. Rather, in his view, nature itself is indeterminate to some extent[55]; moreover, in his (and his contemporaries') view, although man's biological constitution is causally determined by celestial influences, that physiological constitution determines only one's *propensity to act* in a certain way, but not one's *actual behavior* (which may, and indeed should, be subject to the control of the intellect). These two elements of indetermination obviously undermine the possibility of predicting man's conduct and hence judicial astrology.

We may thus conclude that astrology, which was incompatible with Maimonides' philosophy, interested him first and foremost as a social, not as a scientific or intellectual, issue. Maimonides knew that although Peripatetic natural philosophy could seem to provide astrology with a rational basis, which indeed the astrologers claimed for themselves, the great Peripatetic philosophers had refuted the claims of judicial astrology which went beyond natural astrology. Nonetheless, judicial astrology was popular, in Egypt as well as in Yemen and in France, and this popular belief in astrology, which had nothing to do with its alleged theoretical foundation, appeared to him to present serious religious-social threats. To counter those threats was the goal of Maimonides' admonitions against astrology, and this aim accounts for the character of his treatments of the subject. We can, I believe, safely discard the idea that Maimonides had an esoteric doctrine that was closer to judicial astrology than his professed statements suggest.

We have seen that Maimonides had little room in which to maneuver. The received doctrine of the celestial influences on sublunar substances and processes, including man, provided the foundation for the philosophic view of God's governance of the world. This is why Maimonides dwells on this doctrine even in the "Letter on Astrology," where at first glance it may appear as counterproductive. Maimonides had to acknowledge these physical influences, to concede that the heavenly bodies are the "lords of every body," while still denying the possibility of judicial astrology. Indeed, other medieval thinkers interpreted the same evidence differently and held that judicial astrology is a perfectible art whose theoretical foundations are sound. It would be misguided and unjust to characterize them as irrational; some of these thinkers, as for instance Gersonides (see below), were perfectly rational. Nor is Maimonides' rejection of astrology a product of an ahistorical rationality. For Maimonides, astrology presented a social and religious, rather than a scientific, problem, and his views of it followed the *opinio communis* of the great Peripatetic Muslim philosophers. We can, therefore, in sum, appreciate this: Maimonides was fortunate that, at least on this issue, his scientific considerations (pertaining to *truth*) and his pragmatic ones (pertaining to social-religious *utility*) were in harmony.

Maimonides' rejection of astrology is congruent with (although not a consequence of) an essential stance of his philosophy, namely his view of the purpose *(telos)* of material existents generally, and of the celestial bodies in particular. Maimonides vehemently rejects the popular view that "the finality of all that exists is solely the existence of the human species so that it should worship God, and . . . that even the heavenly spheres only revolve in order to be useful to it and to bring into existence that which is necessary for it." According to him, all existents have only one final cause, namely, "God has wished it so, or: His wisdom has required this to be so." The consequence is this: "It should not be believed that all the beings exist for the sake of the existence of man. On the contrary, all the other beings too have been intended for their own sakes and not for the sake of something else. . . . Just as He has willed that the human species should come to exist, He also has willed that the spheres and their stars should come to exist." Therefore, the stars "do not exist for our sake and so that good should come to us from them."[56] Maimonides' stance is decidedly *anti-anthropocentric*.[57] It is absurd to suppose that the most perfect beings exist for the sake of infinitely inferior ones, the heavenly bodies and the separate intellects for the sake of man;[58] rather, they exist for their own sake, although some good overflows from them, from which man benefits.

This theological cosmology seems to have a bearing on the question of astrology. Maimonides' world picture is unfavorable to, although not strictly incompatible with, a science predicting the effects upon man of the celestial bodies. Rather, astrology is much more naturally imbedded in a worldview in which the heavenly bodies exist for the sake of man. This can be perceived from a brief comparison with the views of one of Maimonides' most brilliant followers, R. Levi ben Gershom, *Ralbag*, also known as Gersonides (1288–1344).

Gersonides explicitly rejects Maimonides' view that the spheres and their stars exist only for their own sake. Chapter 3 of Treatise V, Part II, of his *Sefer Milhamot ha-Shem* is devoted precisely to the demonstration that "the stars exist in the spheres for the sake of the things [down] here," in the sublunar world.[59] In Gersonides' view, when God created the celestial bodies, He endowed them with such natures as to allow them to perfect, as far as possible, all sublunar existents. The heavens were designed with utmost wisdom for the benefit of the sublunar existents, and particularly for that of man, the most perfect among them. Gersonides does not doubt that every particularity of each planet and of its motions follows a specific intention, for only the extreme diversity of the planetary actions can bring about the extreme diversity, and indeed great perfectness, of their effects down here.[60]

Now this anthropocentric optimism – Gersonides' view of a divine providence exercised through an "astral determinism" (as Charles Touati has fittingly called it) – is in much greater congruity with astrology than Maimonides' view. If the heavenly bodies exist, nay were *created, in order to* govern the sublunary world and particularly to exert providence over man, it is reasonable to expect, as Gersonides did, that their actions can be studied and, eventually, become known and predicted.[61] Here, obviously, Gersonides' epistemological optimism (the antipode of Maimonides' skepticism) comes into play too, for Gersonides does not doubt that the motions of the stars and the precise mode of their exercising providence can (and indeed should) be investigated empirically and that ultimately they will become known to man. It is in fact precisely this idea that underscores the value of the science of the heavens and that supplied Gersonides with the motivation to devote his days (and nights) to it.[62]

I thus suggest that Maimonides' view that the celestial bodies do not exist for the sake of the sublunar world and his rejection of astrology are related. Although there is no logical necessity here, a common tendency nonetheless seems to underlie Maimonides' views on both issues. Maimonides' is a *theo-* rather than *anthropo-*centric worldview, in which the sublunar world, including man, is of negligible significance. But if the celestial bodies do not exist for the sake of anything, if their influence on man is unrelated to their purpose – they govern the world accidentally and not essentially, so to speak – then there is little wonder that man cannot, by gazing at the skies, know his future.[63]

NOTES

1. This subject has received increasing attention in recent years. Cf., e.g., J. D. North, "Medieval Concepts of Celestial Influence: A Survey," in P. Curry, ed., *Astrology, Science and Society. Historical Essays* (Woodbridge/Wolfeboro: Boydell Press, 1987), pp. 5-17; E. Grant, "Medieval and Renaissance Scholastic Conceptions of the Influence of the Celestial Region on the Terrestrial," *Journal of Medieval and Renaissance Studies* 17(1987):1-23. Cf. also T. Litt, *Les corps célestes dans l'univers de Saint Thomas d'Aquin* (Louvain: Publications universitaires; Paris: Béatrice-Nauwelaerts, 1963).

2. Aristotle, *De gen. et corr.*, II.10; F. Solmsen, *Aristotle's System of the Physical World* (Ithaca, NY: Cornell University Press, 1960), pp. 379-89.

3. Aristotle, *De gen. anim.* IV.10, 777b26 ff. Aristotle was unaware that the moon causes the tides.

4. The argument is found in Ptolemy's *Tetrabiblos* I.2 and is repeated by most medieval proponents of astrology. Cf., e.g., R. Lemay, *Abu Ma'shar and Latin Aristotel-*

ianism in the Twelfth Century (Beirut: American University of Beirut, 1962), pp. 50 ff., 55 ff. All natural philosophers, including those who rejected judicial astrology, endorsed this argument and in fact invoked the astrologers as supplying corroborating evidence for it. Consider Ibn Rushd's exposition. After reporting Aristotle's view that the sun's motion along the ecliptic is the efficient cause of generation and corruption (it brings about the four seasons), he goes on to argue:

> This latter movement is not limited to the sun alone, but is also that of the moon and of all the planets, although actually it is more apparent in the case of the sun. For the effect of the sun upon the alteration of the four seasons in its course along its inclined circle is precisely that of *each star* in its course along its specific circle. For although the particular affect which each one of the stars exerts upon things about us is hidden from us, a comprehensive account would reveal that they do have a bearing upon generation and corruption to the extent that if we were to imagine the removal of any star or its movement, generation as a whole or the generation of some beings could not be completed. Indeed, it may be observed that some things are attributed to the action of this or that star. We therefore find that those who have observed the stars since ancient times [the reference is doubtlessly to the astrologers] divided all things into kinds according to them and characterized one thing as stemming from the nature of one star and another thing as stemming from the nature of another star.

Averroes, *Epitome on Aristotle's 'De generatione et corruptione',* English translation by S. Kurland (Cambridge, MA: The Medieval Academy of America, 1958), p. 133.

5. All quotations are taken from M. Maimonides, *The Guide of the Perplexed,* translated with an Introduction and Notes by S. Pines (Chicago: University of Chicago Press, 1963); here 2:9, p. 269.

6. Ibid., 2:10 pp. 270 f.

7. I will argue these theses in detail in a forthcoming book on Aristotle's "chemistry" and its aftermath, to be published by Oxford University Press.

8. Cf. Maimonides, *Guide* 2:12.

9. For a masterly history of these ideas, cf. H. A. Davidson, "Alfarabi and Avicenna on the Active Intellect," *Viator* 3(1972):109–78; *idem,* "Averroes on the Active Intellect as a Cause of Existence," *Viator* 18(1987):191–225.

10. Lemay, pp. 66 f., 74–85.

11. Maimonides, *Guide* 2:5, p. 260.

12. This specific claim was controversial, for some thinkers held that minerals too receive their forms from the "Giver of Forms." Maimonides' view is shared by Averroes: cf. H. Blumberg, ed., *Averrois Cordvbensis Compedia librorum Aristotelis qui Parva naturalia vocantvr,* Arabic version (Cambridge, MA: The Medieval Academy of America, 1972), p. 76; Hebrew version, ed. by *idem* (Cambridge, MA: The Medieval Academy of America, 1954), p. 50; English translation by *idem:* Averroes, *Epitome of Parva naturalia* (Cambridge, MA: The Medieval Academy of America, 1961), p. 44 f. Since Maimonides elsewhere says that *all* forms originate in a separate intellect (*Guide* 2:12, p. 278), his view is apparently that minerals do not have properties beyond those they have by virtue of their components. As Shem-Tov, in his commentary on the *Guide,* points out, Avicenna and al-Ghazâlî were of the opposite opinion. Cf. *Sefer More Nebukhim . . . im Shlosha Perushim . . . Efodi, Sem-Tov, Ibn Qresqas . . .* (Vilna, 1902; reprinted Jerusalem, 1960), p.

112. Their view was also that of Gersonides. Cf. my "Human Felicity and Astronomy: Gersonides's Revolt Against Ptolemy" (in Hebrew), *Daat* no. 22(Winter 1989):58. H. A. Wolfson compares the views of J. Halevi and Maimonides on this issue in his "Halevi and Maimonides on Design, Chance and Necessity" (1941), reprinted in I. Twersky and G. H. Williams, eds., *Studies in the History of Philosophy and Religion*, vol. 2 (Cambridge, MA: Harvard University Press, 1977), pp. 26–34.

13. Maimonides, *Guide* 1:72, pp. 186 f.

14. Ibid.

15. Ibid., 2:6, p. 261.

16. Ibid., 2:10, pp. 269 f. Qalonymos ben Qalonymos, writing at the beginning of the fourteenth century, ironizes on precisely this saying: "*. . . nikkar sho'a lifney dal, lefi she-yesh lo mazzal mi-le-mala, makke oto pizey oheb we-omer lo: gedal.*" A. M. Habermann, ed., *Eben Bohan* (Tel Aviv: Mahbarot Le-Sifrut, 1956), p. 48.

17. Maimonides, *Guide* 2:10, pp. 269 f.

18. Ibid., 2:12, p. 280.

19. Cf. C. Sirat, "Moïse de Narbonne et l'astrologie," *Proceedings of the Fifth World Congress of Jewish Studies*, vol. 3 (Jerusalem: World Union of Jewish Studies, 1972), pp. 61–72.

20. Moshe Narboni, *Be'ur le-Sefer Moreh Nebukhim*, ed. Y. Goldenthal (Vienna: K. K. 1852 Hof- und Staatsdruckerei, 1852), p. 29^r.

21. Cf., e.g., R. Sorabji, "Body and Soul in Aristotle," in J. Barnes, M. Schofield, and R. Sorabji, eds., *Articles on Aristotle*, vol. 4: *Psychology and Aesthetics* (London: Duckworth, 1979), pp. 42–64; F. Nuyens, *L'évolution de la psychologie d'Aristote* (Louvain: Publications Universitaires, 1948).

22. H. H. Biesterfeldt, "Galens Traktat 'Dass die Kräfte der Seele den Mischungen des Körpers folgen' in arabischer übersetzung," *Abhandlungen für die Kunde des Morgenlandes*, vol. 40 (Wiesbaden: F. Steiner, 1973), p. 4. On the great currency of the theory in both Arabic medicine and Arabic philosophy cf. *idem*, "Gâlînûs Quwâ n-nafs, Zitiert, adaptiert, korrigiert," *Der Islam* 63(1986):119–36. The entire theory received its classical exposition in R. Klibansky, E. Panofsky, and F. Saxl, *Saturn and Melancholy* (Cambridge: Cambridge University Press, 1964).

23. Maimonides, *Guide* 2:40, p. 381.

24. Galen devotes his book *Peri Ethon* precisely to these kinds of "accidents," which he denotes by that name, for example, anger, desire, fear, love. Cf. the English translation of the Arabic version of the work: J. M. Mattock, "A Translation of the Arabic Epitome of Galen's Book *Peri Ethon*," in S. M. Stern, A. Hourani, and V. Brown, eds., *Islamic Philosophy and the Classical Tradition: Essays presented . . . to Richard Walzer* (Oxford: Cassirer, 1972), pp. 235–60.

25. Maimonides, *Guide* 2:38, p. 376.

26. Ibid., 2:36.

27. This argument is adduced, e.g., by Ptolemy, *Tetrabiblos*, I.2, ed. and trans. by F. E. Robbins (London: Heinemann; Cambridge, MA: Harvard University Press [The Loeb Classical Library], 1940), p. 13: "Why can we not, too, with respect to an individual man, perceive the general quality of his temperament from the ambient at the time of his birth . . . and predict occasional events, by the use of the fact that such and such an ambient is attuned to such and such a temperament and is favorable to prosperity, while another is not so attuned and conduces to injury." Cf. also 3:1, pp. 223 f.

28. Cf., e.g., such an opponent of judicial astrology as al-Fârâbî. Referring to the effects of the rays of the stars, he states:

On en vient ensuite à examiner l'influence que ceux-ci excercent sur les humeurs des hommes. On peut alors connaître l'influence que ces humeurs excercent sur les actes volontaires. Si on prédit de quelque façon en fonction des corps célestes quelque chose de ce qui concerne un acte volontaire, il se peut qu'on fasse une prédiction [i.e., a correct prediction!] à propos d'actions volontaires dépendant d'humeurs.

Al-Fârâbî then argues that some voluntary actions entirely proceed from deliberation; on these the heavenly bodies have no influence. Quoted from T.-A. Druart, "Le second traité de Fârâbî sur la validité des affirmations basées sur la position des étoiles," *Bulletin de philosophie médiévale* 21(1979):51. The argument is also put forward in Levi ben Gershom, *Sefer Milhamot ha-Shem* 2:2 (Leipzig, 1866), p. 96, quoted in G. Freudenthal, "Levi ben Gershom as a Scientist: Physics, Astrology and Eschatology," *Proceedings of the Tenth World Congress of Jewish Studies,* Division C, vol. 1: *Jewish Thought and Literature* (Jerusalem: World Union of Jewish Studies, 1990), pp. 67–8.

29. This consequence is judiciously drawn by R. Levi ben Gershom who, in accordance with his general philosophy, transforms it into yet another argument from design. Cf. his *Milhamot ha-Shem* 2:2, p. 97 and Freudenthal, "Levi ben Gershom as a Scientist."

30. Maimonides, *Shemonah Perakim,* chap. 8. Cf. also the chapter "Hanhagat Beriut ha-Nefesh," in *Hanhagat ha-Beriut* in S. Muntner, ed., *Moshe ben Maimon (Maimonides), Medical Works* (in Hebrew), vol. 1 (Jerusalem: Mosad Harav Kook, 1987), p. 58 ff. On Maimonides' differing conceptions of what the "healthy" state of the soul in fact is, and consequently of the aims of the "corrective therapy," cf. H. A. Davidson, "The Middle Way in Maimonides' Ethics," *Proceedings of the American Academy for Jewish Research* 54(1987):31–72.

31. More precisely: an environment that is not perfectly balanced will always cause the equilibrium within a composite substance to shift; this is why states of equilibrium do not last. Cf. G. Freudenthal, "The Theory of the Opposites and an Ordered Universe: Physics and Metaphysics in Anaximander," *Phronesis* 31(1986):197–228.

32. The classical study of this doctrine is E. Honigmann, *Die Sieben Klimata* (Heidelberg: C. Winter, 1929).

33. *Kuzari* 1.1; H. Hirschfeld's translation quoted from I. Heinemann, ed., *Jehuda Halevi, Kuzari,* in *Three Jewish Philosophers* (New York: Atheneum, 1977), p. 28.

34. Ibid., 2.10 ff. Cf. the excellent article by the late A. Altmann, "The Climatological Factor in Yehudah Hallevi's Theory of Prophecy" (Hebrew), *Melilah* 1(1944): 1–17. This paper gives a concise but broad overview of the issues to which I briefly allude here in a single paragraph. Cf. also A. Melamed, "The Land of Israel and Climatology in Jewish Thought" (Hebrew), in M. Hallamish and A. Ravitzky, eds., *The Land of Israel in Medieval Jewish Thought* (Jerusalem: Yad Izhak Ben-Zvi, 1991), pp. 52–78.

35. Maimonides, *Pirké Moshe* 25:57–58, *Medical Works* 2:361 f. Maimonides, as he himself indicates, here follows Galen and al-Fârâbî.

36. Maimonides, *Hanhagat ha-Beriut* 4:1, *Medical Works* 1:66.

37. The argument is put forward in Ptolemy, *Tetrabiblos* II.2.

38. Al-Fârâbî, *Sefer ha-Hat'halot*, ed. Z. Filipowski (Leipzig, 1849), p. 33, quoted after, Melamed, "The Land of Israel and Climatology in Jewish Thought," p. 55 f. n. 11.

39. A. Marx, "The Correspondence Between the Rabbis of Southern France and Maimonides About Astrology," *Hebrew Union College Annual* 3(1926):351; Y. Shailat, ed., *Iggerot ha-Rambam*, vol. 2, (Jerusalem, 1988), p. 481; translation cited from R. Lerner, "Maimonides, Letter on Astrology," in R. Lerner and M. Mahdi, eds., *Medieval Political Philosophy: A Sourcebook* (Ithaca, NY: Cornell University Press, 1963), pp. 229 f.

40. A. S. Halkin and B. Cohen, ed. and trans., *Moses Maimonides' Epistle to Yemen* (New York: American Academy for Jewish Research, 1952), pp. 64–65 (texts), p. xiii (translation).

41. Maimonides, *Mishneh Torah, Hilkhot Akum*, 11.

42. L. Strauss, "The Literary Character of *The Guide for the Perplexed*," *Persecution and the Art of Writing* (1952) (reprinted: London and Chicago: The University of Chicago Press, 1988), pp. 84, 94.

43. One of them was quoted above; another is at *Guide* 3:37.

44. Marx, "The Correspondence," p. 351; Shailat, *Iggerot*, p. 481 f.; translation cited from Lerner, "Maimonides, Letter," p. 230.

45. For an overview and bibliography cf. B. R. Goldstein and D. Pingree, *Levi ben Gerson's Prognostication for the Conjunction of 1345*, in *Transactions of the American Philosophical Society* 80, pt. 6, 1990, particularly p. 3 ff.

46. Marx, "The Correspondence" p. 351; Shailat, *Iggerot*, p. 480 f.; translation quoted from Lerner, "Maimonides, Letter," p. 229 f.

47. For the same reason Maimonides postulates that the members of the Sanhedrin must be knowledgeable, among other things, in astrology and "other stupidities of idolatry." This is necessary "in order that they be able to sit to judgment over them." Maimonides, *Mishneh Torah, Sefer Shoftim, Hilkhot Sanhedrin* 2:1. This is not sufficiently appreciated in R. Lerner, "Maimonides' Letter on Astrology," *History of Religions* 8(1968):148.

48. S. Pines, "Excursus. Notes on Maimonides' Views Concerning Free Will," included in his "Studies in Abu'l-Barakât al-Baghdâdî's Poetics and Metaphysics," *Scripta Hierosolymitana* 6(1960):195–8; A. Altmann, "The Religion of the Thinkers: Free Will and Predestination in Saadia, Bahya and Maimonides," S. D. Goitein, ed., *Religion in a Religious Age* (Cambridge, MA: Harvard University Press, 1974), pp. 25–51, reprinted in Altmann, *Essays in Jewish Intellectual History* (Hanover and London: University Press of New England and Brandeis University Press, 1981), pp. 35–64.

49. T.-A. Druart, "Astronomie et astrologie selon Fârâbî," *Bulletin de philosophie médiévale* 20(1978):43–47; *idem*, "Le second traité de Fârâbî sur la validité des affirmations basées sur la position des étoiles"; A. F. Mehren, "Vue d'Avicenne sur le rapport de la responsabilité humaine avec le destin," *Le Muséon* 3(1884):383–403. For a recent overview cf. B. Radtke, "Die Stellung der islamischen Theologie und Philosophie zur Astrologie," *Saeculum* 39(1988):259–67.

50. This point has been well taken by A. Halkin. Cf. *Moses Maimonides' Epistle to Yemen*, Hebrew Introduction, p. xxiii.

51. For this point I am grateful to the late Professor S. Pines.

52. Maimonides' perception of the connection between astrology, star worship, and fatalism is elaborated in Y. T. Langermann, "Maimonides' Repudiation of Astrology," forthcoming in *Maimonidean Studies 2*. The fatalism to which astrology is prone to give rise is delightfully described by al-Fârâbî: if all events were predetermined, hope

and fear would cease and all incentive to do anything on behalf of one's future would be thwarted. For want of fear, no one would anymore obey his superiors and social order would be subverted. Cf. F. Dietirici, ed., *Al-Fârâbî's philosophische Abhandlungen aus Londoner, Leidner und Berliner Handschriften* (Leiden: Brill, 1890), p. 105, translated in *idem, Al-Fârâbî's philosophische Abhandlungen aus dem Arabischen übersetzt* (Leiden: Brill, 1892), p. 174. Al-Fârâbî did not anticipate that some six centuries after his time, Calvinistic and Puritan doctrines of predestination would, according to M. Weber's and R. K. Merton's analyses, play a major role in giving rise to capitalism and modern science. Cf. my review of F. Klein-Franke, *Vorlesungen über die Medizin im Islam* (= *Sudhoffs Archiv,* Beiheft 23) (Wiesbaden: Franz Steiner, 1982), in *History and Philosophy of Life Sciences* 9(1987): 119–22.

53. Cf. *Guide* 2:24, and, generally, S. Pines, "The Limitations of Human Knowledge According to Al-Farabi, Ibn Badja, and Maimonides," in I. Twersky, ed., *Studies in Medieval Jewish History and Literature* (Cambridge, MA: Harvard University Press, 1979), pp. 82–109.

54. Maimonides, *Guide* 2:37.

55. Cf. A. Funkenstein, "Maimonides: Political Theory and Realistic Messianism," *Miscellanea Mediaevalia* 11(1977):88–90; *idem, Maimonides: Nature, History and Messianic Beliefs* (Hebrew) (Tel Aviv: Ministry of Defense, 1983), pp. 30–33; and *idem, Theology and the Scientific Imagination* (Princeton: Princeton University Press, 1986), pp. 227–39.

56. Maimonides, *Guide* 3:15, pp. 451–2.

57. This crucial point has been forcefully and repeatedly stated by Y. Leibowitz, notably in his *The Faith of Maimonides* (New York: Adama, 1987), chap. 3.

58. Cf. Maimonides, *Guide* 3:15, p. 455.

59. Cf. Levi ben Gershom, *Milhamot ha-Shem,* p. 194 ff.; Ch. Touati, *La pensée philosophique et théologique de Gersonide* (Paris: Editions de Minuit, 1973), p. 306.

60. *Milhamot ha-Shem,* bk. 5, pt. 2, chaps. 7–9.

61. Ibid., bk. 2, chap. 2, pp. 95–98; Freudenthal, "Levi ben Gershom as a Scientist."

62. Cf. G. Freudenthal, "Human Felicity and Astronomy."

63. For helpful remarks on a draft of this paper I am grateful to Professors Bernard R. Goldstein (Pittsburgh) and Samuel Kottek (Jerusalem). To Dr Y. Tzvi Langermann (Jerusalem) I am grateful for a prolonged exchange of views on the issues discussed here and for having sent me prior to its publication a copy of his paper, "Maimonides' Repudiation of Astrology," forthcoming in *Maimonidean Studies* 2.

IV

Gersonides:
Levi ben Gershom

❧ INTRODUCTION ❧

Rabbi Levi ben Gershom, or Gersonides (1288–1344), is one of the most original medieval Jewish thinkers, whose interests and writings spanned philosophy, biblical exegesis, astronomy, mathematics, natural science, logic and medicine. Like most contemporary Jewish philosophers in southern France, Gersonides wrote in Hebrew and drew almost solely on sources available to him in that language. But since most of these were translations from Arabic, Gersonides can be viewed as an innovative continuer of the Arabic philosophical tradition that had culminated in Ibn Rushd (Averroes), indeed as someone who developed his own philosophical ideas through a critical dialogue mainly with two major thinkers who had written in Arabic: Maimonides and Ibn Rushd, as well as, to a lesser extent, the astronomer al-Biṭrūjī.

Although Gersonides greatly admired Maimonides and embraced the latter's programme of creating a synthesis of Judaism and Peripateticism (in one of its versions), he was yet in sharp opposition to cardinal Maimonidean positions. In a nutshell, Gersonides upheld, *contra* Maimonides, that (1) it can be demonstrated that God purposefully created the world in time; that (2) God has designed the world so as to suit perfectly the sublunar creatures living in it, particularly humans; that (3) humans are capable of knowing the world and indeed human perfection consists in acquiring such knowledge; and that (4) knowledge about the created world in fact bears upon the Creator, who therefore to some extent is knowable by man. Thus, Maimonides' uncompromising anti-anthropocentrism, his epistemological scepticism and the associated negative theology, as well as his elitism and esoterism, are all emphatically rejected by Gersonides: on both the cosmological and the epistemological planes, Gersonides' world-view is decidedly optimistic. Gersonides' commitment to the idea of scientific progress and his life-long scientific practice are the consequences of this confidence in the privileged position of humankind in God's world.

➳ LIFE AND WORKS[1] ➳

Our knowledge of Gersonides' life is very scanty. He was born in 1288 and lived most of his life in Orange in southern France, which had a middle-size Jewish community.[2] We do not know anything definite about the course of his studies or about who his teachers were, although a few references in his writings to opinions held by his father suggest that the latter was a scholar too.[3] Gersonides' knowledge of Arabic and Latin has been the subject of some controversy. In his writings Gersonides mentions only works available in Hebrew, although on a few occasions he remarks that he checked the Arabic version of a problematic passage;[4] it seems certain however that he could not read entire works in Arabic.[5] The same presumably holds with respect to Latin: although, as the late Shlomo Pines has shown, Gersonides' doctrine of divine attributes reveals similarities to contemporary Scholastic doctrines, this possible influence was presumably oral.[6]

When Gersonides was eighteen years old (1306), Philippe the Fair expelled all Jews from the Kingdom of France. Yet this historic catastrophe (which did not hit Orange) left no definite traces in Gersonides' writings, although it may perhaps be accountable for the fact that Gersonides began writing relatively late in his life. He set on writing his major philosophic work, the *Sefer Milḥamot ha-Shem*[7] ("The Wars of the Lord"), in 1317, at the age of twenty-nine, and was to pursue it during the following twelve years. In parallel, however, he composed two series of works. The first series, written between *c.* 1319 and 1324, consists of specialized scientific treatises: an innovative work on logic, *The Book of the Correct Syllogism* (1319); a treatise in arithmetic comprising an original chapter on combinatorial theory (1321); and a set of super-commentaries on many of Ibn Rushd's epitomes of, or middle commentaries on, Aristotle's treatises in natural philosophy (1321–4).[8] Subsequently, Gersonides set out to write a series of commentaries on various biblical books: Job (1325); Song of Songs (1326); Ecclesiastes (1328); Esther (1329); Ruth (1329); Genesis (1329); Exodus (1330); most of Leviticus (1332). After an interruption of a few years, Gersonides pursued the series with commentaries on Isaiah; the remaining books of the Torah (Pentateuch; completed 1338); the First Prophets (1338); Daniel (1338); Ezra, Nehemiah and the Books of Chronicles (1338); and the Proverbs (1338).[9]

Concomitantly with these philosophic–theological writings, Gersonides most intensively pursued an astronomical research programme. In fact, book 5, part 1 of the *Wars of the Lord* (which comprises six books) is a fully fledged technical astronomical treatise, whose 136 chapters (mostly still in manuscript[10]) are about equal in length to the rest of the *Wars*. This work, often considered as independent and referred to as Gersonides' *Astronomy*, contains the results of Gersonides' own astronomical observations (begun at least in 1320 and continued throughout his life), tables, an incisive criticism of Ptolemy's astronomy and the descriptions of Gersonides' own astronomical models for the different planets.

Gersonides' accomplishments in astronomy and mathematics made him into a highly respected figure, even outside the Jewish community. Whereas most surprisingly we know next to nothing about contacts (intellectual or other) Gersonides presumably had with Jewish contemporaries,[11] we have some information about his continued connections with high-ranking Christians. Early in his career he composed his astronomical tables "at the request of many great and noble Christians"[12] and in 1342 the influential Philippe de Vitry, the future Bishop of Meaux, asked him for advice on a mathematical theorem connected with his own *ars nova* in musical theory.[13] Also in 1342, Gersonides dedicated to Pope Clement VI the Latin version of a trigonometrical treatise, drawn from his *Astronomy*: since, as has recently been shown, this translation is a part of the (incomplete) Latin translation of Gersonides' astronomical work, presumably the translation (in which Gersonides collaborated actively) was done at the behest and under the patronage of the Papal court.[14] Lastly, Gersonides on at least two occasions composed astrological predictions at the request of two popes. The last of them, a prognostication for the great conjunction of 1345, was composed by Gersonides on his deathbed, and through this circumstance we know the time of his decease with unusual precision: the Latin translator of the *Prognostication* informs us that "Master Leo, prevented by death in the year of Christ 1344 on the 20th day of April about noon, put nothing more in order concerning this conjunction".[15]

GERSONIDES' VIEWS ON HUMAN KNOWLEDGE, GOD, CREATION AND THE IMMORTALITY OF THE SOUL

The bedrock on which rests Gersonides' entire philosophic and scientific endeavour is perhaps his unlimited confidence in the power of human reason to attain ever more knowledge of the world and, hence, of God. Maimonides had argued for sceptical positions on a series of questions, not the least important being the question of the createdness or eternity of the world. Gersonides unambiguously rejects Maimonides' stance at the very beginning of the *Wars*:

> Many people will deem it to be arrogance and audacity on our part that we inquire into [the question of] the eternity or createdness [of the world]. For they may perhaps think that the intellect of the wise man is wanting of means to attain the truth on this problem, except if he be a prophet. All the more so since they see that the earlier perfect [men] of our nation, and among them the crown of the glory of the sages of the Torah, our Master Rav Moshe ben Maimon, may he rest in peace, did not pursue an inquiry on such a topic. They may conclude that it is impossible to attain [knowledge] on this question through the means of [philosophical] inquiry. For if this were possible, it would not have escaped the earlier [sages].
>
> Yet this is a very weak argument. For that which had escaped the early [sages] need not necessarily escape their successors as well. For time suffices to bring forth the truth, as the Philosopher said in Book Two of the *Physics*.[16] Indeed, were it otherwise, then there would be no one who, investigating one of the sciences, would know anything but what he had learnt from others. But, should this be assumed to be the case, then there would be no science at all, and this is patently false.[17]

Attaining new knowledge of the world by means of rational, scientific inquiry is possible, Gersonides holds. This confidence in man's capability to know the world has momentous consequences also for Gersonides' view of man's knowledge of God. Gersonides construes the Active Intellect as comprising the *nomos* – in fact the entire natural order – of the created world. This implies that every bit of knowledge about the world is at the same time knowledge of the Active Intellect and hence (as will be seen) of the divine plan of creation. Consequently, if one apprehends an empirical fact or even a mathematical theorem, one has thereby apprehended an intelligible that is a constitutive part of the Active Intellect: one can therefore attain an adequate, if partial, knowledge of the Active Intellect. Now the *nomos* of the created world, which makes up the Active Intellect, is the object of God's thought: it is in fact through God's thinking the *nomos* that it has come into existence, an idea Gersonides borrowed from Themistius.[18] It follows that God's knowledge and man's have the same object, viz. the *nomos*, and that they differ only by degree: "it is clear that the sole and only difference between the knowledge of God, may He be blessed, and our knowledge is that His knowledge is exceedingly more perfect".[19] One can thus attain some positive knowledge of God: Gersonides in fact rejects the Maimonidean thesis that predicating attributes of God would introduce in Him a multiplicity.

Gersonides' optimistic epistemology provides the basis for his heartening theory of the immortality of the soul. To the commonplace view that one's perfection and afterlife depend on the knowledge one had acquired during one's lifetime Gersonides gives a personal twist. Contrary to Maimonides, he holds that the knowledge that is conducive to felicity is not only, and not even mainly, metaphysical, bearing on the separate entities, but rather knowledge of the material world (being in fact knowledge of the Active Intellect).[20] Further, Gersonides shares the received view that eternal felicity belongs to the acquired intellect – to that part of the rational soul that has been actualized by apprehending intelligibles. But whereas most Jewish philosophers, apparently including Maimonides, followed Ibn Bājjah and Ibn Rushd in holding that after death the acquired intellect loses its individuality by being fused into the Active Intellect,[21] Gersonides upholds the survival of the *individual* acquired intellect[22]: acquiring knowledge, specifically empirical knowledge, thus is the supreme good in life. This view gave Gersonides both a theological legitimation for his scientific research and a forceful motivation to invest himself in it. On the question of one's route to eternal felicity too, then, Gersonides and Maimonides parted company.

One cardinal question that can be submitted to scientific inquiry is whether the world is created or eternal. Gersonides, as already noted, believes, *pace* Maimonides, that he can adduce *proofs* for the createdness of the world. These proofs are largely based on what Gersonides takes to be empirical evidence, namely to the effect that the entire cosmos is perfectly designed. For instance:

> in the foregoing it has been conclusively established that whatever is found in the substance of the heaven is of the utmost possible perfection with a view to perfecting these [sublunar] beings.
> Indeed, were that [heavenly] order corrupted even slightly, these beings would be corrupted [i.e. destroyed] too.[23]

Heavens which are so perfectly designed *with a view to* endowing sublunar existence with the utmost possible perfection cannot but be intentionally, and hence "newly", created, whence it follows that the entire world was created by the volition of a wise Creator.[24]

Gersonides' original cosmogony seeks to give a scientifically sound explanation of creation which is in conformity with the account given in the Torah, reconciling at the same time the thesis of creation in time with the impossibility, postulated by Aristotelian science, of any coming-to-be *ex nihilo*.[25] Gersonides posits a pre-existing "body devoid of all forms", and affirms that the act of creation consisted in God's imprinting upon it the elemental forms: thus ensued the four sublunar elements and the heavenly bodies.[26] The Creator, Gersonides further maintains, conceived these supra- and sublunar forms in such a way that through their influences the celestial bodies would continually control the generation and corruption in the sublunar realm. This is of primary importance. Gersonides gave great prominence to the received medieval physical theory on which the sublunar world is not a closed system: the forms of substances (the vegetative souls of plants and animals, notably, but also the specific forms of some minerals such as the magnet) would not come to be, nor would they subsist, without the informing and sustaining influences continuously issuing from the heavenly bodies. Gersonides repeatedly stresses, drawing on Aristotle's *Meteorologica*, 4, that the equilibrium of the opposite qualities (hot/cold; dry/humid) constituting any sublunar substance is inherently unstable and precarious: left to itself, any substance would soon perish, because one of the qualities would overpower the others:[27] the fact that sublunar substances usually persist over certain periods of time is thus due to the "preserving" influences of the heavenly bodies. (These "influences" were held to consist of "efficient causes" transmitted by the stars' rays, and of "formal causes" emanating from the separate intellects moving the stars.[28]) Gersonides sees the perfection of the world as a whole as consisting precisely in the fact that these celestial influences are faultlessly conceived so as to endow sublunar substances – particularly humanity, the most perfect among them – with maximal perfection and perseverance, a sure indication of a divine plan. Gersonides is here in diametrical opposition to Maimonides' radical anti-anthropocentric stance.[29]

The combination of the influences of the heavenly bodies with the aptitude of the sublunar matter to be suitably affected by them, all "programmed" at the creation, constitute the natural order: once the formless quasi-matter received its forms, the universe became autonomous, functioning solely according to the *nomos* resulting from the interactions of the natures which God has given to its different parts. (More precisely: each separate intellect controls the influences – formal and efficient – emanating from "its" planet; the synthesis of the partial knowledges of all the intellects is the Active Intellect and is in fact the *nomos* of the world.[30]) The consequence is that all events and processes which have taken place after the first act of creation, including the sequel of creation and the events reputed to be miracles, are subsumable under *naturalistic* explanations. Here, as in most of his natural science, Gersonides is obviously a rigorous follower of Ibn Rushd.

Gersonides saw no contradiction whatsoever between his belief that the natural order was autonomous and his commitment to the authoritative texts of Judaism: he rather saw them as fundamentally compatible and complementary. To him, truth could be attained either through scientific inquiry or through a hermeneutic inquiry into the Torah – but both routes were equivalent, necessarily leading to the same single truth. Thus, Gersonides stresses that it is not the case that religious belief *constrained* him to accept the traditional view of temporal creation as found in the Torah: rather, since the Torah is "a *nomos* perfected to the utmost" guiding one to one's ultimate felicity, its statements are necessarily true and in fact directed him in his scientific inquiry.[31] Gersonides would certainly have endorsed the later metaphor according to which Scripture and the book of nature were written by the same hand: revelation and reason are perforce equivalent.

As postulated by Gersonides, the celestial bodies' control over all generation and corruption "down here" naturally encompasses living beings, including humankind. This doctrine, however, does not imply determinism. Gersonides holds each of the celestial bodies to exert its influences only on one *general* aspect of the sublunar physical reality (e.g. the sun "fortifying" the quality of heat, the moon that of humidity, etc.). Consequently, even the Active Intellect, and God too, can have no knowledge of *singular* events. Specifically, while at any time the astral influences give one a *disposition* to act in a certain way (as when one's "heat" is increased and one tends to behave hot-headedly), one can, by following one's intellect rather than one's passions, extricate oneself from the effect of these influences. Gersonides, indeed, forcefully upheld human free will.

The theory of astral influences upon sublunar processes to some extent opens the door for astrology: this is recognized even by Maimonides.[32] Gersonides in fact accords astrology a role, albeit a limited one, in keeping with his view that only general aspects of sublunar occurrences are determined by the heavenly bodies, and with the associated notion of human free will. Unlike Maimonides, who mainly for religious reasons opposed astrology vehemently, Gersonides believed that by being able to predict dispositions to certain types of behaviour, astrologers occasionally succeed in their forecasts of singular events, a feat that is all the more remarkable if one considers that the knowledge of the celestial movements and of astral influences are both (still) wanting.[33] But astrologers cannot, in Gersonides' view, possibly foresee with certainty singular events concerning a given individual. Indeed, Gersonides' only preserved prognostication predicts events involving entire nations, i.e. a great number of individuals: in a large mass, only a few individuals extract themselves from natural determination by following reason using their intellects; the great majority continue to belong to the realm of nature, and so their conduct is largely predictable. Therefore, great upheavals in history (natural and human) can be foreseen by astrologers, although the final and crucial upheaval, namely the establishing of the eternal messianic

Kingdom, will be due to God's special providence and His intervention in the course of history, not to natural necessity.[34] Gersonides ascribes foreknowledge of singular events not to astrologers but notably to prophets, who receive "revelations" from the active intellect: the latter communicates to the prophet "information" pertaining to the general order of reality, which the prophet then applies to the concrete reality, thereby arriving at concrete true predictions.

Traditionalist thinkers of later generations castigated Gersonides for his naturalism, which seemed to belittle miracles. Similarly, his view that God has no knowledge of individuals (because the pre-programmed celestial influences determine only general aspects of the occurrences in the sublunar world and the associated doctrine of free will) implied a denial of individual providence and seemed to leave God no place within human, specifically Jewish, history: this was another stance for which later Jewish thinkers were to disparage him. Within the history of Jewish thought Gersonides' image is that of an audacious freethinker.[35]

GERSONIDES' EMPIRICISM: NATURAL SCIENCE AND ASTRONOMY

The foregoing will have made clear that for Gersonides empirical knowledge of the material world is of crucial importance. It is therefore not surprising that Gersonides himself engaged in science. In natural science his main theoretical paradigms are borrowed from Ibn Rushd: Gersonides' supercommentaries on Ibn Rushd's epitomes and commentaries reveal a profound agreement, although Gersonides very often interjects personal statements to dissent on specific points.

The most remarkable feature of Gersonides' science, of both the sub- and the supralunar realms, is the pronounced empirical attitude it displays. Gersonides apparently conducted botanical experiments: on the occasion of a statement by Ibn Rushd concerning the relationship between the germination of seeds and the type of the soil, Gersonides remarks briefly: "we have tested this [affirmation] for all the [kinds of] seeds and found that the matter is always as stated by Ibn Rushd".[36] Gersonides also envisaged the use of a parabolic mirror as a sort of microscope in order to examine the parts of animals which are too small to be observed with the naked eye: this very impressive idea, which presumably remained unrealized, apparently has no parallel at the time, at least in Europe.[37] Indeed, for Gersonides even the most humble empirical fact was an intelligible, a component of the world's divine *nomos* continuously thought by God, so that by apprehending whatever component of it one shares in His knowledge; apprehending whatever fact was conducive to the immortality of one's soul.

Yet it is the celestial bodies that were at the focus of Gersonides' scientific research programme. God exerts His providence over the created world through the celestial bodies: therefore, by studying their design one can gauge the perfection of the Creator. What counts for Gersonides and what he finds so remarkable is not only the perfection of the celestial realm *per se* – the constancy and regularity of the heavenly motions – but above all the supposed fact that the heavenly realm is perfected so as to bring about and constantly maintain the (relative) perfection of the sublunar realm as well. Gersonides in fact holds that the most tiny details of each and every of the heavenly motions and of the influences emanating from them are indispensable for the preservation of the ordered world: this is why their study reveals God's divine plan, bearing witness to the Creator's wisdom and goodness. Astronomy therefore emerges as the divine science par excellence:

> The prophets and those who spoke by virtue of the Holy Spirit made us aware that it is appropriate to expand this [astronomical] investigation because from it we are led to understand God, as will become evident in this study. Indeed, the orbs and the stars were created by the word of God, as will become clear from our treatise, God willing, by making evident the ampleness of God's wisdom and the ampleness of His power [as manifest] in His bringing into existence these noble bodies in this wondrously wise way and in His endowing them with heterogeneous emanations – even though [the heavenly bodies] are all of one single nature, devoid of the qualities that emanate from them – by virtue of which this lowly [sublunar] existence is perfected.[38]

Gersonides' motivation to study the heavens was thus theological and philosophical; indeed he accorded little value to knowledge whose finality is practical, and on one occasion even adduces astronomy as an example of a science devoid of practical utility.[39] This outlook profoundly shaped the astronomical theory he was to elaborate. Many, presumably most, medieval astronomers approached the study of the heavenly motions with an "instrumentalist", or "fictionalist", image of science: they took their job to consist in "saving the phenomena", i.e. in devising mathematical models and in calculating tables from which stellar positions could be determined with sufficient accuracy; it did not matter to them that the models they used were incompatible with the received (Aristotelian) physics.[40] But Gersonides obviously could not accept this position: his immodest aim was to uncover the blueprint of creation, not to tinker with merely useful computational models. Necessarily, therefore, his epistemology was bound to be realist. Consequently, since he wanted to know the configuration of the supralunar realm as it really was, he set out to construct a theory of the heavens that would accord both with calculation and with physical theory, explicitly rejecting the instrumentalist construal of astronomy.[41]

The awareness of the problem posed by the incompatibility of (Aristotelian) physics and mathematical (Ptolemaic) astronomy was not new: Gersonides is an heir to an Andalusian tradition which goes back at least to Ibn Bājjah and Ibn Ṭufayl and is echoed by Ibn Rushd and by Maimonides;[42] the latter indeed qualified the problem as "the true perplexity", whose resolution was presumably beyond human ken, but which must not preoccupy the astronomer, who should confine himself to calculations.[43] In rejecting the received instrumentalism of the astronomers, Gersonides in fact walked in the footsteps of the astronomer al-Biṭrūjī, whose astronomical treatise had been translated into Hebrew in 1259.[44] Yet, although sympathetic to al-Biṭrūjī's goal, Gersonides found that the latter's system was unsatisfactory: it was refuted by observation and, in addition, was incompatible with the principles of physics and metaphysics.

Gersonides' goal in studying astronomy – to achieve immortality by acquiring some knowledge of the *nomos* of the world – implied that precision was of the highest value: every error in apprehending an intelligible would be fatal to the soul's survival. This is what presumably incited Gersonides to undertake astronomical observations, which he used to test the planetary models – both very rare procedures in the Middle Ages.[45] In order to ensure precision, Gersonides devised two instruments. One, called "Jacob's Staff", allows the determination of the angular distance between planets. The second combines the Jacob's Staff with a *camera obscura* and is used to determine the apparent sizes of the planets.[46] The invention of this instrument depended on Gersonides' philosophical concerns, because the apparent sizes of planets were a relevant parameter for astronomical theory only from a realist stance. (From an instrumentalist perspective only the positions of the planets are of fundamental importance.) Thus, underlying Gersonides' astronomical innovations is his astronomical realism, which in turn depends on his global philosophy.

❧ CONCLUSION ❧

Looming behind Gersonides' variegated cognitive quests was a threefold confidence: firstly, that knowledge of the world was also knowledge of God; secondly, that such knowledge was attainable; and thirdly that knowledge was the guarantor of the immortality of the individual soul. On all these pivotal points Gersonides' views are antithetic to those of Maimonides, just as, generally, Gersonides opposes Maimonides on most crucial issues, while at the same time he follows globally the Maimonidean programme of creating a synthesis of Torah and philosophy.

The distinctive quality of Gersonides' intellectual endeavour seems to be its quest for consistency and coherence: Torah and philosophy, mathematical astronomy and physical theory, theory and observation, all had to match. Knowledge had different, equally valid sources – sense experience, theory and revelation (transmitted through tradition) – and, if properly understood, they could not but lead up to identical results. "It is the hallmark of truth that it agrees with itself from all aspects", Gersonides repeats time and again after having shown that different methods of inquiry yielded one and the same conclusion.

This search for coherence (again the converse of Maimonides) had different consequences in philosophy and in science. Gersonides' philosophical positions are constructed from materials he found in the writings available to him, most notably those of Ibn Rushd: in philosophy Gersonides' drive for consistency results in an "instinct for originality [that] expresses itself in manoeuvring among the texts at his disposal".[47] Gersonides' philosophy indeed remained thoroughly medieval. By contrast, in his scientific work, the quest for coherence resulted in a scientific practice which is entirely modern in outlook (although, to be sure, not in content): Gersonides made his own astronomical observations and criticized and revised mathematical models in their light. It is here, in his science which is wholly out of tune with the norms of the age, that Gersonides' originality bore its best fruits.

❧ NOTES ❧

1 In what follows only few indications about editions of Gersonides' writings, translations and secondary literature are given. Full information can be found in Kellner (1992).

2 The fact that Gersonides was referred to in Latin as "magister Leo de Balneolis" gave rise to the persistent error that he lived in Bagnols-sur-Cèze in the Département du Gard. In point of fact, "de Balneolis" was the name of an extended family living in Orange. Cf. Shatzmiller (1972).

3 It has been repeatedly conjectured that Gersonides' father was Gershom ben Shlomo, the author of the well-known encyclopedic work Sha'ar ha-Shamayim; cf. Shatzmiller (1992).

4 Cf. Lévy (1992).

5 Touati (1973): 38f.; Feldman (1984): 5ff.

6 Pines (1967): 31ff.; Touati (1973): 38; Pines (1986a).

7 There are two editions of the Hebrew text: Gersonides (1560); Gersonides (1866).

8 Ibn Rushd's works on which Gersonides wrote supercommentaries are notably the following: the Epitome of, and the Middle Commentary on the *Physics* (1321); the Epitome of *De generatione et corruptione* (1321); the Epitome of *De caelo* (1321); the Epitome of the *Meteorologica* (1322); the Epitome of books

11 to 19 of the so-called *Book of Animals* (= *The Parts of Animals* and *The Generation of Animals*; 1323); the Middle Commentary on the first seven books of the *Organon* (1323); the Epitome of the *De anima* (1323); the Epitome of the *Parva naturalia* (1324); the Middle Commentary on the *Metaphysics* (written before 1328; lost).

9 For the works and their dates cf. Touati (1973): 49–82; Feldman (1984): 8–30; the works are chronologically arranged in Weil-Guény (1992).

10 Only chapters 1–20 have been published (with an English translation) in Goldstein (1985).

11 Only very recently it has been discovered that Gersonides taught philosophy to a group of students, none of whom however rose to any distinction. Cf. Glasner (1995).

12 Goldstein (1974): 20.

13 Cf. Chemla and Pahaut (1992) for a study of this work.

14 Mancha (1992).

15 Goldstein and Pingree (1990): 34.

16 For references cf. Touati (1973): 87–8, including note 28.

17 *Wars*, Introduction; Gersonides (1560): 2va; Gersonides (1866): 4.

18 Gersonides knew Themistius' Commentary on book Lambda of Aristotle's *Metaphysics*, which had been translated into Hebrew in 1255. Cf. Pines (1987): 199f.; Davidson (1992).

19 *Wars*, 3.3; Gersonides (1560): 22vb; Gersonides (1866): 133.

20 For Maimonides, not the apprehension of natural entities, composed of matter and form, but rather the intellection of separate – divine – entities, results in the survival of the soul. Man's true perfection is in studying metaphysics, not physics. Cf. the famous parable in *Guide of the Perplexed*, 3.51 and Harvey (1977). Yet, as is well known, Maimonides in fact paradoxically holds that the separate entities, whose knowledge alone he holds to lead to salvation, are in fact unknowable, with the consequence that finally happiness is to be sought in the practical-political realm; cf. e.g Pines (1979); Stern (1995). The late Shlomo Pines suggestively argued that this contradiction is the result of a dramatic change of mind on Maimonides' part: cf. Pines (1986b).

21 Pines (1963b): ciiif. For an overview of the background cf. Leaman (1985): 87–107.

22 Cf. Feldman (1978).

23 *Wars* 6.1.7; Gersonides (1560): 51va, Gersonides (1866): 310.

24 Cf. also Feldman (1967); Davidson (1987): 209–12.

25 Cf. Freudenthal (1986).

26 Gersonides believed he could empirically confirm the existence of the primeval quasi-matter. Medieval physical astronomy postulated the existence of rotating spheres carrying the planets; these spheres had to turn independently, so as not to perturb one another's motion. To "isolate" the motions of the spheres, Gersonides argued, there must be a fluid matter filling the inter-spherical spaces, and this is none other than the rest of the "formless" quasi-matter, out of which all celestial and sublunar matter was created.

27 Cf. Freudenthal (1995).

28 Cf. Freudenthal (1993).

29 Maimonides holds that the stars "do not exist for our sake and so that good

should come to us from them"; cf. *Guide*, 3.15. His unbending anti-anthro-pocentrism has forcefully and repeatedly been highlighted by the late Yeshaiahu Leibowitz; cf. notably Leibowitz (1987): chapter 3. Gersonides, by contrast, maintains that the "stars are in the spheres not for their own sake, but in order to exert influence on this sublunar existence" so as to perfect it to the utmost; *Wars*, 5.2.3; Gersonides (1560): 32; (1866): 196.

30   The notion that the Active Intellect is a "synthesis" of the other intellects is one of the few innovative points in Gersonides' theory of the intellect. For an exhaustive comparison of Gersonides' views with his sources cf. Davidson (1992).

31   Cf. *Wars*, 6.2.1; Gersonides (1560): 69; Gersonides (1866): 419.

32   Cf. Maimonides, *Guide*, 2.12; Freudenthal (1993).

33   *Wars*, 2.2; Gersonides (1560): 17; Gersonides (1866): 95.

34   Cf. Goldstein and Pingree (1990); Freudenthal (1990).

35   Touati (1973): 541ff.; Kellner (1976).

36   Gersonides, *Supercommentary on* [Ibn Rushd's] *Epitome to the "Book of Animals"*, MS Vatican Urb. 42, fol. 44; quoted after Freudenthal (1989): 62.

37   *Ibid.*, fol. 9f.; quoted after Freudenthal (1989): 62.

38   *Astronomy*: chapter 2 (= *Wars*, 5.1.2); quoted (with modifications) after Goldstein (1985): 24 (English), 303 (Hebrew).

39   For what follows cf. Freudenthal (1989); Freudenthal (1992b); Freudenthal (1992c).

40   Cf. Duhem (1908); Jardine (1984): 225–57; Hugonnard-Roche (1992).

41   *Astronomy*: chapter 1 (= *Wars*, 5.1.1), Goldstein (1985): 305 (Hebrew), 22 (English); cf. also Goldstein's introductory remarks in *ibid*.: 2–9.

42   Cf. Gauthier (1909); Sabra (1984).

43   Maimonides, *Guide*, 2.24. For a different interpretation of Maimonides' views cf. Langermann (1991).

44   Goldstein (1971).

45   Gersonides recorded forty-five observations of planetary longitudes and latitudes; cf. Goldstein (1988).

46   Cf. Goldstein (1991).

47   Davidson (1992): 195.

REFERENCES

Chemla, K. and Pahaut, S. (1992) "Gersonide et la théorie des nombres", in Freudenthal (1992a): 149–91.

Davidson, H. A. (1987) *Proofs for Eternity, Creation, and the Existence of God, in Medieval Islamic and Jewish Philosophy* (New York).

—— (1992) "Gersonides on the Material and Active Intellects", in Freudenthal (1992a): 195–265.

Duhem, P. (1908) *Sozein tà phainomena. Essai sur la notion de théorie physique de Platon à Galilée* (reprinted Paris, 1982).

Feldman, S. (1967) "Gersonides' Proofs for the Creation of the Universe", *Proceedings of the American Academy for Jewish Research*, 35: 113–32.

—— (1978) "Gersonides on the Possibility of Conjunction with the Agent Intellect", *American Jewish Studies Review*, 3: 99–120.

—— (1984) Levi ben Gershom (Gersonides), *The Wars of the Lord*. Volume One: *Book One: Immortality of the Soul*, trans. with introduction and notes by Seymour Feldman (Philadelphia).

—— (1987) Levi ben Gershom (Gersonides) *The Wars of the Lord*. Volume Two: *Book Two: Dreams, Divination, and Prophecy; Book Three: Divine Knowledge; Book Four: Divine Providence*, trans. and with an appendix and notes by Seymour Feldman (Philadelphia).

Freudenthal, G. (1986) "Cosmogonie et physique chez Gersonide", *Revue des études juives*, 145: 295–314.

—— (1989) "Human Felicity and Astronomy: Gersonides' Revolt Against Ptolemy" (Hebrew), *Da'at*, 22: 55–72.

—— (1990) "Levi ben Gershom as a Scientist: Physics, Astrology and Eschatology", *Proceedings of the Tenth World Congress of Jewish Studies*, Division C, 1, *Jewish Thought and Literature* (Jerusalem): 65–72.

—— (ed.) (1992a) *Studies on Gersonides – a Fourteenth-Century Philosopher-Scientist* (Leiden).

—— (1992b) "Sauver son âme ou sauver les phénomènes: sotériologie, épistémologie et astronomie chez Gersonide", in Freudenthal (1992a): 317–52.

—— (1992c) "Rabbi Lewi ben Gerschom (Gersonides) und die Bedingungen wissenschaftlichen Fortschritts im Mittelalter: Astronomie, Physik, erkenntnistheoretischer Realismus, und Heilslehre", *Archiv für Geschichte der Philosophie*, 74: 158–79.

—— (1993) "Maimonides' Stance on Astrology in Context: Cosmology, Physics, Medicine, and Providence", in Fred Rosner and Samuel S. Kottek (eds), *Moses Maimonides: Physician, Scientist and Philosopher* (Northvale and London): 77–90.

—— (1995) *Aristotle's Theory of Material Substance. Form and Soul, Heat and Pneuma* (Oxford).

Gauthier, L. (1909) "Une réforme du système astronomique de Ptolémée, tentée par les philosophes arabes du XIIe siècle", *Journal asiatique*: 483–510.

Gersonides (Levi ben Gershom) (1560) *Sefer Milḥamot ha-Shem* (Riva di Trento).

—— (1866) *Milchamot Ha-schem. Die Kämpfe Gottes. Religionsphilosophische und kosmische Fragen, in sechs Büchern abgehandelt von Levi ben Gerson* (Leipzig).

Glasner, R. (1995) "Levi ben Gershom and the Study of Ibn Rushd in the Fourteenth Century: Historical Reconstruction", *Jewish Quarterly Review*, forthcoming.

Goldstein, B. R. (1971) *Al-Bitrûjî: On the Principles of Astronomy*, 1: *Analysis and Translation*; 2: *The Arabic and Hebrew Versions* (New Haven).

—— (1974) *The Astronomical Tables of Levi ben Gerson* (= *Transactions of the Connecticut Academy of Arts and Sciences*, 45) (New Haven).

—— (1985) *The Astronomy of Levi ben Gerson (1288–1344): A Critical Edition of Chapters 1–20 with Translation and Commentary* (New York and Berlin).

—— (1988) "A New Set of Fourteenth-Century Planetary Observations", *Proceedings of the American Philosophical Society*, 132(4): 371–99.

—— (1991) "Levi ben Gerson: On Astronomy and Physical Experiments", in Sabetai Unguru (ed.) *Physics, Cosmology and Astronomy, 1300–1700: Tension and Accommodation* (= *Boston Studies in the Philosophy of Science*, 126) (Dordrecht, Boston and London): 75–82.

Goldstein, B. R. and D. Pingree (1990) *Levi ben Gerson's Prognostication for the Conjunction of 1345* (= *Transactions of the American Philosophical Society*, 80(6)) (Philadelphia).

Harvey, W. Z. (1977) "R. Hasdai Crescas and His Criticism of Philosophical Felicity" (Hebrew), in *Proceedings of the Sixth World Congress of Jewish Studies* (Jerusalem), 3: 143–9.

Hugonnard-Roche, H. (1992) "Problèmes méthodologiques dans l'astronomie au début du XIVe siècle", in Freudenthal (1992a): 55–70.

Jardine, N. (1984) *The Birth of History and Philosophy of Science* (Cambridge).

Kellner, M. (1976) "Gersonides and his Cultured Despisers: Arama and Abravanel", *Journal of Medieval and Renaissance Studies*, 6: 269–96.

—— (1992) "Bibliographia Gersonideana: An Annotated List of Writings by and about R. Levi ben Gershom", in Freudenthal (1992a): 367–414.

Langermann, Y. T. (1991) "The 'True Perplexity': The *Guide of the Perplexed*, Part II, Chapter 24", in Joel L. Kraemer (ed.) *Perspectives on Maimonides: Philosophical and Historical Studies* (Oxford): 159–74.

Leaman, O. (1985) *An Introduction to Medieval Islamic Philosophy* (Cambridge).

Leibowitz, Y. (1987) *The Faith of Maimonides* (New York).

Lévy, T. (1992) "Gersonide commentateur d'Euclide: traduction annotée de ses gloses sur les *Eléments*", in Freudenthal (1992a): 83–147.

Mancha, J. L. (1992) "The Latin Translation of Levi ben Gerson's *Astronomy*", in Freudenthal (1992a): 21–54.

Pines, S. (1963a) *Moses Maimonides, The Guide of the Perplexed*, trans. S. Pines (Chicago).

—— (1963b) "Translator's Introduction", in Pines (1963a): lvii–cxxxiv.

—— (1967) *Scholasticism after Thomas Aquinas and the Teachings of Hasdai Crescas and his Predecessors* (= *Proceedings of the Israel Academy of Sciences and Humanities*, 1(10)) (Jerusalem).

—— (1975) "Maimonides, Rabbi Moses ben Maimon", *Dictionary of Scientific Biography* (New York), 9: 27–32.

—— (1979) "The Limitations of Human Knowledge According to Al-Farabi, ibn Bajja, and Maimonides", in I. Twersky (ed.) *Studies in Medieval Jewish History and Literature* (Cambridge, Mass.): 82–109.

—— (1986a) "Problems Concerning Gersonides' Doctrine", appendix to his "Some Views Put Forward by the 14th-Century Jewish Philosopher Isaac Pulgar, and some Parallel Views Expressed by Spinoza" (Hebrew), in J. Dan and J. Hacker (eds) *Studies in Jewish Mystics, Philosophy and Ethical Literature, Presented to Isaia Tishby on his Seventy-Fifth Birthday* (Jerusalem): 395–457, on pp. 447–57.

—— (1986b) "Le Discours théologico-philosophique dans les oeuvres halachiques de Maïmonide comparé avec celui du *Guide des égarés*", in *Délivrance et Fidélité, Maïmonide: Textes du colloque tenu à l'Unesco en décembre 1985 à l'occasion du 850e anniversaire du philosophe* (Toulouse): 119–24.

—— (1987) "Some Distinctive Metaphysical Conceptions in Themistius' Commentary on Book Lamda and Their Place in the History of Philosophy", in J. Wiesner (ed.), *Aristoteles Werk und Wirkung Paul Moraux gewidmet*, 2 (Berlin and New York): 177–204.

Sabra, A. I. (1984) "The Andalusian Revolt against Ptolemaic Astronomy: Averroes and al-Bitrûjî", in E. Mendelsohn (ed.) *Transformation and Tradition in the Sciences: Essays in Honor of I. Bernard Cohen* (Cambridge): 133–53.

Shatzmiller, J. (1972) "Gersonides and the Jewish Community of Orange in His Day" (Hebrew), *Studies in the History of the Jewish People and the Land of Israel*, 2: 111–26.

—— (1992) "Gersonide et la société juive de son temps", in G. Dahan (ed.), *Gersonide en son temps: science et philosophie médiévales* (Louvain and Paris): 33–43.

Stern, J. (1995) "Maimonides on Language and the Science of Language", *Maimonides and the Sciences* (= *Boston Studies in the Philosophy of Science*) (Dordrecht): forthcoming.

Touati, Ch. (1973) *La Pensée philosophique et théologique de Gersonide* (Paris).

Weil-Guény, A.-M. (1992) "Gersonide en son temps: un tableau chronologique", in Freudenthal (1992a): 355–65.

V

SAUVER SON ÂME OU SAUVER LES PHÉNOMÈNES: SOTÉRIOLOGIE, ÉPISTÉMOLOGIE ET ASTRONOMIE CHEZ GERSONIDE

I. Introduction

Le but de cette étude est de tenter de répondre à la question suivante: quelles sont les sources de l'originalité de la pensée scientifique, surtout astronomique, de Gersonide? Qu'est-ce qui a motivé et structuré l'œuvre gersonidienne, si singulière dans le contexte de l'époque?

Cette question est étroitement liée, me semble-t-il, à une autre. A la lecture des études sur Gersonide, on reste parfois avec l'impression que nous avons affaire à deux personnes au moins: d'une part le philosophe et l'exégète biblique, d'autre part le scientifique— l'astronome, le mathématicien et le commentateur des traités scientifiques d'Aristote (tels que présentés par Averroès). On n'a guère tenté de voir si et comment toutes les préoccupations de Gersonide sont liées entre elles. Dans ce qui suit je m'adresse précisément à cette question: l'œuvre gersonidienne, avec ses recherches dans des domaines si variés, possède-t-elle une unité, une cohérence interne, et si oui laquelle? Ou encore, pour reprendre un terme cher à l'historienne et philosophe des sciences Hélène Metzger: quel est *le ressort profond* de la pensée scientifique, philosophique et théologique de Gersonide[1]? La réponse à cette question permettra également, nous le verrons, de rendre compte de l'originalité de Gersonide dans le domaine scientifique.

La réponse que je propose est, en bref, celle-ci: Gersonide a pu innover dans le contenu même de la science—et je m'intéresserai ici surtout à la science astronomique—puisqu'il avait une conception neuve, hétérodoxe pour l'époque, de ce que la science *doit* être. Pour reprendre les termes de Yehuda Elkana: une nouvelle *image de la*

---

[1] Cf. Hélène Metzger, *De la Méthode philosophique en histoire des sciences. Textes 1914-1939* réunis par Gad Freudenthal (Paris, Fayard [Corpus des œuvres de philosophie en langue française], 1987).

science, image étant à entendre ici tant dans un sens descriptif que dans un sens normatif, a donné lieu à une *science* nouvelle[2].

De quoi s'agit-il? Je propose un aperçu de mon argumentation, avant d'entrer dans les détails.

Le très regretté professeur Shlomo Pinès a récemment souligné l'importance qu'il faut accorder à *l'ordre de présentation* que suit Gersonide dans *Milḥamot ha-Shem*[3]. Or le premier sujet qu'aborde Gersonide dans son *magnum opus*, le sujet du premier traité, est celui de la béatitude de l'âme (*shlemut ha-nefesh*) et son immortalité: l'âme humaine (ou une partie d'elle) subsiste-t-elle après la mort physique? Autrement dit: la béatitude est-elle possible et si oui, comment? Contre Averroès, et peut-être Maïmonide, entre autres, Gersonide soutient que: 1° l'âme humaine subsiste dans son individualité et 2° la félicité dépend de la *science* que l'intellect humain avait acquise durant la vie ici-bas. Cette conception que se fait Gersonide du chemin menant à la félicité éternelle a pour conséquence une vision, elle aussi peu commune, du but et de la valeur de la connaissance. Selon Gersonide, en effet, l'acquisition de la science ou, pour employer le langage du moyen-âge: l'acquisition d'intelligibles, devient un *but en soi*; bien plus: elle devient le but primordial de la vie humaine. Et il convient de souligner qu'il s'agit bien de *toute* science du réel, portant sur le monde sublunaire (minéraux, plantes, et animaux) et supralunaire (corps célestes). *Toute* science mène au salut de l'âme.

Conséquence de l'idée évoquée, la position épistémoloqique de Gersonide, son image de la science, est décidément *réaliste*: Gersonide refuse la conception traditionnelle, "instrumentaliste", selon laquelle l'astronomie a pour tâche de "sauver les phénomènes"; la science des cieux, comme toute autre science, est, pour lui, une science du réel—en fait: du *créé*—à laquelle il incombe de chercher la connaissance de la véritable structure du monde céleste. Pour Gersonide cela implique que les modèles mathématiques de

---

[2] Cf. Y. Elkana, "A Programmatic Attempt at an Anthropology of Knowledge" *in* Everett Mendelsohn et Yehuda Elkana (éd.), *Sciences and Cultures* [= *Sociology of the Sciences*, t. 5 (1981)] (Dordrecht, Reidel, 1981), p. 1-76.

[3] Sh. Pines, "Some Views Put Forward by the 14th-Century Jewish Philosopher Isaac Pulgar, and some Parallel Views expressed by Spinoza" (en hébreu) *in* J. Dan et J. Hacher (éd.) *Studies in Jewish Mystics, Philosophy and Ethical Literature, Presented to Isaia Tishby on his Seventy-fifth Birthday* (Jérusalem, Magnes Press, 1986), p. 395-457; voir l'Appendice sur les "Problèmes concernant la doctrine de Gersonide" aux p. 447-457.

l'astronomie doivent satisfaire aux principes de la physique et de la métaphysique aristotéliciennes. Aussi rejette-il l'astronomie ptolé-méenne de son temps, qu'il s'efforce de remplacer par un autre système astronomique. Son système, nous le verrons, tient compte de données qui, ayant une signification uniquement dans le cadre d'une épistémologie réaliste, ont été ignorées de l'astronomie traditionnelle. L'image réaliste de la science se révélera ainsi avoir eu une incidence directe sur le contenu même de la théorie astronomique de Gersonide. Le "ressort profond" de l'originalité de l'œuvre astronomique de Gersonide s'avérera donc être son épistémologie réaliste, elle-même une conséquence de sa sotériologie, qui l'a conduit à tenter une synthèse de la physique arisotélicienne et de l'astronomie mathématique de Ptolémée.

II. L'EUDÉMONIE DE LA CONNAISSANCE

Le problème que Gersonide souhaite élucider dans le premier traité des *Milḥamot ha-Shem* est énoncé dans les termes suivants:

"L'âme raisonnable ayant reçu une certaine mesure de perfection, subsiste-t-elle [après la mort]? Et si elle subsiste, y a-t-il une différence de degré entre les hommes en ce qui concerne la façon de subsister [de leurs âmes respectives]? Cette question est extrêmement importante et semée de doutes; une erreur à son égard éloigne l'homme énormément de sa véritable félicité."[4]

Pour Gersonide, le thème de l'immortalité de l'âme—il s'agit toujours de la partie raisonnable de l'âme—est donc crucial, et ceci précisément à cause de la doctrine que soutient Gersonide à ce sujet.

Certaines théories aristotéliciennes précédentes, celles d'Alexandre d'Aphrodise, de Thémistius, d'Ibn Bājja et d'Averroès notamment, nient l'immortalité de l'âme individuelle[5]. Ces théories, suivies

---

[4] *M.H.*, Introduction, p. 2ᵃ. Toutes les références à *Milḥamot ha-Shem* (= *M.H.*) sont données d'après l'édition de Riva di Trento, 1560. J'indique successivement le traité, la partie (lorsqu'il y a lieu) et le chapitre; les numéros indiquent les feuillets; les lettres, les colonnes.

[5] La bibliographie sur le sujet traité dans les paragraphes qui suivent est très abondante. Citons, à titre d'indication: Sh. Pines, "Translator's Introduction", *in* Moses Maimonides, *The Guide of the Perplexed*, translated by S. Pines (Chicago, University of Chicago Press, 1963), p. ciii sqq.; H.A. Davidson, "Alfarabi and Avicenna on the Active Intellect", *Viator* 3 (1972), p. 109-178; *idem*, "Averroes on the Material Intellect", *Viator* 17 (1986), p. 91-137; *idem*, "Averroes on the Active Intellect as a Cause of Existence", *Viator* 18 (1987), p. 191-225; A. Ivry, "Averroes

peut-être même par Maïmonide, soutiennent, chacune à sa façon, qu'après la mort physique, l'âme devient immortelle en s'unissant à l'intellect agent. Or cette conjonction, l'âme l'acquiert au prix de la perte de son individualité: "dans" l'intellect agent, toutes les âmes sont numériquement une.

Gersonide, lui, soutient (à l'instar, d'ailleurs, d'al-Fārābī et d'Avicenne) une théorie autrement plus optimiste. Selon lui, les âmes raisonnables, ou les intellects humains, peuvent atteindre l'immortalité individuelle. Son raisonnement est le suivant[6]: en acquérant des connaissances, l'intellect humain se transforme d'un intellect hylique (matériel) en un intellect acquis, possédant des formes, des intelligibles. Or ces intelligibles se trouvent—nous verrons plus loin pourquoi—dans l'intellect agent et, de fait, sans le concours de celui-ci l'homme ne saurait point les acquérir. Or puisque l'intellect acquis consiste en formes qui, par leur nature même, sont impérissables, il est incorruptible, donc éternel, immortel; il subsiste après la mort. Dans ce sens, concède Gersonide, l'acquisition d'intelligibles, puisqu'ils se trouvent dans l'intellect agent, constitue effectivement en quelque façon une conjonction avec l'intellect agent[7]. Néanmoins, dans la mesure où les individus acquièrent, durant leur vie terrestre, des ensembles d'intelligibles différents, chaque individu aura un intellect acquis singulier. Aussi l'immortalité est-elle individuelle[8].

---

on Intellection and Conjunction", *Journal of the American Oriental society* 86 (1966), p. 77-85; A. Hyman, "Aristotle's Theory of the Intellect and Its Interpretation by Averroes", *in* D.J. O'Meara (éd.), *Studies in Aristotle* (Washington, The Catholic University of America Press, 1979), p. 161-190. Sur les idées de Gersonide sur ce sujet cf. Ch. Touati, *La Pensée philosophique et théologique de Gersonide* (Paris, Éditions de Minuit, 1973), p. 395-442; S. Feldman, "Gersonides on the Possibility of Conjunction with the Agent Intellect", *AJS Review* 3 (1978), p. 99-120.

[6] *M.H.* 1.11.; cf. égalemnt l'article d'H.A. Davidson dans ce volume, §4.

[7] *M.H.* 1.12, p. 15[d]. Gersonide explicite son idée dans son *Commentaire du Livre de Job* (fin de Chapitre 38; *Miqra'ot Gedolot*, p. 190[b]), où il formule l'idée que l'homme parvient à une certaine *union* avec Dieu du fait que Dieu, en qui les intelligibles existent tous en acte, aide l'homme, en qui ces mêmes intelligbles existent en puissance, à "actualiser" en lui, c'est-à-dire à appréhender, des intelligibles.

[8] "Il s'ensuit que celui qui a appréhendé un seul intelligible des intelligibles géométriques vivra éternellement, puisque ces intelligibles se trouvent dans l'intellect agent. Ceci est cependant une illusion et un faux", devait s'exclamer le philosophe Hasdaï Crescas. Voir Z. Harvey, "R. H. Crescas et sa critique de la félicité philosophique" (en hébreu), *Proceedings of the 6th World Congress of Jewish Studies* (Jérusalem, 1977), t. 3, p. 143-149. Un problème se pose cependant au sujet de la critique de Crescas. La raison pour laquelle la connaissance mène au salut est que les intelligibles se trouvent dans l'intellect agent, c'est-à-dire qu'ils sont des parcelles

On voit la valeur suprême que confère la sotériologie de Gersonide à l'acquisition des connaissances[9]. Puisque chaque intelligible,
une fois acquis par l'intellect, est une forme et donc incorruptible,
il confère à l'âme, si j'ose dire, une "parcelle d'immortalité". Il va
de soi que cette *eudémonie du savoir*, pour reprendre un terme de
Georges Vajda[10], crée chez celui qui l'accepte une motivation puissante à se consacrer à la science: c'est donc dans la sotériologie de
Gersonide que se trouve le fondement de sa vocation pour la science.
Le sérieux avec lequel Gersonide soutient cette conception se révèle,
entre autres, dans une remarque de son *Commentaire du Livre de Job*.
A la question éternelle, comment il se fait que des justes souffrent,
Gersonide répond que, souvent, les hommes pèchent sans le savoir.
Son exemple: quelqu'un qui, sans en être conscient, ne s'applique
pas à acquérir des intelligibles au plus haut niveau qui lui est
possible.

*

La doctrine gersonidienne selon laquelle l'acquisition de toute
connaissance, de tout intelligible, contribue à l'immortalité de
l'âme comporte encore un autre aspect important. Il nous faut,
pour le saisir, considérer brièvement la conception gersonidienne de
l'intellect agent. Ce sujet reçoit, dans ce volume même, un traite-
ment exhaustif par M. H.A. Davidson, ce qui me permet de me
limiter à de brèves remarques sur quelques aspects du sujet ayant
des implications directes pour notre problématique.

Dans la tradition de la philosophie aristotélicienne de la nature,
la notion de l'intellect agent s'insère dans le cadre de la théorie qui
complète la physique sublunaire en postulant des influences célestes
sur le monde de la génération et de la corruption. Cette théorie com-

---

du plan de la création et de la Providence. Cela s'applique-t-il effectivement,
comme le suppose la critique de Crescas, aux intelligibles mathématiques? Ces
intelligibles décrivent-ils le réel? Gersonide donne une réponse positive à cette question, soutenant que l'intellect agent possède les intelligibles mathématiques, comme
cela se manifeste dans maintes structures du monde sublunaire qui révèlent des
régularités mathématiques (*M.H.* 1.7, p. 9^d-10^a). Gersonide va jusqu'à affirmer
que les intelligibles mathématiques, y compris les notions générales telles que "le
tout est plus grand que la partie", nous parviennent, à l'instar des autres, par l'ex-
périence (*ibid.* 1.9, p. 10^c; également *Commentaire sur le Cantique des cantiques*, in:
Be'ur Hamesh Megillot [Königsberg, 1860], p. 14^c). La critique de Crescas est valide
uniquement si l'on tient compte de cette position de Gersonide.

[9] Cf. Touati, *op. cit.* (n. 5), p. 83-86.
[10] Georges Vajda, *Introduction à la pensée juive du moyen âge* (Paris, Vrin, 1947),
p. 143.

porte deux composantes: le postulat d'une influence des *corps* célestes, agissant en tant que *cause efficiente*, sur les éléments sublunaires; celui que des *intellects séparés*, en particulier l'intellect agent,
qui sont associés aux astres en tant que leurs moteurs, agissent, en
tant que *causes formelles*, sur les êtres qui sont sujets à la génération
et à la corruption. Les deux composantes de cette théorie viennent
répondre à deux ensembles de problèmes inhérents à la physique
aristotélicienne dont le point commun est que leurs solutions respectives font intervenir dans le monde sublunaire des influences célestes, révélant ainsi que, dans la description aristotélicienne, celui-
ci n'est pas un système autonome.

Un premier problème qui se pose dans le cadre de la physique
aristotélicienne est pourquoi les quatre éléments n'ont pas, depuis
longtemps, gagné chacun leur place naturelle, formant un monde
immobile, constitué de quatre sphères concentriques. De même,
comment expliquer l'alternance régulière de périodes de génération
et de corruption? Aristote avait déjà soutenu que, par ses mouvements, le Soleil agit sur les quatre éléments: le mouvement journalier
mélange et transforme continuellement les éléments, les empêchant
de s'éterniser dans leur lieux naturels; le mouvement annuel, lui, est
la cause de la génération et de la corruption[11]. L'aristotélisme médiéval complète cette théorie en soutenant que *toutes* les planètes
agissent sur les éléments sublunaires: si le Soleil chauffe (c'est-à-dire:
fortifie l'élément feu) et si la Lune exerce une influence certaine sur
l'eau, n'est-on pas en droit de penser que les autres planètes agissent
pareillement, même si nous ignorons le détail de leurs actions[12]?
Cet argument justifie aux yeux des aristotéliciens, y compris les
adversaires les plus résolus de l'astrologie, cette ''concession'' aux
thèses astrologiques[13]. Ainsi, nonobstant sa démarcation très ri-

---

[11] Aristote, *De gen. et corr.* II, 10.

[12] Cf. *Averrois Cordubensis Commentarium medium & epitome in Aristotelis De generatione et corruptione libros*, version hébraïque, éd. S. Kurland (Cambridge, Mass., The
Medieval Academy of America, 1958), §§56-62, p. 88-94 (commentaire moyen),
p. 121-123 (épitomé, v. particulièrement lignes 68-76); traduction anglaise: *Averroes
on Aristotle's* De generatione et corruptione: *Middle Commentary and Epitome*, trad.
S. Kurland (Cambridge, Mass., The Medieval Academy of America, 1958),
p. 101-106 et p. 132-135, respectivement.

[13] Cf. par exemple Maïmonide, *Guide des égarés*, I.72, II.10 et II.12. L'argument se trouve déjà dans Ptolémée, *Tetrabiblos* I.2 et il est répété par la plupart des
astrologues médiévaux: cf. p. ex. Richard Lemay, *Abu Maʿshar and Latin Aristotelianism in the Twelfth Century* (Beirut, American University of Beirut, 1962), p. 50 sqq.,

goureuse entre les mondes sub- et supralunaire, la physique aristo-télicienne se voit dans la nécessité d'admettre que le monde sub-lunaire n'est pas un système clos et que son équilibre dynamique est maintenu par des influences des corps célestes, agissant en tant que cause efficiente sur les éléments sublunaires.

Un deuxième problème concerne la génération et la perdurabilité des *formes* dans la matière sublunaire. En effet, la matière—les quatre éléments—ne saurait se doter de formes par elle-même, ni, une fois dotée de formes, préserver celles-ci. Comment, en effet, expliquer que les quatre éléments dans un corps composé ne re-gagnent pas leurs lieux naturels respectifs? De même, comment ex-pliquer que l'équilibre précaire entre les quatre qualités contraires, dont chacune cherche à l'emporter sur les autres, est maintenu? Certes, la perdurabilité d'un corps naturel composé, notamment d'une plante ou d'un animal, est attribuée par Aristote à sa forme, c'est-à-dire à son âme végétative[14]. Mais Aristote n'explique pas la provenance de cette cause active qu'est l'âme et le moyen âge postule que les formes des êtres composés dépendent, pour leur genèse ainsi que pour leur préservation, du concours d'un don-nateur de formes, le plus souvent identifié avec l'intellect agent, notion qui a son origine dans la psychologie d'Aristote[15].

En somme, les influences célestes se révèlent indispensables pour le maintien de l'ordre naturel. Le fait que chez Aristote se trouve déjà (ou plutôt: encore) une localisation du premier moteur trans-cendant ''en haut''[16] a facilité l'association entre la Cause première et un amalgame des influences des corps célestes en tant que cause efficiente avec celles, de type ''formel'', provenant des intellects. Pour Gersonide, nous le verrons dans un instant, ces influences ap-

---

55 sqq. J'étudie comment la physique aristotélicienne fondait certaines des thèses de l'astrologie dans mon article ''Maimonides' Stance on Astrology in Context: Cosmology, Physics, Medicine, and Providence'', *in* Fred Rosner et Samuel S. Kottek (éd.) *Moses Maimonides as Physician, Scientist and Philosopher* (Northvale, N.J. et Londres, Jason Aronson, à paraître en 1993).

[14] Cf. par exemple M.L. Gill, *Aristotle on Substance. The Paradox of Unity* (Prince-ton, Princeton University Press, 1989), notamment chapitre 7.

[15] Voir les travaux de H.A. Davidson cités note 5 ainsi que Gad Freudenthal, ''The Theory of the Opposites and an Ordered Universe: Physics and Metaphysics in Anaximander'', *Phronesis* 31 (1986), p. 197-228.

[16] H. Happ, ''Kosmologie und Metaphysik bei Aristoteles. Ein Beitrag zum Transzendenzproblem'', *in* K. Flasch (éd.), *Parusia. Studien zur Philosophie Platons und zur Problemgeschichte des Platonismus. Festgabe für Johannes Hirschberger* (Frankfurt, Minerva, 1965), p. 155-87.

portent, en effet, la démonstration de la Providence. Néanmoins sa version de ces théories assigne à l'intellect agent une place différente de celle qui fut la sienne chez les autres philosophes; la divergence de Gersonide de l'*opinio communis* sur ce point a des conséquences immédiates pour notre sujet.

Selon Gersonide, chaque astre et chaque intellect séparé (l'intellect étant le moteur de l'astre, l'astre l'instrument matériel de l'intellect) contrôlent une partie, ou un aspect, des processus physiques dans le monde sublunaire. Cette idée est à la base de sa cosmogonie et de sa cosmologie: en fait, la Providence sur le monde sublunaire actuel n'est que le prolongement, dans le présent, de sa création initiale[17]. Selon Gersonide, en effet, une fois les astres et les intellects mis en place, la création du monde s'est poursuivie de façon autonome, selon les mêmes lois naturelles en vigueur dans le monde constitué et par lesquelles s'exerce la Providence. Or bien que l'ensemble des corps célestes et des intellects structurent et gouvernent tous les processus physiques du monde sublunaire, pourtant chaque corps céleste ne contrôle et ne connaît qu'un segment des processus sublunaires; encore faut-il souligner que le contrôle porte uniquement sur les aspects généraux de ces processus. La sagesse et la bonté divines se manifestent précisément dans le fait que, dès la création, toutes ces actions des corps célestes, séparées et indépendantes, sur le monde sublunaire ont été coordonnées et programmées de façon cohérente dans une sorte d'harmonie préétablie. Cette harmonie des actions multiples, ignorée par chacun des intellects séparés qui y participent, n'est connue que par l'intellect agent (et par Dieu). Gersonide, se démarquant par là des autres philosophes du moyen-âge, conçoit en effet l'intellect agent comme une sorte de *synthèse* des connaissances partielles des autres intellects séparés; il est "le *nomos* de tout ce qui existe" (*nimus kol 'elu ha-nimṣa'ot*)[18] et donc le seul intellect à posséder une connaissance de la Providence et de la bonté divines, qui gouvernent notre monde ici-bas. Ainsi conçu, l'intellect agent devient pour Gersonide le

---

[17] Cf. Touati, *op. cit.* (n. 5), p. 161-298; Gad Freudenthal, "Cosmogonie et physique chez Gersonide", *Revue des études juives* 145 (1986), p. 295-314.

[18] Sur l'historique de cette notion cf. Sh. Pinès, "Some Distinctive Metaphysical Conceptions in Themistius' Commentary on Book Lamda and Their Place in the History of Philosophy", *in* J. Wiesner (éd.), *Aristoteles Werk und Wirkung Paul Moraux gewidmet*, tome 2 (Berlin et New York, W. De Gruyter, 1987), p. 177-204.

réceptacle de toute science possible et, partant, la fin de toute recherche.

L'idée que toutes les formes des êtres existants, tant sub- que supralunaires, se trouvent dans l'intellect agent implique évidemment que chaque intelligible qu'acquiert l'homme contribue à sa connaissance de l'intellect agent. Une telle connaissance est d'autant plus précieuse que l'intellect agent est, en fait, le plan, ou, pour recourir encore une fois à cette métaphore: le programme, selon lequel Dieu a créé le monde et exerce sa Providence. La connaissance scientifique de chaque détail du monde supra- et sublunaire contribue donc à notre connaissance de la Providence divine, d'où, encore une fois, sa signification sotériologique[19]. Dans son *Commentaire du Cantique des cantiques*, Gersonide est on ne peut plus explicite:

> "La félicité suprême de l'homme (*ha-haṣlaḥa ha-takhlitiyit la-adam*) est d'appréhender et de connaître Dieu, béni soit-Il, pour autant que cela lui est possible. [L'homme] y parvient en observant tout ce qui a trait aux êtres existants, leur ordre et leur régularité, ainsi que la manière dont s'exerce la sagesse divine en les disposant tels qu'ils sont"[20].

De fait, insiste Gersonide, ce n'est que par ces actions que nous pouvons connaître Dieu[21]; l'on parvient à l'immortalité en appréhendant, à partir des substances accessibles aux sens, des intelligibles, ceux-là même qui se trouvent dans l'intellect agent.

Mais les formes dans le monde sublunaire proviennent-elles *toutes* de l'intellect agent, de sorte que la connaissance de chacune d'elles contribue à la félicité? Un examen rapide de ce sujet controversé montrera quel enjeu important pouvait se cacher, au moyen âge, derrière un débat qui ne touchait qu'à une question à l'apparence anodine de la théorie physique. La controverse portait, bien évidemment, uniquement sur la provenance des formes du règne minéral[22]. Averroès, de même d'ailleurs que Maïmonide, est d'avis

---

[19] Cf. Touati, *op. cit.* (n. 5), p. 86.

[20] Lévi ben Gershon, *Commentaire sur le Cantique des cantiques, op. cit.* (n. 8), fol. 2^b; une édition critique de cette partie du *Commentaire* a été publiée *in*: M. Kellner, "Gersonides' Introduction to his Commentary of Song of Songs" (en hébreu), *Da'at* 23 (été 1989), p. 15-32; le passage cité se trouve p. 18.

[21] *Ibid.*

[22] S'agissant des plantes et des animaux, il y avait des discussions sur la provenance exacte de la cause formelle, mais il n'était pas question de soutenir que l'âme végétative (et encore moins les âmes plus élevées) puisse être générée sans l'intervention d'une telle cause. Cf. l'article de H.A. Davidson dans ce volume.

326

que toutes les propriétés d'un minéral sont la conséquence du mélange spécifique des éléments qui le compose: les minéraux n'ont pas de formes qui seraient dues à une cause formelle[23]. Gersonide, par contre, fait valoir que les vertus de certains minéraux, celle de l'aimant par exemple, ne sauraient être expliquées par la seule référence à leur composition matérielle et il en conclut que les minéraux reçoivent leurs formes, eux aussi, des intellects, à l'instar des végétaux et des animaux[24].

Cette divergence entre Gersonide et Averroès au niveau de la théorie physique entraîne une différence quant à leurs conceptions respectives de la science. En effet, l'intellect agent de Gersonide connaît les formes des minéraux, que celui d'Averroès ou de Maïmonide ignore. Aussi, les formes des minéraux, quoique parmi

---

[23] *Averrois Cordubensis Compendia librorum Aristotelis qui Parva naturalia vocantur*, version arabe éd. H. Blumberg (Cambridge, Mass., The Medieval Academy of America, 1972), p. 76; version hébraïque, éd. *idem* (Cambridge, Mass., The Medieval Academy of America, 1954), p. 50; traduction anglaise: *Averroes, Epitome of Parva naturalia*, H. Blumberg trad. (Cambridge, Mass., The Medieval Academy of America, 1961), p. 44 sq.; Maïmonide, *Guide des égarés*, I.72: "Il faut savoir que les facultés [ou: forces; *quwwa*] qui arrivent de la sphère céleste à ce monde-ci sont [...] au nombre de quatre, savoir: une faculté qui produit le mélange et la composition et qui suffit indubitablement pour la production des minéraux [...]" (traduction de S. Munk). Cf. le commentaire de Shem-Tov *ad loc*: "le Maître [Maïmonide] est d'avis qu'il n'est pas nécessaire [de postuler] un agent à part donnant les formes aux minéraux, comme le pensent Ibn Sīnā et Abu Ḥamid [al-Ghazālī], qui sont d'avis que les préparations [de la matière] sont le fait des sphères, mais que tout existant acquiert sa forme de l'intellect agent". (*Sefer More Nebukim ... ʿim Shlosha Perushim... Efodi, Sem-Tov, Ibn Qresqas...* [Vilna, 1902; réimprimé Jérusalem, 1960], p. 112.) Maïmonide affirme ailleurs: "Quant à ce que nous voyons naître sans que ce soit la simple conséquence du mélange—et ce sont toutes les formes—il faut pour cela un efficient, je veux dire quelque chose qui donne la forme" (*Guide* II.12, trad. S. Munk): il faut en conclure que, pour Maïmonide, les minéraux ne possèdent pas de formes—toutes leurs propriétés sont dues aux seules composantes du mélange. Ajoutons qu'Al-Ghazālī souscrit effectivement à l'opinion que lui attribue Shem-Tov: cf. *Maqāṣid al-falāsifa (Opinions des philosophes)*, traduction hébraïque d'Isaac Albalag, Paris, Bibliothèque nationale, ms. héb. 956, fol. 179ᵛ sq. Harry A. Wolfson compare les attitudes de Judah Halévi et de Maïmonide à propos de ce sujet dans son article: "Halevi and Maimonides on Design, Chance and Necessity" (1941) réimprimé dans ses *Studies in the History of Philosophy and Religion*, tome 2, I. Twersky et G.H. Williams éd. (Cambridge, Harvard University Press, 1977), p. 1-59, notamment p. 26-34.

[24] Gersonide, *Supercommentaire sur l'Épitomé [d'Averroès] des Parva naturalia*, Berlin, Staatsbibliothek preussischer Kulturbesitz, ms. Orient. Fol. 1055, fol. 145ᵇ sqq.; Gersonide, *Supercommentaire sur l'Épitomé [d'Averroès] des Météorologiques*, Paris, Bibliothèque nationale, ms. héb. 949, fol. 57ᵃ, cité en partie *in* G. Freudenthal, *op. cit.* (n. 17), p. 308, notamment n. 30. Cf. également l'article de Y.T. Langermann dans ce volume.

les plus basses, se trouvent, elles aussi, dans l'intellect agent, de sorte que leur connaissance contribue également, à l'instar de celle des formes plus élevées, à la connaissance de Dieu et ainsi à l'immortalité de l'âme. Du reste, la théorie de Gersonide implique qu'une forme qui ne proviendrait pas des intellects séparés serait ignorée de l'intellect agent: soutenir que les formes des minéraux ne proviennent pas de l'intellect agent équivaudrait donc à affirmer que le règne minéral est soustrait à la Providence.

Les idées gersonidiennes sur le nexus de causalités régissant le monde physique et sur l'information qu'en possède l'intellect agent ont ainsi des conséquences au niveau de l'épistémologie: elles déterminent les vues de Gersonide quant à ce que l'on peut savoir sur le monde et, partant, sur ce que l'on doit s'efforcer de savoir. Ainsi l'image de la science chez Gersonide dépend-t-elle de ses idées cosmologico-métaphysiques, en même temps qu'elle contribue, comme nous le verrons, à déterminer sa théorie physique du monde.

*

Les idées gersonidiennes que nous venons d'exposer font de la connaissance théorique une valeur en soi. Par cette conception du rôle de la connaissance, Gersonide s'oppose à Averroès, qui soutient une conception plutôt utilitariste de la connaissance. Ainsi, dans son *Epitomé du De anima*, Averroès explique-t-il que l'homme assure son existence au moyen des "arts utiles à sa survie", dont les sciences théoriques[25]. Gersonide rejette résolument cette notion utilitariste de la science. Dans son commentaire du texte d'Averroès il dit:

---

[25] Contrairement aux animaux, dit Averroès, l'homme ne saurait assurer sa survie au moyen des sens et de l'imagination seulement. Effectivement, dit-il, l'homme

"Possède encore une [autre] puissance [de l'âme] par laquelle il appréhende les choses abstraites de la matière, les compose l'une avec l'autre et, par le raisonnement, il fait ressortir une chose d'une autre. Ainsi cette puissance lui permet de perfectionner plusieurs arts qui sont *utiles à sa survie*. Certains de ces arts sont nécessaires, par exemple les arts pratiques préparant la nourriture de l'homme, son vêtement, son habitat, ses médicaments et les autres choses nécessaires à sa survie. [...] D'autres [arts] sont plus élevés, par exemple les arts pratiques qui ne sont pas absolument nécessaires, si ce n'est afin de [produire] le luxe, comme la décoration des vêtements ou de l'habitat. [...] Il en est de même en ce qui concerne les sciences théoriques dont la finalité est la seule recherche théorique (ʿiyyun), comme la science de la nature, la géométrie, etc."

Averroès, *Epitomé du De anima*, cité d'après la traduction hébraïque que commente Gersonide *in* Berlin, Staatsbibliothek preussischer Kulturbesitz, ms. Orient. Fol. 1055, fol. 171ᵃ.

"Ce [raisonnement] appelle une réflexion ayant trait à la question de l'immortalité de l'âme. En effet, pour les chercheurs en cette science, il est évident que la nature ne fait rien en vain. Et puisqu'il en est ainsi, pour quel but a été placée en nous la capacité d'appréhension des sciences théoriques, si ce n'est en raison d'une utilité quelconque? Or il n'y a pas lieu de dire que ces sciences théoriques sont les fondements [lit. les prémisses] des arts pratiques, car il y a des sciences théoriques qui n'ont aucune intention pratique, comme par exemple l'astronomie (*hokhmat ha-kokhabim*) et d'autres."[26]

Dans *Milḥamot ha-Shem* Gersonide explicite sa pensée davantage encore:

"Non seulement [les choses théoriques] n'ont aucune utilité pour la vie corporelle, mais l'effort de les acquérir est même au détriment de la bonne vie. Ainsi nous trouvons que les chercheurs, qui s'efforcent d'acquérir les intelligibles [théoriques], ne s'occupent pas de l'agrément de la vie corporelle; ils ne s'emparent même pas des choses matérielles qui leur sont absolument nécessaires. Et si quelqu'un disait que le pouvoir d'acquérir les intelligibles a été placé en nous parce que [ces intelligibles] sont les fondements des arts pratiques dont nous avons besoin pour nous fournir ce qui est nécessaire pour la vie, de sorte que [les sciences théoriques] sont utiles de cet aspect-là, nous lui répondrions que cela est faux. Il y a en effet des sciences [lit. arts; *melakhot*] théoriques qui n'ont aucune vocation pour la pratique. De plus, si on le supposait ainsi, il s'ensuivrait que les arts pratiques, étant la finalité des sciences [lit. arts] théoriques, seraient plus nobles qu'elles; or cela ne s'accorde pas avec la façon dont nous avons été constitués. Nous trouvons en effet que, pour tout homme, les sciences théoriques ont une grande préséance sur les pratiques; le petit peu que nous pouvons acquérir de ces sciences théoriques nous réjouit davantage que la grande quantité de ce que nous pouvons atteindre dans les arts pratiques[27]. [...] De même, nous avons trouvé que la nature a placé en nous, la communauté des hommes, un désir de la recherche théorique (*'iyyun*) qui l'emporte sur notre désir de la pratique. Tout cela concourt à montrer que la partie théorique en nous n'existe pas en vue de la partie pratique."[28]

La science théorique, dont Gersonide se fait ici le protagoniste, constitue le véritable but de l'existence humaine; c'est par elle que l'on peut parvenir à la félicité:

---

[26] Gersonide, *Supercommentaire du commentaire [d'Averroès] du De anima, ibid.*
[27] Gersonide répète souvent ce credo; cf. Touati, *op. cit.* (n. 5), p. 85, n. 15. Il s'agit, bien entendu, d'une citation d'Aristote, *Parties des animaux* I.5, 644b31.
[28] *M.H.* 1.4, p. 6^{a-b}. Ce texte reprend, en l'abrégeant un peu, un texte parallèle du *Supercommentaire de l'Épitoné du De anima*, cité plus haut.

"La félicité (*haṣlaḥa*) humaine s'accomplit lorsque l'homme sait, autant que possible, une chose d'entre les choses existantes. [La félicité] s'obtient davantage par l'appréhension des choses élevées que par l'appréhension des choses d'un rang et d'une noblesse inférieurs. Pour cette raison nous désirons l'appréhension limitée qui nous est possible des choses élevées davantage que celle, complète, que nous avons des choses qui leur sont inférieures.''[29]

*

La conception gersonidienne de la connaissance a deux autres composantes encore qu'il convient de relever. D'abord, Gersonide manifeste une confiance dans le pouvoir de l'intellect humain et dans sa capacité à atteindre la connaissance. Dès l'introduction à *Milḥamot ha-Shem*, il s'oppose sur ce sujet à Maïmonide, pour qui il a pourtant la plus grande vénération. Maïmonide, on le sait, est, en ce qui concerne l'épistémologie, plutôt sceptique: s'agissant de la question de l'éternité du monde, par exemple, il affirme que les limites de son pouvoir cognitif interdisent à l'homme d'atteindre une certitude à ce sujet.[30] N'est-il pas démesuré, dès lors, que de vouloir entreprendre une recherche jugée impossible par un si illustre prédécesseur, demande Gersonide. Dans sa réponse, il ne mâche pas ses mots:

"Ceci est une raison bien faible. Il n'est pas nécessaire que ce qu'ont ignoré les premiers [savants], leurs successeurs l'ignorent également. Car le temps suffit pour faire ressortir la vérité, comme le dit le Philosophe [...][31]. En effet, s'il en était autrement, toute personne menant des recherches dans quelque science que ce soit n'en [saurait] que ce qu'elle aurait appris d'autrui. Et si on le supposait ainsi, il n'y aurait aucune science, ce qui est manifestement faux.''[32]

Pour souscrire à l'opinion de Maïmonide, poursuit Gersonide, il aurait fallu *démontrer*, positivement, que cette recherche est effectivement impossible. A l'évidence, Gersonide a une notion très claire du progrès scientifique, concept dont l'élaboration, au cours de la

---

[29] *M.H.*, Introduction, p. 2^d.
[30] L'attitude de Maïmonide à l'égard de cette question fait partie intégrante de ses vues générales sur les limites de la connaissance humaine. Cf. Sh. Pinès, "The Limitations of Human Knowledge According to Al-Farabi, Ibn-Bajja, and Maimonides", *in* I. Twersky (éd.). *Studies in Medieval Jewish History and Literature* (Cambridge, Mass., Harvard University Press, 1979), p. 82-109.
[31] Cf. Touati, *op. cit.* (n. 5), p. 87-88, y compris les références données dans la note 28.
[32] *M.H.*, Introduction, p. 2^c.

330

Renaissance, a été si importante pour le renouveau des sciences au XVIIe siècle[33].

Donc, faire progresser la connaissance et, ce faisant, assurer l'immortalité de son âme, est possible. Cette doctrine, Gersonide la complète par une autre idée d'une grande portée: elle concerne la question de la diffusion de la connaissance. On connaît les idées bien élitistes d'un Maïmonide ou d'un Averroès à ce sujet. Gersonide les rejette avec véhémence:

"Il ne serait pas approprié que quelqu'un, ayant appréhendé quelque chose des choses théoriques, s'abstienne d'épancher sur autrui ce qu'il a appréhendé. Ceci serait complètement méprisable. De même que la totalité de ce qui existe a émané de Dieu, béni soit-Il, sans que Lui en tire profit, de même il convient que chacun, étant parvenu à une part de perfection, s'efforce à en parfaire autrui. Ceci lui permet de ressembler, dans la mesure du possible, à Dieu, béni soit-Il."[34]

Selon Gersonide, la connaissance est assortie d'une obligation éthique: le philosophe a le devoir moral de transmettre ses connaissances et de contribuer ainsi à la félicité des autres.[35] Gersonide lui-même tâchait d'agir en conformité avec ces exigences. Dans un rare moment de confidence personnelle, il écrit:

"Je jure par la vérité que je ne cessais de craindre d'écrire ce genre de choses dans un livre, connaissant bien les habitudes des ignorants qui, à leurs propres yeux, sont des sages. [...] Ce qui nous y a [pourtant] amené est notre puissant désir de déblayer du chemin des philosophes (baᶜaley ha-ᶜiyyun) les obstacles concernant ces grandes questions, où toute erreur éloigne fortement l'homme de la béatitude."

Et Gersonide de se féliciter:

"Il est évident qu'il ne serait pas approprié qu'un des savants accomplis (ha-shlemim) nous condamne pour être entré dans ce genre de recherches difficiles. Bien au contraire, il est approprié que nous soyons loué pour nos efforts dans la recherche de ces choses profondes, et ceci même dans

---

[33] R. Nisbet, *History of the Idea of Progress* (New York, Basic Books, 1980).

[34] *M.H.*, Introduction, p. 2ᵈ.

[35] De fait, dans chaque génération, la Providence place dans la Cité quelques "hommes accomplis" (*shlemim*)

"afin de nous diriger vers notre félicité. [...] [Ils le font] soit par la parole, soit—en poursuivant un désir naturel—par l'écrit. Il est en effet naturel de ressembler, autant que possible, à Dieu, béni soit-Il, duquel a émané cette parfaite existence, sans qu'Il en ait la moindre utilité."

Gersonide, *Commentaire du Cantique des cantiques*, *op. cit.* (n. 8), p. 6ᵈ.

le cas où nous ne serions pas parvenu à aller au delà du seul effort. Ayant parachevé, comme on va le voir, autant que possible, l'exposé de ces recherches, c'est à plus forte raison que nous devons être loué. Il est également clair qu'il n'est pas approprié que celui qui étudie nos propos nous persécute en échange de notre amour pour lui et notre volonté de lui être utile.''[36]

Ces idées sur les devoirs du savant de divulguer ses connaissances éclairent, me semble-t-il, la composition du corpus gersonidien avec la coexistence en son sein d'éléments apparemment dissemblables: commentaires philosophiques et bibliques, ouvrages de philosophie et de science. S'agissant des supercommentaires sur les commentaires d'Averroès, il est évident que les écrits du philosophe de Cordoue renferment, pour Gersonide, des intelligibles: aussi est-il de son devoir d'en répandre la connaissance, d'où la longue série de supercommentaires, issus peut-être d'une activité d'enseignement. Il en est de même des textes bibliques: "La Tora, dit Gersonide, n'est pas un *nomos* nous contraignant à croire des choses fausses; non, elle nous dirige, pour autant que cela est possible, à l'appréhension de la vérité''[37]. En fait, la Tora, donnée en un temps où les hommes ignoraient encore tout de la cause formelle, poursuit l'intention de "diriger le comportement humain vers la véritable perfection'';[38] elle est "un *nomos* conduisant ceux qui le suivent vers la véritable félicité''[39]. La Tora et la philosophie expriment donc, dans des langages différents, les mêmes vérités: à maintes reprises, lorsqu'il montre, à propos d'un problème particulier, que le résultat de la recherche est identique à ce qui est (selon lui) énoncé dans la Tora, Gersonide s'exclame: "C'est de la nature de la vérité que d'être en accord avec elle-même de tous les côtés!''. Aussi, les commentaires bibliques poursuivent-ils le même but que les recherches philosophiques. Puis, il y a les ouvrages originaux: l'œuvre philosophique, *Milḥamot ha-Shem*, bien entendu, qui, plus que tout autre, est supposée indiquer à l'homme le chemin de sa félicité, et les ouvrages spécialisés, notamment en logique, en mathématiques et en astronomie: quel que soit le domaine, dès lors qu'une nouvelle idée éclairait l'horizon, Gersonide considérait qu'il était de son

---

[36] *M.H.*, Introduction, p. 2[d].
[37] *M.H.*, Introduction, p. 2[d] sq. Cf. également Touati, *op. cit.* (n. 5), p. 92-97.
[38] Gersonide, *Perush ha-Tora* (Venise, 1547), introduction, p. 2[a].
[39] *Ibid.*, p. 9[a].

devoir éthique incontournable de la diffuser; ces intelligibles, non appréhendés auparavant, une fois trouvés, il lui incombait de les ''épancher'' (pour reprendre son terme) sur le public, ressemblant par là, pour autant que cela est possible, à Dieu.

III. Réalisme et Théorie Astronomique

Manifestement, la sotériologie de Gersonide suscite une motivation puissante à la recherche scientifique: toute bribe de connaissance, chaque intelligible, contribue à la félicité. Et pour Gersonide l'effort pour acquérir des intelligibles doit porter en premier lieu sur le monde matériel (sub- et supralunaire): c'est à partir des êtres créés que l'homme peut appréhender les formes que possède l'intellect agent. La recherche métaphysique et théologique aboutit donc à la conclusion—quelque peu paradoxale, mais rappelant en ceci précisément la conclusion à laquelle parvient Wittgenstein dans le sixième chapitre de son *Tractatus*[40]—qu'il faut se tourner vers la recherche *empirique*, d'où l'impossibilité d'acquérir des intelligibles après la mort. La sotériologie gersonidienne implique donc une attitude décidément *innerweltlich*; nous dirons quelques mots de la signification historique d'une telle attitude dans notre conclusion.

La valeur qu'attribue Gersonide aux sciences détermine non seulement le fait même qu'il faut s'y consacrer, mais encore que la science la plus noble est l'astronomie, dont la sotériologie gersonidienne conditionne jusqu'à certains aspects du contenu.

Tout d'abord, les substances du monde réel et, partant, les sujets de la recherche, ne sont pas tous du même rang. Bien au contraire, Gersonide soutient, avec Averroès, une *hiérarchie* stricte des substances et des formes. Selon Averroès, en effet, cette hiérarchie a à sa base les formes des quatre éléments; puis viennent successivement les formes des homéomères, les formes végétatives, les formes perceptives, les formes imaginaires, les intelligibles théoriques et, enfin, les formes des corps célestes, c'est-à-dire les intellects séparés, ces derniers ayant ainsi le statut ontologique le plus élevé. La même hiérarchie se retrouve au niveau des matières respectives: ''L'eau,

---

[40] L. Wittgenstein, *Tractatus Logico-Philosophicus*, §6.54: "Meine Sätze erläutern dadurch, dass sie der, welcher mich versteht, am Ende als unsinnig erkennt, wenn er durch sie—auf ihnen—über sie hinausgestiegen ist. (Er muss sozusagen die Leiter wegwerfen, nachdem er auf ihr hinaufgestiegen ist.)"

dit Averroès, est la perfection de la terre, l'air celle de l'eau, le feu celle de l'air; enfin le corps céleste est la perfection du feu et les formes séparées sont la perfection du corps céleste"[41]. Cette conception, entièrement partagée par Gersonide,[42] établit évidemment une hiérarchie des intelligibles correspondants, au sommet de laquelle se trouvent les corps célestes. (Au-dessus de ces intelligibles accessibles à une investigation au moyen des sens se trouve bien sûr la Forme première.) La science des corps célestes est l'astronomie, d'où la place priviligiée qu'elle occupe dans la pensée de Gersonide:

"Nous avons considéré qu'il serait approprié de nous étendre longuement sur cette recherche, car cette science est très précieuse, tant en elle-même que par la direction qu'elle donne au sujet des autres sciences. Qu'elle soit précieuse en elle-même est manifeste, puisque le rang du *quaesitum* [*derush*] est selon le rang de l'objet sur lequel porte la recherche. Or il est manifeste que l'objet sur lequel porte cette recherche, à savoir le corps céleste, est le plus noble parmi tous les corps naturels et que la forme le mouvant est la plus noble parmi toutes les formes naturelles."[43]

L'importance de la recherche astronomique découle encore et surtout du fait que c'est par les corps célestes que Dieu exerce sa Providence sur le monde sublunaire. Si l'on veut connaître les lois gouvernant, dès la création, l'ordre naturel parfait ici-bas, il faut tourner les yeux vers le ciel. La science de l'astronomie

"est précieuse [...] à cause des indications qu'elle fournit à l'adresse des autres sciences. [...] Elle donne une direction admirable pour la science

---

[41] *The Epistle on the Possibility of Conjunction with the Active Intellect by Ibn Rushd with the Commentary of Moses Narboni. A Critical Edition and Annotated Translation* by Kalman P. Bland (New York, The Jewish Theological Seminary of America, 1982), 14.18-29. Cf. également les passages d'Averroès se trouvant dans Gersonide, *Supercommentaire de l'Épitomé [d'Averroès] du De caelo*, Berlin, Staatsbibliothek preussischer Kulturbesitz, ms. Orient. Fol. 1055, fol. 18ᵃ et idem, *Supercommentaire de l'Épitomé [d'Averroès] du De anima*, ibid., fol. 145ᵃ⁻ᵇ, 149ᵃ⁻ᵇ. Averroès développe une idée se trouvant en germe déjà chez Aristote; cf. *De caelo* IV.3, 310b15 et M.L. Gill, *Aristotle on Substance, op. cit.* (n. 14), p. 239.

[42] *M.H.* 1.12, p. 15ᵇ.

[43] *M.H.* 5.1.2, 1-2. On sait que la première partie (astronomique) du cinquième livre de *Milḥamot ha-Shem* n'a pas été incluse dans les éditions de cet ouvrage. Nous disposons d'une édition des seuls vingt (sur 136!) premiers chapitres de cette partie: Bernard R. Goldstein, *The Astronomy of Levi ben Gerson (1288-1344). A Critical Edition of Chapters 1-20 with Translation and Commentary* (Berlin, Springer, 1985). Toutes les références ultérieures aux vingt premiers chapitres de l'*Astronomie* de Gersonide se font d'après cette édition. Les chiffres derrière la virgule indiquent le numéro de la phrase, selon la numérotation de M. Goldstein.

de la nature et pour la science divine. En effet, la recherche portant sur ses formes et sur leurs rangs par rapport à la Forme première [montre que ces formes sont] le fruit de la sagesse divine et sa finalité. [...] Car ces sphères et ces astres ont été créés par la parole de Dieu, béni soit-Il. [...] Ils font apparaître la grandeur de la sagesse et de la puissance qu'a manifestées Dieu, béni soit-Il, en produisant ces corps célestes nobles [...] desquels Il fait émaner des influences produisant des choses différentes, qui perfectionnent l'existence ici-bas.''[44]

Si bien que certains commandements bibliques ont été donnés afin d'orienter l'homme vers la contemplation des cieux[45].

Gersonide est douloureusement conscient des carences de la science des cieux de son temps. Son imperfection se manifeste, d'après lui, dans la faiblesse de l'astrologie judiciaire. A l'instar de tous ses contemporains, y compris les adversaires de l'astrologie, Gersonide ne met jamais en doute le principe même d'influences astrales et il accepte, jusqu'à un certain degré, les thèses des astrologues[46]. Il reconnaît cependant que ces derniers se trompent trop souvent, même si l'on tient compte du libre arbitre et du fait que le déterminisme astral n'est que partiel. Comment expliquer ces erreurs? Elles tiennent, avance-t-il, à l'imperfection des connaissances de la physique céleste—des mouvements des astres et de leurs différentes influences sur le monde sublunaire[47]. D'où, encore une fois, l'importance des recherches astronomiques.

Ainsi, c'est par leur rang ontologique et par leur place centrale dans l'économie du monde physique, où ils jouent le rôle d'instrument de la Providence, que les corps célestes sont, pour Gersonide, les objets privilégiés de la recherche. Ceci explique la grande attention que porte Gersonide à l'astronomie, science à laquelle il a probablement consacré plus de temps (de nuits notamment!) qu'à toutes les autres sciences mises ensemble.

*

Nous avons, jusqu'ici, vu comment la sotériologie gersonidienne détermine le fait même que notre philosophe considère la science

---

[44] *M.H.* 5.1.2, 2 sq., 6.

[45] I. Heinemann, *Ta'amey ha-Miṣwot be-Sifrut Yisrael*, t. 1, (Jérusalem, 3e éd., 1954), p. 99.

[46] Cf. Gad Freudenthal, ''Levi ben Gershom as a Scientist: Physics, Astrology and Eschatology'', in: *Proceedings of the Tenth World Congress of Jewish Studies*, Division C, vol. 1: *Jewish Thought and Literature* (Jérusalem, World Union of Jewish Studies, 1990), p. 65-72.

[47] *M.H.* 5.2.1, p. 31d.

comme le chemin menant au salut; nous avons également constaté que les idées cosmologico-métaphysiques de Gersonide font des corps célestes l'objet d'étude par excellence, le site priviligié de la recherche. Nous allons maintenant préciser cette influence des idées philosophiques sur la science en montrant que la sotériologie de Gersonide détermine le biais original par lequel il aborde l'étude de l'astronomie, influençant ainsi directement le contenu même de ses théories astronomiques.

Gersonide, nous avons pu nous en convaincre, n'étudie pas l'astronomie à cause de son utilité potentielle: ne va-t-il pas jusqu'à la citer comme l'exemple même d'une science purement théorique? Ce que cherche Gersonide, c'est connaître les intelligibles du rang suprême, déchiffrer, si peu que ce soit, le secret de la Providence, la sagesse divine ayant établi le *nomos* du monde physique. Pour Gersonide, cela signifie: appréhender des intelligibles se trouvant dans l'intellect agent. Or cela implique que, sur le plan épistémologique, sa position ne peut être que *réaliste*. Le but que poursuit Gersonide dans l'étude de l'astronomie lui interdit l'adoption de la position *instrumentaliste* (ou, si l'on préfère, fictionaliste), qui avait le plus souvent prévalu en astronomie: cette position épistémologique consiste à postuler que le rôle de l'astronome est de "sauver les phénomènes", c'est-à-dire de proposer des modèles mathématiques permettant de produire des tables et ainsi de calculer les positions des astres, sans nullement s'occuper de savoir si ces modèles correspondent à la structure réelle du monde[48]. Gersonide ne saurait

---

[48] Sur la signification du terme "instrumentalisme", cf. par exemple K.R. Popper, "Three Views Concerning Human Knowledge", dans ses *Conjectures and Refutations* (London, Routledge and Kegan Paul, 1963), p. 97-119; E. Nagel, *The Structure of Science* (London, Routledge and Kegan Paul, 1961), p. 129-140. Il y a lieu de souligner qu'en tant que conception épistémologique, l'instrumentalisme n'implique nullement une attitude utilitariste à l'égard de la science. Le terme "fictionalisme" est utilisé par N. Jardine dans *The Birth of History and Philosophy of Science* (Cambridge, Cambridge University Press, 1984), cf. notamment p. 225-257. L'étude historique la plus détaillée des attitudes épistémologiques envers l'astronomie demeure P. Duhem, *Sózein tà phainómena. Essai sur la notion de théorie physique* (1908) (rééd. Paris, Vrin, 1982); cf. également R.S. Avi-Yonah, "Ptolemy vs Al-Bitruji: A study of Decision-Making in the Middle Ages", *Archives internationales d'histoire des sciences* 35 (1985), p. 124-147 et, pour la scolastique latine, Edward Grant, "Eccentrics and Epicycles in Medieval Cosmology", *in* Edward Grant et John E. Murdoch (éds.) *Mathematics and Its Applications to Science and Natural Philosophy in the Middle Ages* (Cambridge, Cambridge University Press, 1987), p. 189-214, ainsi que l'article d'H. Hugonnard-Roche dans le présent volume.

partager cette position épistémologique: sa recherche astronomique est commandée par sa sotériologie, pour laquelle rien n'a de signification sauf la connaissance du monde tel qu'il est véritablement, ce qui, seul, peut éventuellement déboucher sur une connaissance (partielle) de l'intellect agent. "Nous avons constaté, dit-il en effet au début du premier chapitre de son *Astronomie*, que [même] ceux parmi les mathématiciens qui ont fait des recherches appropriées en cette science se sont contentés de trouver un système [ou: modèle; *tekhunah*] astronomique duquel les observations peuvent être inférées approximativement. Ils n'ont pas tâché d'élaborer le système astronomique nécessaire selon la vérité"[49]. Aussi Gersonide se propose-t-il d'élaborer une astronomie nouvelle, qui décrira le monde céleste tel qu'il est en vérité. L'écartement de Gersonide par rapport au consensus *épistémologique* prévalant dans sa discipline aura des répercussions déterminantes sur le *contenu* même de ses théories. Le biais par lequel Gersonide aborde la recherche scientifique implique qu'il sera un astronome "pas comme les autres".

Les tentatives pour élaborer une astronomie réaliste—ou, ce qui revient au même, pour concilier l'astronomie mathématique avec la physique aristotélicienne—sont peu nombreuses. Chez Aristote lui-même, le rapport entre 1° le modèle astronomique d'Eudoxe; 2° l'idée que, de par leur matière, les astres ont un mouvement circulaire naturel; et 3° l'idée que les astres sont mus par des intellects, est bien obscur. Ptolémée tente une explication physique des mouvements célestes dans ses *Hypothèses*, mais sa théorie demeure vague et ne prend pas en compte les détails des mouvements tels qu'ils sont analysés dans l'*Almageste*. La première tentative d'une conciliation entre l'astronomie ptoléméenne et la physique aristotélicienne est due, semble-t-il, à Sosigène, le maître d'Alexandre d'Aphrodise, dans la deuxième moitié du IIe siècle. Sosigène tente d'adapter la philosophie de la nature d'Aristote à l'astronomie nouvelle, celle de Ptolémée, qui l'emporte par la précision de ses prévisions. Simplicius, à qui nous devons notre information sur Sosigène, a adopté l'approche inverse: il exclut l'astronomie du champ de la philosophie et du domaine où l'on cherche la vérité[50].

Parmi les auteurs arabes, Thābit ibn Qurra et, notamment, Ibn

---

[49] *M.H.* 5.1.1, 3-4.
[50] Matthias Schramm, *Ibn al-Haythams Weg zur Physik* (Wiesbaden, F. Steiner, 1963), p. 16-59; cf. également P. Duhem, *op. cit.* (n. 48), p. 3-27.

al-Haitham se sont intéressés au problème d'une explication physique de la mécanique céleste. Dans le présent contexte, c'est cependant une autre tradition qui nous intéresse, une tradition jadis
décrite par Léon Gauthier comme "une réforme du système astronomique de Ptolémée, tentée par les philosophes arabes du XIIe
siècle"[51]. Plus récemment, A.I. Sabra a parlé de "La révolte an-
dalouse contre l'astronomie ptoléméenne"[52]. A l'origine de cette
"révolte" se trouvent, pour autant que nous le sachions, Ibn Bājja
(mort 1138) et Ibn Tofail (mort en 1185); les acteurs principaux en
sont Averroès (mort en 1198) et al-Biṭrūjī; certains échos s'en font
entendre chez Maïmonide. Sans entrer dans les détails, exposés
maintes fois, disons que dans son *Grand commentaire du De caelo* et,
plus tard, dans le *Grand commentaire de la Métaphysique*, écrit après
1186, Averroès affirme sans ambages que l'existence postulée de
sphères excentriques ou épicycliques serait contre nature. Il explique que de tels mouvements créeraient plusieurs centres de
mouvement circulaire, en contradiction avec la physique d'Aristote,
selon laquelle l'existence de tels centres impliquerait celle d'autres
terres que la nôtre. Averroès conclut que "l'astronomie de nos jours
est en accord avec les calculs seulement, mais non avec ce qui
existe" et de lancer un appel à la recherche d'une "vraie astronomie
qui soit possible du point de vue des principes physiques". Averroès
croit d'ailleurs qu'une telle astronomie avait été connue des
anciens[53]. Maïmonide expose en détail le même type de considérations mais, en accord avec son attitude philosophique générale,[54] il
se dégage du problème en trouvant refuge dans l'épistémologie
instrumentaliste[55].

---

[51] "Une réforme du système astronomique de Ptolémée, tentée par les
philosophes arabes du XIIe siècle", *Journal asiatique*, 1909, p. 483-510.

[52] A.I. Sabra, "The Andalusian Revolt against Ptolemaic Astronomy. Averroes
and al-Biṭrūjī", in Everett Mendelsohn (éd.), *Transformation and Tradition in the
Sciences. Essays in Honor of I. Bernard Cohen* (Cambridge, Cambridge University
Press, 1984), p. 133-153.

[53] L. Gauthier, *op. cit.* (n. 51), p. 502-504; A.I. Sabra, *op. cit.* (n. 52), p. 142;
Ch. Genequand, *Ibn Rushd's Metaphysics. A Translation with Introduction of Ibn Rushd's
Commentary on Aristotle's Metaphysics, Book Lām* (Leiden, Brill, 1984), p. 178.

[54] Voir Sh. Pinès, "The Limitations of Human Knowledge", *op. cit.* (n. 30),
notamment p. 93-94.

[55] Maïmonide, *Guide des égarés* II.24: "Je t'ai déjà expliqué de vive voix que tout
cela ne regarde pas l'astronome; car celui-ci n'a pas pour but de nous faire connaître sous quelle forme les sphères existent, mais son but est de poser un système
par lequel il soit possible d'admettre des mouvements circulaires, uniformes et con-

Écrivant autour de 1200, al-Biṭrūjī s'efforce d'édifier une astronomie mathématique selon les principes énoncés par Averroès (qu'il ne mentionne cependant pas)[56]. Le système d'al-Biṭrūjī, qui se veut en accord avec le calcul tout en se conformant aux principes de la physique, est, dans la tradition astronomique médiévale, une véritable révolution. Ainsi Gersonide se réfère toujours à al-Biṭrūjī comme à "l'auteur de la nouvelle astronomie" et son contemporain Isaac Israeli, l'auteur de *Yesod Olam*, l'appelle "l'homme qui, par sa théorie, a mis en émoi le monde entier"[57].

Dans la tradition hébraïque, l'astronomie d'al-Biṭrūjī a été très rapidement connue. Yehuda b. Salomon Kohen ibn Matqa de Tolède, le correspondant de l'empereur Frédéric II, en donne un résumé dès 1247 environ dans son encyclopédie *Midrash ha-Ḥokhmah*, ouvrage cependant resté sans grand écho[58]. Une traduction hébraïque complète de l'ouvrage d'al-Biṭrūjī a été terminée par Moïse ibn Tibbon en 1259[59]. Il semblerait pourtant que sa percée soit restée limitée. Aussi tard que 1318, un auteur pourtant averti en mathématiques et arabisant de surcroît, Qalonymos b. Qalonymos, semble l'ignorer complètement[60].

---

formes à ce qui se perçoit par la vue, peu importe que la chose soit réellement ainsi, ou non". (Traduction de S. Munk.)

[56] Les textes arabe et hébreu, une traduction en anglais et une interprétation de l'*Astronomie* d'al-Biṭrūjī se trouvent *in* B.R. Goldstein, *Al-Biṭrūjī: On the Principles of Astronomy* (New Haven, Yale University Press, 1971), 2 tomes.

[57] Cf. S. Munk *Mélanges de philosophie juive et arabe* (Paris, 1857), p. 521; Goldstein, *Al-Biṭrūjī, op. cit.* (n. 56), t. 1, p. 40-45.

[58] Goldstein, *Al-Biṭrūjī, op. cit.*, t. 1, p. 40.

[59] *Ibid.*, p. 3, 47, 155 (n. 58).

[60] Dans son épître à Joseph ibn Kaspi, Qalonymos critique ce dernier pour avoir donné à un de ses ouvrages le titre prétentieux de "Livre du secret" (*Sefer ha-sod*). Il se gausse de son correspondant en posant, en cascade, une série de questions rhétoriques, un peu dans le style du Livre de Job 38-39. Parmi ces questions:

"T'a-t-on dit et as-tu entendu dire qu'un système astronomique a été trouvé, ne postulant ni épicycle, ni excentrique, ni un mouvement des pôles, de sorte que les principes naturels soient respectés? En effet, nous trouvons que les calculs qui sont en accord avec les observations sont uniquement ceux faits selon une de ces méthodes, qui, toutes, sont en dehors du système de la physique. Le Maître [Maïmonide] a évoqué cela dans le chapitre 24 de la deuxième partie [du *Guide*]. [...] De nombreux jours sont passés et je suis resté perplexe, cette perplexité s'ajoutant à tant d'autres. Enfin, j'ai trouvé qu'Ibn Rushd, dans son *Commentaire* au deuxième traité du *De caelo*, dit que puisqu'Aristote a été attentif à ce problème mais qu'il n'en a rien dit, il y avait, peut-être, de son temps, un système astronomique sans épicycles ni excentriques; ce système astronomique, on l'appelait hélicoïdal (*lulaviyit*), mais de nos jours on l'ignore toujours".

Cf. *Kalonymos ben Kalonymos' Sendschreiben an Joseph Kaspi*, éd. Joseph Perles (München, Theodor Ackermann, 1879), p. 26; quelques corrections à cette édition

Au moment où Gersonide entre en scène, le problème est donc présent aux esprits, du moins aux esprits philosophiques: ces derniers connaissent l'exposé fourni de Maïmonide sur le sujet, ainsi que les textes d'Averroès où l'incompatibilité de l'astronomie et de la physique est constatée. Astronome, Gersonide connaît de surcroît la nouvelle astronomie d'al-Biṭrūjī. Aussi est-il pleinement conscient de la problématique: quelques physiciens et philosophes, dit-il, ont remarqué que les modèles astronomiques, puisqu'ils ne se conforment pas aux lois de la nature, ne sauraient correspondre à la vérité.

"Ils se sont cependant déchargés du fardeau en disant que c'est le mathématicien qui devrait mener la recherche dans ce domaine, cela leur étant impossible en tant que physiciens ou philosophes. De même, le mathématicien soutient que, en tant que mathématicien, il n'a aucune vocation pour mener cette recherche: il lui suffit de postuler un système [ou: modèle] astronomique duquel on peut inférer ce qui est observable des mouvements des astres [...], sans prêter attention si ce système [ou: modèle] est conforme à la nature. [...] Ainsi, poursuit Gersonide, un sceptique pourra dire: 'si la vérité dans cette recherche n'appartient ni au mathématicien, ni au physicien, ni au philosophe, à qui appartient-elle?'"[61]

La position de Gersonide est la suivante:

"La vérité dans cette recherche n'appartient pas à la métaphysique [lit. la science compréhensive; *ha-ḥokhmah ha-kolelet*], qui examine l'être en tant qu'existant, [...] ni à la science de la nature, [...] et elle n'appartient pas non plus, en sa totalité, à la science mathématique. Non, en sa totalité elle appartient à l'ensemble de ces sciences."[62]

Et Gersonide d'énoncer l'idée d'une science d'un type nouveau, interdisciplinaire, mathématique et physique à la fois:

"Puisqu'il en est ainsi, il est impossible qu'une recherche, pour autant qu'elle est une, appartienne en partie à un chercheur dans une science, la partie restante appartenant à un chercheur dans une autre science. [...] Il s'ensuit nécessairement que cette recherche appartient, en sa

---

ont été apportées par M. Steinschneider *in: Hebräische Bibliographie* 19 (1879), p. 115-118. Le texte d'Averroès évoqué par Qalonymos est indiqué par Gauthier, *op. cit.* (n. 51), p. 502 sq., n. 2. Ibn Kaspi lui-même, un contemporain de Gersonide, fait une allusion, discrète et peu developpée, au problème dans son commentaire au *Guide des égarés* I, 72; cf. ses ʿ*Amudey Kesef u-Maskiyot Kesef*, éd. S. Werbluner (Frankfurt, 1848), p. 73.

[61] *M.H.* 5.1.1, 5-8.
[62] *M.H.* 5.1.1, 10-13.

totalité, à quelqu'un qui est à la fois, et mathématicien, et physicien, et philosophe: en effet, quelqu'un qui est ainsi sera à même de compléter cette recherche en s'appuyant sur chacune de ces sciences."[63]

Le lecteur ne sera pas surpris d'apprendre que Gersonide s'estime posséder les qualifications souhaitées.

En élaborant son système astronomique, Gersonide poursuit donc l'intention de satisfaire aux exigences de l'astronomie mathématique, de la physique et de la métaphysique à la fois. Ces deux dernières sciences impliquent l'impossibilité physique de plusieurs centres de mouvement circulaire et Gersonide, comme al-Biṭrūjī avant lui, écarte les épicycles de son système astronomique. Gersonide ne peut pas non plus accepter le système d'al-Biṭrūjī qui, explique-t-il en détail, est contredit par la physique et la métaphysique, aussi bien que par l'observation[64]. Al-Biṭrūjī, dit-il, et les historiens modernes partagent son avis, "ne s'est pas du tout référé, ni à l'observation, ni aux considérations mathématiques. Il les a ignorées et il a établi [son système] selon ce qui lui a semblé être en conformité avec la physique"[65]. De fait, le système astronomique de Gersonide est fondé sur une idée qu'al-Biṭrūjī rejette, à savoir que tous les astres se trouvent sur des sphères excentriques par rapport au centre de la Terre.

*

Le système astronomique de Gersonide porte les marques de la démarche de son auteur. Dans ce qui suit, je voudrais relever deux ou trois caractéristiques de l'astronomie gersonidienne, qui mettent en évidence que sa spécificité est une conséquence directe de l'épistémologie réaliste de son auteur.

Signalons tout d'abord que, pour Gersonide, bien évidemment, il ne peut y avoir qu'un seul système astronomique, le *vrai*. Gersonide est, bien entendu, conscient du fait que des modèles astronomiques

---

[63] *M.H.* 5.1.1, 14-16.

[64] *M.H.* 5.1.40, 5.1.44. Ces chapitres non publiés de la partie astronomique de *M.H.* se trouvent notamment dans les manuscrits suivants: Biblioteca Nazionale, Naples, ebr. III F.9; Bibliothèque nationale, Paris, héb. 724 et 725. Dans les références ultérieures je les désigne, en suivant B.R. Goldstein [*The Astronomy of Levi ben Gerson, op. cit.* (n. 43), p. x], par les lettres N, P et Q, respectivement. Je remercie M. le professeur B.R. Goldstein d'avoir aimablement mis à ma disposition une transcription préliminaire du texte.

[65] *M.H.* 5.1.43, début (N, fol. 151[b]; P, fol. 75[b]; Q, fol. 55[a]); et cf. Goldstein, *Al-Biṭrūjī, op. cit.* (n. 56), p. 44-45, pour une opinion moderne semblable.

peuvent être équivalents et que des modèles différents peuvent donner lieu à des prévisions sensiblement identiques. Il ne perd cependant jamais sa confiance dans la possibilité d'atteindre la vérité. Gersonide examine les propriétés spécifiques (*segulot*) de chaque modèle et c'est sur la base de cet examen qu'il le juge[66]. "Nous avons montré, sans laisser subsister le moindre doute, que, pour chacune des planètes, le modèle astronomique que nous avons postulé est vrai et qu'il ne peut y avoir un autre modèle qui soit en accord avec l'observation des mouvements de la planète", dit Gersonide au début de son *Astronomie*[67].

La confiance qu'a Gersonide de pouvoir distinguer le vrai du faux, même lorsqu'il s'agit des corps célestes et par laquelle il s'oppose de façon si éclatante à Maïmonide, pour qui la vérité sur les mouvements célestes n'est connue que de Dieu seul[68], cette confiance n'est nullement gratuite et ne se limite pas à une déclaration sans suite. En effet, le désir de vérifier les caractéristiques par lesquelles se distinguent les modèles astronomiques amène Gersonide à effectuer des *observations astronomiques* à une époque où elles ne se pratiquent guère et, qui plus est, à inventer et à utiliser de façon intensive des *instruments astronomiques*, notamment son "bâton de Jacob".

S'agissant des observations, entre 1325 et 1345, Gersonide a, d'après M. B.R. Goldstein, effectué quelque quarante-cinq observations de planètes, auxquelles s'ajoutent une dizaine d'observations d'éclipses solaires et lunaires: il les utilise, fait rarissime au moyen âge, afin de mettre à l'épreuve différents modèles astronomiques disponibles[69]. Gersonide fait la plupart de ces observations en utilisant des instruments, dont le bâton de Jacob qu'il a lui-même inventé dans le but d'améliorer leur précision. Dans sa version "standard", le bâton de Jacob permet de déterminer, avec plus d'exactitude que cela n'était possible auparavant, la distance angulaire entre deux étoiles. "Nous nous sommes efforcé d'inventer un instrument dont la construction soit sans faute et dont les observations soient exemptes d'erreurs, écrit Gersonide. Et nous avons

[66] *M.H.* 5.1.20.
[67] *M.H.* 5.1.3, 51.
[68] Maïmonide, *Guide des égarés* II.24.
[69] Bernard R. Goldstein, "A New Set of Fourteenth-Century Planetary Observations", *Proceedings of the American Philosophical Society*, 132 (1988), p. 371-399 et l'article du même auteur dans le présent volume.

commencé à l'employer pour faire des observations, très utiles dans cette recherche. [Cet instrument] nous dirigera, si la Providence le veut bien, jusqu'à la vérité du système astronomique''[70]. Le désir de Gersonide de trouver le système astronomique *vrai* est donc à l'origine de l'invention du bâton de Jacob[71]. C'est l'épistémologie réaliste qui a donné un *sens* à l'augmentation de l'exactitude des observations astronomiques. Dans la perspective d'une épistémologie instrumentaliste, en effet, disposer d'observations plus exactes n'a de sens que pour autant que la différence de précision ait une signification pratique. Dans la mesure où des prévisions astronomiques plus fiables n'étaient pas demandées par la vie pratique en ce début du XIVe siècle (sauf, peut-être, en astrologie), à une époque où effectivement les astronomes ne faisaient guère d'observations, une amélioration de la précision des observations n'avait donc qu'une faible importance. Pour Gersonide, par contre, la précision, la conformité absolue avec la réalité, est primordiale: il s'agit de connaître l'agencement divin du monde et une différence même minime dans les résultats des observations risque de mettre en échec le salut de l'âme. Cette considération explique également la grande préoccupation de Gersonide pour les problèmes de l'erreur observationnelle, sujet sur lequel il revient maintes fois[72].

Bien évidemment, le bâton de Jacob, permettant d'améliorer l'exactitude des observations de la distance angulaire entre deux étoiles, a pu être inventé aussi dans le cadre d'une épistémologie instrumentaliste, quoique probablement plus tard. Mais Gersonide a également développé une deuxième version de cet instrument: il s'agit d'une combinaison du bâton de Jacob et de la *camera obscura*, instrument permettant de mesurer le diamètre apparent des astres[73]. Or les observations de cet instrument ont une signification

---

[70] *M.H.* 5.1.3, 44-45.

[71] Gersonide lui-même appelle son instrument *megalle ʿamuqot*, ''révélateur des profondeurs''. Cette dénomination peut s'entendre comme ayant délibérément un sens double, évoquant à la fois des vérités profondes et la profondeur au sens mathématique, c'est-à-dire la hauteur relative d'un astre par rapport à un autre.

[72] Cf. N. Rabinovitch, ''Early Antecedents of Error Theory'', *Archive for the History of Exact Sciences* 13 (1974), p. 348-358, notamment p. 356-358; B.R. Goldstein, ''Levi ben Gerson: On Instrumental Errors and the Transversal Scale'', *Journal for the History of Astronomy* 8 (1977), p. 102-112.

[73] Cf. B.R. Goldstein, ''Levi ben Gerson: On Astronomy and Physical Experiments'', *in* Sebetai Unguru (éd.), *Physics, Cosmology and Astronomy, 1300-1700: Tension and Accomodation* (= *Boston Studies in the Philosophy of Science*, vol. 126) (Dordrecht / Boston / London, Kluwer, 1991), p. 75-82.

uniquement si l'on considère le système astronomique comme une représentation du réel. Gersonide remarque, en effet, que si les planètes se trouvaient sur des épicycles, leurs diamètres apparents devraient varier considérablement: le diamètre apparent de la Lune dans le rapport 1:2, celui de Mars dans le rapport 1:6, etc., ce qui ne correspond pas à l'observation. Ptolémée, dit Gersonide, a écarté ce problème en disant que notre vue est trop faible pour observer ces changements de luminosité: or, rétorque-t-il,

"Cet argument n'a aucun poids, car nous avons déterminé le diamètre de la Lune en mesurant, par une règle, son rayon passant par l'ouverture de l'instrument et non pas par l'observation du corps même de la Lune. Il est donc évident qu'il [Ptolémée] ne peut attribuer ce fait—à savoir que le diamètre apparent ne correspond pas à ce qu'implique son modèle—à la faiblesse de la vue"[74]

Le diamètre apparent est, pour Gersonide, précisément une de ces caractéristiques par lesquelles on peut distinguer un modèle planétaire d'un autre:

"S'il y a plus d'un système [ou: modèle] astronomique duquel s'ensuit l'ordre [des mouvements planétaires], alors, en prenant en compte ce qui est en accord avec l'observation de la variation du diamètre apparent de l'astre, la recherche nous indiquera qu'un des systèmes [modèles] astronomiques est juste, et aucun autre."[75]

Dans la tradition ptoléméenne, on n'a guère prêté attention aux diamètres apparents. Effectivement, dans une perspective instrumentaliste, seules les positions des astres intéressent. La démarche de Gersonide est donc révolutionnaire: il *redéfinit l'ensemble des paramètres dont l'astronome doit tenir compte.* Pour employer un langage à la mode: Gersonide invente un nouveau discours astronomique. Le nouveau paramètre qu'introduit Gersonide dans son astronomie, ainsi que l'instrument qu'il invente afin de le mesurer, n'ont de sens possible que dans le cadre d'une épistémologie réaliste.

*

Nous terminons par un bref examen de la théorie astronomique de Gersonide, fondée sur l'idée de sphères excentriques par rapport au centre du monde (qui est, bien entendu, identique au centre de la Terre). Tout d'abord, les mesures des diamètres apparents, celles

---

[74] *M.H.* 5.1.43 (N, fol. 158ᵇ; P, fol. 79ᵃ; Q, fol. 58ᵃ).
[75] *M.H.* 5.1.19, 3.

mêmes qui avaient réfuté les modèles épicycliques, confirment le modèle excentrique[76] et permettent de déterminer l'excentricité[77]. De fait, la confiance de Gersonide dans ce modèle est si grande, qu'à plusieurs reprises, confronté à une difficulté concernant les implications physiques ou métaphysiques du modèle, il remarque que le modèle ayant déjà été établi, le raisonnement physique doit s'y accommoder. Mentionnons une seule de ces difficultés. Est-il possible que le mouvement circulaire des sphères se fasse autour d'un centre qui n'est pas le centre du monde? Pourquoi pas, répond Gersonide:

> "Il a déjà été démontré en physique que c'est le corps qui se meut en rotation qui produit le centre. Ainsi, s'il était possible qu'il y ait plus d'un seul monde, les mouvements simples des parties des différents mondes seraient nécessairement dirigés vers leurs centres respectifs."[78]

Cette thèse, tout à fait non orthodoxe dans la perspective aristotélicienne, est corroborée par le remarquable argument suivant:

> "Lorsque les corps sont inclus dans une circonférence, leurs mouvements simples se rapportent au centre de cette circonférence. Voici ce qui l'indique. Tu vois que dans certaines choses naturelles de forme sphérique, la partie lourde est au milieu, la partie légère l'entourant. Ainsi tu vois que l'œuf commence par consister en jaune seulement, puis la nature [en?] crée la partie légère, la partie lourde restant au centre et la partie légère l'entourant. De même, tu trouves que certains mouvements naturels ne se rapportent pas au centre du monde, mais à leur propre centre, comme le battement du cœur et [le mouvement de] l'air fondamental dans les artères."[79]

Certes, les arguments de Gersonide n'emportent pas nécessairement la conviction. Mais il serait erroné de les considérer sous cet aspect-là. Ce qui importe c'est que Gersonide a la hardiesse de s'opposer à toute la tradition artistotélicienne sur un point aussi capital. En effet, au seuil du XVIIᵉ siècle encore, quelqu'un comme William Gilbert est confronté pratiquement au même problème: dans une période où la physique est encore aristotélicienne, sa tentative d'élaborer une physique compatible avec l'astronomie copernicienne le confronte également à la question de savoir comment une

---

[76] *M.H.* 5.1.43.
[77] *M.H.* 5.1.33.
[78] *M.H.* 5.1.43 (N, fol. 152ᵇ; Q, fol. 76ᵃ; P, fol. 55ᵇ).
[79] *Ibid.*

pluralité de centres de gravitation est possible[80]. De fait, l'idée d'une telle pluralité est tout à fait incompatible avec les fondements mêmes de la physique aristotélicienne, de sorte que Gilbert, et à plus forte raison Gersonide avant lui, font effectivement preuve d'une très grande audace intellectuelle en l'acceptant.

*

Deux aspects encore de l'astronomie gersonidienne doivent retenir notre attention. Rappelons d'abord que Gersonide tente une explication physique des mouvements célestes. Il postule un corps liquide remplissant l'espace entre les sphères célestes, corps qui est supposé, d'une part, communiquer le mouvement diurne d'une sphère à celle qui l'entoure et, d'autre part, empêcher que les autres mouvements d'une planète ne perturbent ceux des sphères adjacentes. Conséquence de cette théorie, les thèses hardies de Gersonide concernant les dimensions du monde[81]. L'application de la physique à l'explication du "mécanisme" des mouvements célestes fait évidemment corps avec l'épistémologie réaliste: la description du monde céleste que propose Gersonide est mathématique et physique à la fois.

Gersonide ne perd jamais de vue que le but ultime de la recherche astronomique est de connaître la Providence. Maintes fois il observe que telle ou une telle disposition particulière des corps célestes témoigne de la Providence qui l'a créée. Considérons un seul exemple. Les mouvements des sphères ont une vitesse angulaire constante par rapport au centre postulé. Comment un mouvement régulier peut-il produire des effets si divers dans le monde sublunaire? Précisément, répond Gersonide: en soi, le mouvement est régulier, comme doit l'être tout mouvement céleste; mais par rapport à la Terre, puisqu'elle n'est pas au centre, il ne l'est pas. La Providence a donc su utiliser un mouvement régulier afin de produire ici-bas des effets très variés[82].

---

[80] Gad Freudenthal, "Theory of Matter and Cosmology in William Gilbert's *De magnete*", *Isis* 74 (1983), p. 22-37.

[81] Cf. l'article de B.R. Goldstein dans ce volume. Le rapport de cette théorie à l'idée des intellects en tant que moteurs de sphères reste à élucider.

[82] *M.H.* 5.1.45.

IV. Conclusion

En conclusion, quelques réflexions sur la nature, l'origine et la signification de l'originalité de Gersonide.

1. Une première remarque concerne le lien qui existe, dans l'œuvre de Gersonide, entre progrès scientifique et réalisme épistémologique.

L'originalité de Gersonide, le "ressort profond" de ses innovations, tient à son image réaliste de la science, qui, à son tour, découle de sa sotériologie. Pour lui, la recherche scientifique, loin d'avoir une visée pratique, est une fin en soi et elle a pour seul but de parvenir à une connaissance véridique du réel. Ce que cherche Gersonide, c'est de parvenir à l'immortalité de l'âme en acquérant la connaissance d'intelligibles se trouvant dans l'intellect agent. Il ne doute pas de la capacité de l'homme de parvenir à une connaissance adéquate, bien que partielle, du monde tel qu'il est en vérité. Cette attitude fournit le cadre dans lequel s'insèrent les innovations scientifiques, notamment astronomiques, de Gersonide. La plus conséquente en est l'idée que le diamètre apparent des astres est un paramètre dont l'astronomie doit tenir compte. Les mesures de cette variable—et pour les exécuter Gersonide invente un instrument approprié—apportent, pour lui, l'argument observationnel décisif contre l'hypothèse des épicycles. Notons en passant que nous avons là sans doute la raison pour laquelle l'astronomie de Gersonide est restée sans audience[83]: la quasi-totalité des astronomes médiévaux ne partageaient pas l'épistémologie réaliste de Gersonide, de sorte que pour eux les mesures des diamètres apparents des astres n'avaient simplement pas de signification[84].

---

[83] La question: pourquoi l'*Astronomie* de Gersonide, pourtant traduite en latin du vivant de son auteur (cf. l'article de J.L. Mancha dans ce volume), est-elle restée sans influence? a été formulée par B.R. Goldstein; cf. *The Astronomy of Levi ben Gerson, op. cit.* (n. 43), p. 15.

[84] Heinrich von Langenstein (Henricus de Hassia, 1325-1397) critique la théorie ptoléméenne avec des arguments pratiquement identiques à ceux de Gersonide; cf. Ernst Zinner, *Entstehung und Ausbreitung der Copernicanischen Lehre* (München, C.H. Beck, 2ᵉ éd. 1988), p. 82; Claudia Kren, "Homocentric Astronomy in the Latin West: The *De reprobatione eccentricorum et epiciclorum* of Henry of Hesse", *Isis* 59 (1968), p. 269-281; *idem*, "A Medieval Objection to 'Ptolemy'", *British Journal for the History of Science* 4 (1969), p. 378-393. Etant donné qu'il fut professeur à l'Université de Paris à partir de 1360 environ et jusqu'à son émigration à Vienne vers 1384 (suite au grand schisme), la question se pose s'il avait connu l'*Astronomie* de Gersonide, disponible en latin depuis 1344 (cf. l'article de J.L. Mancha dans ce volume). Le fait qu'Oresme connaissait certains résultats

Le fait que, contrairement à ses contemporains, Gersonide s'emploie à faire des observations et, à plus forte raison, son grand souci d'augmenter leur précision par le biais d'instruments, proviennent également de son désir de trouver la structure réelle du monde. Ce désir le conduit à l'invention du bâton de Jacob et à l'utilisation de la *camera obscura*, deux instruments dont il analyse de façon remarquable les conditions d'utilisation. De ces instruments, le bâton de Jacob dans sa version ''standard'' a connu une grande diffusion: contrairement aux *idées* nouvelles de Gersonide, qui pouvaient être acceptées uniquement par quelqu'un qui partageait également l'image gersonidienne de la science, c'est-à-dire son réalisme épistémologique, cet instrument, lui, était utilisable de façon tout à fait indépendante du contexte théorique dont il était issu. En revanche, l'autre version du bâton de Jacob, destinée à mesurer les diamètres apparents, n'a pas connu une telle diffusion, son utilisation étant directement liée à l'épistémologie réaliste qui présidait à son invention.

Autre conséquence: l'attitude de Gersonide est foncièrement empirique. Toute substance, même les minéraux, a, pour Gersonide, reçu sa forme de l'intellect agent. Aussi toute substance, supra- ou sublunaire, mérite-t-elle d'être un objet de recherche, car elle témoigne de la Providence[85]. Le chemin menant vers la connaissance de Dieu passe par la recherche empirique: ''Car dans la création des animaux et des plantes [Dieu] a déployé une sagesse merveilleuse, [...] afin que dans chacun d'eux existent tous les membres [lit. instruments] qu'il lui est possible de posséder et par lesquels son existence devient parfaite. Et Il nous donne également le temps nécessaire pour appréhender à partir de cela quelque chose de précieux concernant Dieu, béni soit-Il''[86].

C'est dans ce même contexte qu'il faut situer une autre innovation, tout à fait remarquable, de Gersonide qui n'a pas encore, semble-t-il, était notée. Dans son *Supercommentaire sur l'Épitomé [d'Averroès] du Livre des animaux* Gersonide envisage l'utilisation d'un

---

mathématiques de Gersonide (voir l'article de K. Chemla et S. Pahaut dans ce volume) tend à rendre plausible cette possibilité.

[85] Cf. de nouveau Touati, *op. cit.* (n. 5), p. 86.

[86] Gersonide, *Commentaire du Livre de Job*, fin de Chapitre 38; *Miqra'ot Gedolot*, p. 190[b]. Sur les attitudes de Judah Halévi et de Maïmonide à l'égard de ce type d'argument (''argument from design'') cf. H.A. Wolfson, *op. cit.* (n. 23).

348

miroir parabolique (dont il se sert ailleurs afin de concentrer les rayons de la flamme d'une bougie et ceux de la lune[87]) comme une loupe, permettant de discerner des détails des corps d'insectes, invisibles à l'œil nu:

> "De nombreuses espèces [animales] parmi celles qu'a mentionnées Aristote ne nous sont pas connues, surtout pas les espèces dont les individus sont extrêmement petits. Car même si nous pouvons les apercevoir, il nous est difficile, à cause de leur finesse, de connaître les formes et les propriétés de leurs membres. Je pense qu'une astuce [ou: un artifice; *taḥbulah*] au moyen de laquelle nous pouvons rendre compte des membres des animaux qui sont si fins qu'ils ne peuvent être aperçus par les sens est de les regarder au moyen des choses qui nous montrent l'objet perçu plus grand qu'il n'est: par exemple, le miroir ardent et d'autres objets semblables, avec lesquels on voit un objet perçu plus grand qu'il n'est. Avec une telle astuce on peut, en effet, constater la vérité concernant ces membres."[88]

Gersonide énonce ici l'idée d'une sorte de *proto-microscope*, une idée qui, du moins en Europe, n'a pas son semblable au moyen âge[89]. Gersonide ne fait pas état d'observations effectivement réalisées au moyen de l'instrument envisagé, ce qui donne lieu à penser qu'il n'a pas mis son idée en œuvre. Il demeure que même si son invention est restée sur le papier, elle témoigne de l'importance qu'il attachait à l'investigation empirique des formes créées, y compris celles des membres les plus fins des animaux les plus petits.

Nous pouvons ainsi conclure que, dans le cas de Gersonide, le réalisme épistémologique était un moteur important du progrès scientifique. Bien plus, en attribuant une signification au diamètre apparent des astres, variable qui dans la perspective instrumentaliste n'avait pas de grande pertinence, l'épistémologie réaliste de Gersonide a largement conditionné le contenu même de ses théories.

2. Une deuxième observation est d'ordre sociologique. L'image que se fait Gersonide de l'astronomie, qui conditionne, comme nous l'avons vu, son originalité, est étroitement liée à son rôle social, ou

---

[87] *M.H.* 5.2.6, p. 34ª.

[88] Gersonide, *Supercommentaire sur l'Épitomé [d'Averroès] du Livre des animaux*, ms. de la Bibliothèque Vaticane, Urb. ebr. 42, fol. 9ᵛ-10ᵛ. Ce passage est absent du commentaire d'Averroès que Gersonide paraphrase et commente: cf. Paris, Bibliothèque nationale, ms. héb. 956, fol. 421ʳ.

[89] Renseignement aimablement fourni par M. le professeur David Lindberg (Université de Wisconsin).

plutôt professionnel. En effet, Gersonide réunit en lui deux rôles professionnels distincts: celui du philosophe et celui de l'astronome[90]. Au moyen âge, les tâches institutionnalisées associées avec ces rôles sont différentes, voire opposées: chacun des rôles est régi par un autre ensemble de normes le définissant et son exercice est contôlé et sanctionné par un groupe de référence (de "pairs") distinct. Au philosophe il incombe de chercher la vérité, tandis que l'astronome a pour tâche de fournir des calculs utilisables. D'ordinaire, ces rôles étaient dissociés: ceux qui, dans différents contextes sociaux (dans les cours royales, notamment) excerçaient l'astronomie en tant qu'"art" pratique n'étaient pas les mêmes que ceux qui, le plus souvent dans les universités, s'occupaient de la physique et de la métaphysique. C'est d'ailleurs probablement la raison sociologique pour laquelle l'incompatibilité entre les modèles utilisés en astronomie, et la physique et la métaphysique aristotéliciennes a pu être si bien supportée pendant des siècles: la division du travail empêchait que les contradictions ne soient perçues trop fortement. Gersonide représente ainsi un cas singulier. Par son intérêt de connaissance, par la motivation qui l'anime, par sa formation et par le type d'ouvrages qu'il rédige, il est très certainemet un philosophe. Mais il est en même temps un astronome techniquement compétent, qui passe ses nuits à faire des observations et ses jours à calculer des tables astronomiques. Sur le plan sociologique, l'originalité de Gersonide résulte donc du fait que dans sa recherche il se conforme aux *normes des deux rôles sociaux à la fois*.

L'originalité de Gersonide, cognitive et sociologique à la fois, est largement en conformité avec un modèle de l'émergence d'innovations scientifiques, et d'innovations intellectuelles en général, décrit par le regretté sociologue des sciences Joseph Ben-David. Sur la base d'une analyse de l'émergence de nouvelles disciplines (la psychanalyse de Freud, la bactériologie de Pasteur et la psychologie expérimentale de Wundt), Ben-David a montré qu'une nouvelle discipline scientifique est souvent le fait d'un savant qui, pour des raisons d'ordre social (par exemple le blocage de carrières), migre d'une discipline à une autre et qui applique, par la suite, les méthodes et les normes qu'il avait adoptées ("internalisées") dans sa discipline

---

[90] La notion du rôle social a été introduite dans la sociologie des sciences par J. Ben-David; cf. surtout son ouvrage *The Scientist's Role in Society* (Chicago, University of Chicago Press, 2ᵉ éd., 1984) ainsi que la note suivante.

d'origine, à l'étude des objets de la seconde. Ben-David a appelé *hybridation de rôles* ce processus social, par lequel émerge un nouveau rôle social lorsque des normes d'un rôle social, appartenant à une discipline, sont introduites dans un deuxième rôle, associé avec une autre discipline. Sa thèse sociologique est alors que l'hybridation des *rôles sociaux* peut conduire à une hybridation d'*idées*, donnant lieu à des innovations cognitives[91].

Les innovations de Gersonide sont manifestement le résultat d'une hybridation d'idées: Gersonide introduit les normes épistémologiques de la philosophie dans l'étude de l'astronomie mathématique, sur un pied d'égalité avec les normes techniques de celle-ci. Sociologiquement parlant, ces innovations sont le résultat de l'hybridation des rôles de l'astronome et du philosophe. Nous ignorons malheureusement comment et pour quelles raisons cette hybridation de rôles s'est produite: est-elle uniquement le fait d'une curiosité intellectuelle d'un homme aux talents peu communs, qui a goûté autant le raisonnement philosophique que la technique mathématique et qui n'éprouvait aucune difficulté à les maîtriser l'un et l'autre? Ou y eut-il des raisons d'ordre social qui incitèrent Gersonide à se consacrer aux deux domaines, que, par la suite, il considéra comme formant une seule et même science? En l'absence d'éléments concernant la vie de Gersonide et son développement intellectuel, cette question reste sans réponse.

L'analyse sociologique éclaire encore davantage les raisons pour lesquelles l'astronomie de Gersonide est restée sans lendemain. Au moyen âge, Gersonide demeure un des rares à réunir en lui le rôle social de l'astronome et celui du philosophe de la nature. Il fallait attendre Copernic, Kepler et la synthèse newtonienne pour qu'une telle unité de rôles sociaux s'établisse durablement[92]. Or, comme

---

[91] Thèses développées surtout dans les deux travaux suivants: J. Ben-David, "Roles and Innovations in Medicine", *American Journal of Sociology* 65 (1960), p. 557-568; J. Ben-David et Randall Collins, "Social Factors in the Origins of a New Science: The Case of Psychology", *American Sociological Review* 31 (1966), p. 451-465. Ils sont réimprimés *in* J. Ben-David, *Scientific Growth*: *Essays on the Social Organization and Ethos of Science*, éd. Gad Freudenthal (Los Angeles / Berkeley / Oxford, University of California Press, 1991), p. 33-48 et 49-70, respectivement. Une analyse des idées de Ben-David se rapportant à la sociologie de la connaissance se trouve *in*: Gad Freudenthal, "Joseph Ben-David's Sociology of Knowledge", *Minerva* 25 (1987), p. 135-149; Gad Freudenthal et Ilana Löwy, "Ludwik Fleck's Roles in Society: A Case Study Using Joseph Ben-David's Paradigm for a Sociology of Knowledge", *Social Studies of Science* 18 (1988), p. 625-651.

[92] Cf. l'article d'H. Hugonnard-Roche dans ce volume.

nous l'avons vu, les innovations de Gersonide sont intimement liées
à la fusion en lui des deux rôles: nous pouvons ainsi émettre
l'hypothèse que les innovations qui en sont issues n'ont pas trouvé
d'écho puisque nul autre ne fit sien le même double rôle social.

3. Les historiens des sciences se gardent bien aujourd'hui de
rechercher des ''précurseurs'' et c'est une banalité que de dire que
l'histoire des sciences s'efforce de comprendre les penseurs
d'autrefois en les situant dans leurs propres contextes[93]. Il paraît
cependant légitime de relever chez Gersonide une autre sorte de
''modernité'', qui ne se situe pas sur le plan des idées scientifiques
elles-mêmes. Sa théologie l'a amené à porter son attention sur le
monde matériel; elle se solde par une attitude dirigée vers le monde
lui-même (innerweltlich) et non vers le transcendant. Gersonide
cherche la connaissance de Dieu à travers ses œuvres uniquement.
C'est cette théologie qui a légitimé pour lui sa recherche scientifique, en même temps qu'elle lui fournissait, sur le plan psychologique, la motivation de s'y investir. Cette attitude est d'autant plus
remarquable que la quasi-totalité des savants juifs médiévaux se
conformaient à la conception maïmonidienne selon laquelle les
sciences ont pour seul but de préparer l'intellect à l'étude de la
métaphysique, dont dépend la connaissance adéquate de Dieu[94]: en
défendant une théologie légitimant la recherche scientifique en tant
que pratique sociale et en attribuant à la connaissance scientifique
une valeur en soi, Gersonide s'oppose donc au consensus des penseurs juifs médiévaux, qui attribuaient à la science une place subor-
donnée. Or sur ce plan il y a une parenté frappante entre Gersonide
et les penseurs anglais du milieu du XVIIᵉ siècle qui ont fondé la
Royal Society. Là, comme l'avait si bien montré Robert K. Merton,
c'était le puritanisme qui avait suscité un intérêt religieux pour

---

[93] Rappelons qu'Hélène Metzger fut une des premières à énoncer explicitement ce principe méthodologique; cf. notamment ses articles ''L'historien des
sciences doit-il se faire le contemporain des savants dont il parle?'' (1933) et ''Le
rôle des précurseurs dans l'évolution de la science'' (1939), réimprimés in H.
Metzger, op. cit. (n. 1), p. 9-21 et 75-91 respectivement.

[94] Cf. par exemple Maïmonide, Guide, III.51; Maïmonide, Traité des huit
chapitres, chapitre 5; H.A. Davidson, ''The Study of Philosophy as a Religious
Obligation'', in S.D. Goitein (éd.), Religion in a Religious Age (Cambridge, Mass.,
Association for Jewish Studies, 1974), p. 53-68; H.A. Wolfson, ''The Classifica-
tion of Sciences in Mediaeval Jewish Philosophy'', dans ses Studies in the History of
Philosophy and Religion, tome 1, éd. I. Twersky et G.H. Williams (Cambridge,
Harvard University Press, 1973), p. 493-550, aux p. 542-545.

352

l'univers des phénomènes naturels et qui était un facteur important
dans l'institutionnalisation de la science de la nature en tant qu'occupation indépendante et légitime[95]. Du reste, selon une thèse récemment avancée par A.I. Sabra, les grands scientifiques de langue
arabe étaient, eux aussi, motivés dans leurs recherches scientifiques
par une théologie qui attribuait une valeur sotériologique à la connaissance scientifique du monde matériel[96]. Par ce type de lien
entre la théologie et la science—la première légitimant une recherche empirique portant sur les détails des manifestations divines
dans ce monde-ci—Gersonide est certainement un précurseur de la
science moderne[97].

---

[95] Cf. R.K. Merton, *Science, Technology and Society in Seventeenth-Century England*
(1938) (New York, Harper Torchbooks, 1970). Pour une analyse pénétrante de la
thèse de Merton cf. J. Ben-David, "Puritanism and Modern Science: A Study in
the Continuity and Coherence of Sociological Research", *in* E. Cohen, M. Lissak
et U. Almagor (éd.), *Comparative Social Dynamics. Essays in Honor of S.N. Eisenstadt*
(Boulder/Colo. et London, Westview Press, 1985), p. 207-223, réimprimé *in*
J. Ben-David, *Scientific Growth, op. cit.* (n. 91), p. 343-360. Un développement particulièrement perspicace de la thèse de Merton est donné par Michael Heyd, "The
Emergence of Modern Science as an Autonomous World of Knowledge in the Protestant Tradition of the Seventeenth Century", *in* S.N. Eisenstadt et I.F. Silber
(éd.), *Cultural Traditions and Worlds of Knowledge: Explorations in the Sociology of
Knowledge* (= *Knowledge and Society: Studies in the Sociology of Culture Past and Tresent*,
7) (Greenwich/Con. et London, JAI Press, 1988), p. 165-179.

[96] A.I. Sabra, "The Appropriation and Subsequent Naturalization of Greek
Science in Medieval Islam: A Preliminary Statement", *History of Science* 25 (1987),
p. 223-243. Il faut cependant noter que l'"appropriation" des sciences par les
savants arabes n'était pas toujours une étape précédant leur "prolongement créatif"; cf. Roshdi Rashed, "Problems of the Transmission of Greek Scientific
Thought into Arabic: Examples from Mathematics and Optics", *History of Science*
27 (1989), p. 199-209.

[97] Le sujet traité dans cet article a fait l'objet de deux conférences qui ont par
la suite été publiées: "*Haṣlaḥah nafshit we-astronomiya: Milḥamto shel ha-Ralbag neged
Talmay*" ("Perfection de l'âme et astronomie: La révolte de Gersonide contre
Ptolémée"; en hébreu), *Daʿat* n° 22 (1989), p. 55-72 (où l'on trouvera les textes
originaux hébreux de quelques passages cités dans le présent article en traduction
française); "Rabbi Lewi ben Gerschom (Gersonides) und Bedingungen wissenschaftlichen Fortschritts im Mittelalter: Astronomie, Physik, erkenntnistheoretischer Realismus, und Heilslehre", *Archiv für Geschichte der Philosophie* 74 (1992),
p. 158-179. Le présent article est cependant le plus complet.

Pour avoir lu ce texte et m'avoir fait part de leurs critiques et suggestions je
remercie vivement MM. B.R. Goldstein (Pittsburgh), M. Kellner (Haifa), T.Y.
Langermann (Jérusalem), Juliane Lay et J.-P. Rothschild (Paris).

Levi ben Gershom as a Scientist: Physics, Astrology and Eschatology

The heavenly bodies play a major role in R. Levi ben Gershom's (= Ralbag's) view of the world. They are nothing less than the instruments through which God exercises His providence over the sublunary world, and Ralbag in fact holds that at the creation, the influences emanating from the heavenly bodies have been "programmed" so as to make everything down here the best possible one: both in the design of nature and in the arrangement of human social affairs. In view of this paramount role Ralbag ascribes to the heavenly bodies in governing sublunary affairs — Prof. Charles Touati appropriately described his position as one upholding an "astral determinism" — it does not come as a surprise that practically all those who have written about Ralbag, have described him as one subscribing to astrology. Moreover, we know today that Ralbag in fact cast prognostications. My aim in this paper is to try and make explicit in what sense precisely we may say of Ralbag that he believed in astrology and argue that his position was founded on the best contemporary science and was thus as rational as any. I will also show how his vision of mankind as influenced by the stars relates to his ideas on human freedom of choice and to his eschatological views; this will also clarify Ralbag's ideas on general and particular providence.

*

Consider, to begin with, the Peripatetic physics of the sublunary world. It has to be realized, and this point must be emphasized, that all medieval thinkers, without exception, subscribed to *natural astrology*, i.e. the physical theory upholding that the generation and corruption of sublunary substances depend upon the heavenly bodies. Consider a summary of its canonical form, from the pen of such an adversary of astrology as Rambam (Maimonides):

It is known and generally recognized in all the books of the philosophers speaking of governance that the governance of this lower world — I mean the world of generation and corruption — is said to be brought about through the forces overflowing from the spheres. ... [Y]ou will find likewise that the Sages say: "There is not a single herb below that has not a '*mazzal*' in the firmament that beats upon it and tells it to grow" (*Bereshit Rabbah* 10:6). ... Now they also call a star: *mazzal*. ... By means of this dictum they have made it clear that *even individuals subject to generation have forces of the stars that are specially assigned to them*. (*Guide for the Perplexed*, II.10, trans. S. Pines, p. 269 f.; cf. also I.72, p. 186 f.; II.5, p. 260.)

Indeed, Rambam himself recognized that these universally accepted Peripatetic doctrines seemingly provided a suitable basis for astrology: "[T]he stars act at some particular distances; I refer to their nearness to, or remoteness from, the center or their relation to one another. *From here astrology comes in*." (*Guide* II.12, p. 280.) Rambam thus subscribes even to the idea that the actions of the stars depend on their positions with respect to one another, i.e. on what in the technical vocabulary of astrology is called their "aspects" — a distinctively astrological doctrine. And he notes himself that this physical thesis is the foundation of astrology.

Some words must now be said on a further, quite distinct, doctrine, which complemented the theoretical foundation of astrology: This is psycho-physiology, which was also universally accepted in the Middle Ages. The Hippocratics and Aristotle already held that certain physiological characteristics of the body — e.g. whether it is hot or cold — determine the person's character. This idea was elaborated in the doctrine of the four humors, whose equilibrium was held to determine the person's temperament. Whence the idea that the faculties of the soul of an individual, including his moral traits, depend upon the blend of his humors. This idea was elaborated notably by Galen, one of whose treatises, translated into Arabic, bears the title: "That all the faculties of the soul depend upon the mixtures [or: temperaments] of the body."

Let me again take Rambam as my crown witness. In order to explain why men are psychologically so dissimilar — it is this dissimilarity that makes social organization indispensable — Rambam draws on the psycho-physiological doctrine:

[T]here are many differences between the individuals belonging to [the human species], so that you can hardly find two individuals who are in any accord with respect to one of the species of moral habits. ... The cause of this is the difference of the mixtures, owing to which the various kinds of matter differ, and also the accidents consequent to the form [i.e. the soul] in question. (*Guide* II.40, p. 381.)

The theoretical argument underlying astrology comes to the fore in yet another doctrine postulating external physical influences on the constitution of bodies, and therefore on psychical characteristics, namely *climatology*. According to this universally accepted theory, the physiological equilibrium inside a living body is strongly influenced by the physical conditions of the environment; consequently, the psychical faculties of each individual, and indeed of entire nations, strongly depend on the physical and climatic conditions of the place. (Cf. e.g. Yehudah Hallevi, *Kuzari* 1.1; 2.10 ff.; Rambam, *Pirqey Moshé* 25.57.)

We may conclude that medieval philosophy of nature accepted as indisputable the idea that the heavenly bodies to a large extent shape the natural processes in the sublunary world. The heavenly bodies are here construed firstly as *material* bodies acting upon the sublunary matter — they in fact mix the elements and also influence their proportions within each substance; secondly, a separate *intellect*, the so-called agent intellect, imposes corresponding forms upon matter. Now since the psychical qualities of beasts and men depend upon their material constitution, it follows that men's disposition to behave in one way or another depends upon celestial influences. All this, I repeat, is good Aristotelian and medical theory, accepted by all medieval thinkers. Writers on astrology, beginning with Ptolemy in his *Tetrabiblos*, and then, most thoroughly, by the astrologer Abū Ma'shar, in his *Kitāb al-madkhal al-kabīr*, drew on this natural philosophy to provide their art with a theoretical foundation.

*

The universally-accepted tenets of natural philosophy I have just mentioned are the point of departure of Ralbag's doctrine of astral determinism. The general principles of Peripatetic physics imply, according to Ralbag, that

the heavenly bodies preserve the sublunary existents by reinforcing now this contrary, now another. ... As when you say that Mars reinforces the fiery nature, the moon the watery nature. The contraries thus produced by [the heavenly bodies] in the temperament of the human individuals influence the moral qualities [*middot*] and the practical capacities. It follows that in one position the heavenly bodies direct a man toward one quality, and in the contrary position they direct him toward its opposite. (*Milḥamot ha-Shem* 2.2, Leipzig ed., p. 96.)

In other words, Ralbag's thesis is: "the temperament of every person is given him by the spheres and by the stars, as this has already been established in this science" (cf. A. Altmann in *American Academy for Jewish Research*, Proceedings, vol. 46–47 (1979–80), p. 17).

Now this astral determinism is doubly limited. First, as is well known, Ralbag holds that the separate intellects — and consequently divine providence too — know only some *general* traits of the sublunary world: the effects produced by the celestial influences differ according to the matter receiving them, with the resulting forms having a certain "breadth." Thus, men born under a given astral aspect will all share certain common general traits, namely precisely those known by the agent intellect; still, each will have an individual temperament (*mezeg 'ishī*).

The second limitation of the astral determinism is due to the fact that according to the psycho-physical theory, our temperament does not *determine* our behavior; rather it only gives us a *disposition* (or propensity) to act in a certain way. Man, and this is an idea of capital importance to Ralbag, is free to counteract the tendencies due to his biologically-determined temperament: "God," Ralbag urges, "has placed in us an intellect, whose end is to make us achieve something different from what has been determined by the celestial bodies" (*M.H.* 2.2, p. 97). Astral determinism may give you a "hot" nature, but if you wish to, you will succeed in containing your choleric temperament.

For Ralbag, then, the astral determinism of human behavior is inherently limited. It does not entirely determine an individual's temperament, nor does the individual temperament itself wholly determine conduct. This means that the physical laws concerning the influences on man's actions of the celestial bodies are not deterministic laws. Rather they are what in modern philosophy of science is called *probabilistic* laws, laws specifying statistical distributions of certain properties in given sets of individuals. The laws of astral determinism state that when a given astral configuration is dominant, certain characteristics of temperament will appear in the population with a great probability (*me'odiyyi*). These laws are therefore deterministic with respect to the statistical *distribution* which they state; but they are not deterministic with respect to any given *individual* and thus do not conflict with the idea of human freedom of choice.

This interpretation of Ralbag's astral determinism can be illustrated and confirmed by the following example. How, Ralbag asks (*M.H.* 2.2, p. 97 f.), can we explain that in each society we find someone to choose even the most unpleasant work? We already mentioned that Rambam ascribed the diversity of men to the differences of their physiological temperaments. But whereas Rambam did not say a word on the origin of these differences, Ralbag explicitly says that the differences of temperaments, which in turn lead to differences of dispositions and hence of the occupations chosen, are due to astral influences. Ralbag insists that only the overall distribution of the professions is unfailingly brought about by these influences and is thus due to divine providence. But the determinism is not absolute with respect to any given individual: concerning the occupation of a given person, astral influences determine only his *disposition* to behave in one way or another, a disposition which one can, however, counteract by his will to follow his intellect. Ralbag's indeterminism is thus ontological and not epistemological: it is inherent to nature itself, and is not due merely to the limitations of our knowledge of the natural order.

Consider now Ralbag's stance on the possibility to predict human actions. Ralbag (*M.H.* 2.2, p. 95) does not doubt that the astrologers often predict the future correctly. How is this possible? His answer is simple: true, every given person is free to choose his course of action and thus escape the effects of the astral determinism. To be sure, astrology can predict nothing whatsoever of the future of someone who thus makes use of his freedom. Nonetheless, the depressing fact of life is that most people do *not* follow their intellect and thus do not escape astral determinism. *De facto*, therefore, the astral determinism *does* structure the conduct of the great majority of people. Consequently, the astrologer who predicts what the astral determinism implies for the temperament, and consequently the conduct, of a given person, will *in most cases* be right. This is, according to Ralbag, the reason why astrological predictions so often come true.

Another kind of astrological predictions, which are of special interest to us in the present context, concerns *macro-events*. By this term I refer to events involving large social groups, notably nations. Now because the astral determinism is probabilistic, predictions of macro-events can be absolute without running against human liberty. Take the following example: suppose that at a given period Mars is preponderant, thus strengthening in all men the element of fire and, consequently, the belligerent tendencies. One can safely predict that a war will inevitably result. Hence: the war itself, a macro-event involving numerous persons, is determined, indeed *caused*, by the physical influences emanating from the stars. Nonetheless, any individual who will follow his intellect, will be able to avoid succumbing to this combative lust of the time. If, in addition, this person can predict what the astral determinism implies for his nation, he may be able to undertake appropriate steps so as not to be hurt personally.

The logical conclusion of the foregoing is that certain events in history are due to the astral determinism, that is to providence. It should be realized that the providence in question is that which acts through physical influences of the heavenly bodies on *the world of the elements*; it follows *natural* laws. This is not the particular divine providence which, according to Ralbag, watches over the just. The predictions of Ralbag's *Prognostication* concerning the significance of the conjunction of 28 Mars 1345 (to be published by B. R. Goldstein and D. Pingree in the *Transactions of the American Philosophical Society*, vol. 80.6 [1990]) belong precisely to this category. Predicted are macro-events, to begin in 1355, concentrated in the Eastern Mediterranean: the destruction of one nation by another of a different religion; plagues; periods of sufferings for most men; shedding of much blood in wars; periods of dryness and of famine; storms drowning ships; etc. This lengthy period of upheavals, Ralbag says, will end through the rise of a kingdom of truth and justice. In the introduction, Ralbag considers the implications for the individual of the predicted macro-events: having an intellect and thus capable of free choice, man can avoid the consequences of the catastrophes that the global astral determinism will bring about. This means that the just will escape the cataclysms through personal providence: when the stars produce a calamity in a given country, he affirms, the just are not hurt.

We have here, I suggest, Ralbag's *philosophy of history*. For Ralbag, history happens, so to say, on two distinct, if related, planes. There is, first, the history of the entire human species, of men inasmuch as they are physiological entities underlying the laws of nature. Men have bodies whose composition is determined by the heavenly bodies, and these bodies determine their dispositions to act. Since the great majority of men do not use their intellect to make free choices, they belong to the realm of nature, just as the beasts. Consequently, social macro-events such as wars and peace, depend upon the astral determinism *in exactly the same way as natural events* such as periods of draught or storms. This part of history depends upon God's global providence, exercised through the natural, regular and law-like influences of the heavenly bodies; it is an integral part of *the natural history of the world.*

Then, on another plane, there is the history of those on whom God exercises, in addition, His particular providence: this privileged group is not fully under the influence of the astral determinism. To it belong the people of Israel and the "community of the investigators [or: philosophers; *'adat ha-me'ayyenim*]," namely those who escape the astral determinism using their intellect. One may wonder how Ralbag can possibly hold that with respect to history, the philosophers and the people of Israel constitute a single group — indeed this is the group whose history alone is really human. The answer to this puzzle is that Ralbag takes philosophy and the Torah to be tracing two ways to one and the same goal: the Torah, he holds, guides those who follow it toward the very same theoretical and moral truths which philosophers attain through reflection (*'iyyun*). This group alone, then, is partially outside the natural history of the world, including that of the rest of mankind.

Let us now consider Ralbag's view of the history of Israel, comparing the *Prognostication* with his *Commentary on the Book of Daniel.* One important conclusion of the preceding analysis is that Ralbag does not subscribe to the "astrological interpretation" of history as found notably in *Megillat ha-megalleh* of Abraham bar Ḥiyya. For bar Ḥiyya believes that all major historical events, including the history of Israel, are determined by the stars, specifically by different types of conjunctions of the planets. By contrast, for Ralbag, astral determinism structures only *nature*; but Israel, by virtue of the Torah it follows, is not entirely subject to the natural determinism. For Ralbag, there can thus be no question of describing the history of Israel as depending upon the conjunctions of the planets. This indeed explains why — as it seems — he did not cast prognostications for Jews: only gentiles, with the exception of true philosophers, entirely underlie astral determinism. The *Prognostication* concerning the conjunction of 1345 too seems to have been written on the demand of the papal court.

Nonetheless, the history of nature is the framework for the history of the just too: events depending upon the global providence — i.e. those originating in the astral determinism — *may* be significant for the history of Israel too (cf. *Commentary* 12:1). This is the case of the events described in the *Prognostication.* In his *Commentary on the Book of Daniel*, Ralbag argues that the only event Daniel had prophesized which is yet to come is the war between the king of the North — that of the Christians — and the

Moslem king of the South (11:40). This war will end with the victory of the
king of the North (11:30; 11:40–44), who in turn will be defeated because
"he will be fought from heaven" (11:44): this ultimate defeat, Ralbag
explicitly says, will be due not to the star configuration, but to particular
providence, for the sake of the Patriarchs (end of the *Commentary*, eighth
"lesson"; *Commentary* 7:17). This defeat will be followed by the
establishment, in 1358, of the fifth Kingdom, the eternal messianic
Kingdom.

What is the relationship between the events predicted in the
Prognostication for 1355 and those predicted in the *Commentary* for 1358?
Prof. B. R. Goldstein has already argued that the events of the *Prognostication* are not messianic events, for those Ralbag held to begin only
three years later. Nonetheless, there are obvious similarities between the
events described in the two texts. The *Prognostication* predicts that "if
southerners will fight northerners, the southerners will be defeated
according to the [planetary] configuration." This and other predicted events
unmistakably recall those that according to the *Commentary* will precede
the messianic era. Moreover, the period of wars predicted in the *Prognostication* will end by the establishment of a kingdom of truth and rectitude,
traits that are also those of the fifth Kingdom according to the *Commentary*.

I suggest that the relationship between the events described in the two
texts is this: The events predicted in the *Prognostication* for 1355, notably
the war ending with the victory of the northerners which is predicted in the
Commentary too, are events preceding and preparing the establishment of
the fifth Kingdom and of the messianic times, but they do not themselves
begin the messianic era. For these events belong to the history of nature: this
indeed is precisely the reason why they can be foreseen by astrological
predictions. Thus, the establishment of the fifth Kingdom is due to particular
providence and is distinct from the victory predicted in the *Prognostication*,
although this victory is a prior condition for it. This is why the events
brought about by the astral determinism are after all related to the messianic
times: they are nothing else than the *ḥevley Mashi'aḥ*, they mark the
atḥaltah di-ge'ulah (the beginning of deliverance).

SUR LA PARTIE ASTRONOMIQUE DU
LIWYAT ḤEN DE LÉVI BEN ABRAHAM BEN ḤAYYIM *

Lévi b. Abraham b. Ḥayyim, philosophe provençal qui a fleuri dans la dernière partie du XIIIème siècle, nous a laissé, entre autres, deux ouvrages: le premier ,בתי הנפש והלחשים, est une sorte de compendium, écrit en partie en vers, des différentes sciences [1]; le deuxième, לוית חן (= *LḤ*), une grande encyclopédie du savoir philosophique et de la réflexion religieuse, est le sujet de cette note.

Neubauer et Renan [2] et, plus récemment, Mme Colette Sirat [3], ont déjà remarqué que dans l'état actuel des manuscrits, le *LḤ* s'apparente à un puzzle: il n'existe aucun manuscrit complet de l'ouvrage, de sorte qu'il n'est pas aisé de déterminer les places respectives de ses parties qui, pour autant qu'elles existent, sont dispersées dans différentes bibliothèques, et ceci d'autant plus que l'ouvrage a connu au moins deux rédactions [4]. Mme Sirat a établi une liste des chapitres de la fin du cinquième livre (*ma'amar*) — traitant de la métaphysique — et du sixième livre, dont les parties traitent: 1° de la prophétie et des secrets de la *Torah*; 2° des secrets de la religion; 3° de la création [5]. Pour notre part, nous voudrions, dans cette note, apporter quelques lumières sur la structure et sur le contenu des autres parties de l'ouvrage.

La liste des chapitres donnée par Mme Sirat nous apprend que le sixième livre de l'ouvrage entier (הספר הכולל המכונה לוית חן) est le premier livre de

* Ce travail a été soutenu par le Sidney Edelstein Center for the History and Philosophy of Science, Technology and Medicine à l'Université hébraïque de Jérusalem. Je lui exprime ma grande gratitude.

1. I. Davidson, «L'introduction de Lévi b. Abraham à son encyclopédie poétique», *REJ* CV, 1940, pp. 80-94; *idem*, «Levi b. Abraham b. Hayyim. A Mathematician of the Thirteenth Century», *Scripta Mathematica* 4 (1), 1926, pp. 52-66.

2. E. Renan, *Les Rabbins français du quatorzième siècle*, Paris, 1877, pp. 628-647.

3. C. Sirat, «Les différentes versions du *Liwyat Ḥen* de Lévi ben Abraham», *REJ* CXXII (1-2), 1963, pp. 167-177.

4. Cf. la notice de M. Steinschneider in *Hebräische Bibliographie*, IX, 1869, pp. 24-25.

5. Une liste moins complète avait été publiée par A. Geiger in *He-Halutz* 2, 1853, pp. 17-21.

sa deuxième «colonne» (*'amûd*), nommée *Boaz*. Nous pouvons ainsi, sans trop de risque, conclure que la première colonne s'intitulait *Jachin* et qu'elle comportait cinq livres. Quant à *Boaz,* la liste de chapitres de Mme Sirat indique qu'il comporte deux livres dont le premier est divisé en trois parties, le second en deux: מעשה מרכבה et שער ההגדה ⁶. Dans ce qui suit nous nous concentrerons sur *Jachin*.

Le manuscrit de la Bibliothèque Vaticane, ebr. 383, commence ainsi (fol. 1ʳ):

המאמר השלישי מן החבור הכולל הנקרא לוית חן בחכמת התכונה. אמר לוי בר׳
אברהם לוי. אחר שהשלמנו מאמר שני מספרנו והבאנו קצת כוללים מן המדות הנה
נחל המאמר השלישי בחכמת התכונה. ומנין שערי זה המאמר השלישי ארבעים
שערים על סדר החרוזות אשר בניתי בתכונה בספר בתי הנפש והלחשים.

Il s'avère que le troisième livre, portant sur l'astronomie, a été précédé par un livre traitant de la géométrie. Quant au quatrième livre, son sujet est indiqué par le manuscrit de Cambridge, Add. 1563, fol. 81ᵛ:

נשלם המאמר השלישי המדבר בחכמת התכונה ת״ל. ויבוא אחריו המאמר הרביעי
בחכמה הטבעית.

Ainsi, la structure de *Jachin* était: 1° inconnu (l'arithmétique?) ⁷; 2° la géométrie; 3° l'astronomie (y compris, nous le verrons, l'astrologie); 4° la physique; 5° la métaphysique.

Quelles sont les parties de *Jachin* qui existent encore aujourd'hui? En consultant le catalogue de l'Institut des microfilms de manuscrits hébreux de l'Université hébraïque à Jérusalem ⁸, nous avons trouvé uniquement des textes appartenant au troisième livre, ainsi qu'un fragment du deuxième ⁹. Le reste — le premier, la plupart du deuxième, le quatrième et le cinquième livre de *LḤ* — semble ainsi perdu.

Intéressons-nous donc à ce troisième livre de *LḤ*, le ספר תכונה. Comme nous l'avons déjà constaté, le manuscrit Vat. ebr. 383 annonce 40

---

6. Des extraits en furent publiés, d'après le manuscrit de Munich n° 58, fol. 27 sqq., par M. Steinschneider et J. Kobak in *Jeschurun,* VIII, 1871, partie hébraïque, pp. 1-13.

7. Dans le *LḤ* Lévi entendait reprendre, en en augmentant le détail, les sujets de בתי הנפש והלחשים. Or dans cet ouvrage, le septième traité porte sur המספר והמדות et il est suivi des traités d'astronomie, de physique, et de métaphysique (Davidson, *op. cit.*, p. 60-61). Il est donc vraisemblable que, dans *LḤ*, מספר et מדות ont été partagés entre deux livres. (Je remercie M. le professeur B.R. Goldstein pour cette suggestion.) Cf. aussi Renan, *Rabbins*, pp. 638 sqq.

8. Nous remercions l'Institut pour avoir pu y travailler dans les meilleures conditions. Les cotes des microfilms des manuscrits mentionnés dans le texte sont indiquées dans l'Annexe.

9. Il s'agit du manuscrit de Paris, Bibliothèque nationale, héb. 1050, fol. 48ᵛ-50ᵛ, où le texte porte le titre מס׳ לוית חן בחלק חכמת התשברת.

chapitres (שערים) pour ce livre. Ce manuscrit, paraît-il, est le seul à les comporter tous et à donner ainsi le texte complet du livre. De fait, les 39 premiers chapitres d'une part, et le 40ème chapitre d'autre part, ont été transmis de façon largement indépendante, les manuscrits respectifs étant destinés à des publics ayant des intérêts différents. En effet, le quarantième «chapitre» — à lui seul quelque trente pour cent de l'ensemble du traité — porte sur l'astrologie et il est en fait un traité indépendant; les 39 premiers chapitres, en revanche, constituent un traité d'astronomie. Ainsi, un manuscrit de la Bibliothèque nationale de Paris (héb. 1047, fol. 147ᵛ-220ᵛ), comporte une sélection des premiers 39 chapitres astronomiques, puis donne les toutes premières phrases du quarantième chapitre, avant de s'arrêter en notant (fol. 220ᵛ):

והנה זה השער [הארבעים] נמצא אתי יותר באריכות בספר בפני עצמו ושם הספר שער הארבעים ודי לי בו.

Effectivement, en tant qu'ouvrage indépendant, la partie astrologique reçoit dans les manuscrits le titre de שער הארבעים ou, plus explicitement: שער הארבעים בכחות הכוכבים ומפעלם על צד הכלל.

Dans l'Annexe, nous donnons: 1° une liste de manuscrits comportant différentes parties du troisième livre de *LH*; 2° une liste des quarante chapitres de ce livre.

Nous pouvons maintenant mentionner et résoudre une petite énigme littéraire. Dans *Les Écrivains juifs français du XIVème siècle*, A. Neubauer attribue à Lévi b. Abraham un ouvrage intitulé דלוגים, inclus dans le manuscrit de Cambridge Add. 1563 (fol. 92ʳ-104ᵛ), sans traduire ou expliquer ce titre bien singulier[10]. M. Steinschneider a déjà soutenu que «in der That ist hier nicht von einem Titel, am allerwenigsten von einer so benannten Schrift ... die Rede», mais qu'il s'agit là de «Weglassungen» et il a émis l'hypothèse suivante: «Es fragt sich, ob diese Nachträge zu einem, in demselben Ms. vorangehenden Stücke gehören»[11]. Cette hypothèse s'avère exacte. En effet, à sept endroits en marge du premier texte dans ce manuscrit, texte qui comporte le quarantième chapitre (astrologique) du livre d'astronomie de Lévi, on trouve les mentions דלוג ראשון, דלוג שני etc. Ces notes marginales indiquent donc la place où doivent être insérés les sept passages «omis», intitulés דלוג ראשון, דלוג שני etc., se trouvant aux fol. 92ʳ-104ᵛ. En effet, en consultant certains autres manus-

10. E. Renan, *Les Écrivains juifs français du XIVème siècle*, Paris, 1893, p. 262, n° XXIII.
11. M. Steinschneider, *Die Mathematik bei den Juden*, Hildesheim, 1964, p. 133.

crits de ce quarantième chapitre, nous avons pu constater que les passages omis du manuscrit de Cambridge y font corps avec le texte même (p. ex. Paris, Bibliothèque nationale, héb. 1066; Rome, bibl. Vittorio Emanuele 12). Il est donc évident que le manuscrit de Cambridge a été copié sur un manuscrit défectueux (effectivement, le manuscrit Vat. ebr. 383 comporte les mêmes lacunes), mais qu'un jour il a été collationné avec un manuscrit complet.

Effectivement, à la fin des sept passages on trouve la remarque (fol. 104ᵛ):

ע"כ [= עד כאן] דלוגי זה הספר מחלק התכונה שחבר ר' לוי בר' גרשום [צ"ל:
אברהם] בעל בתי הנפש והלחשים.

S'agissant des six premiers passages, il est manifeste que les lacunes sont dues à un hasard (feuilles déchirées, etc.). En effet, ces passages (sauf le sixième) commencent au milieu d'une phrase. Il en va autrement du septième passage: celui-ci a sa place juste après une longue discussion de huit raisons pour lesquelles le futur d'un individu, tel qu'il est indiqué par les astres, se trouve modifié (devenant ainsi imprévisible) à cause de l'appartenance de cet individu à une collectivité, discussion que Lévi conclut par la phrase ואין מזל לישראל (fol. 80ᵛ). Or, le septième passage aborde un nouveau sujet avec les mots suivants (fol. 100ᵛ):

ראיתי לזכור הנה על צד הקצור קצת עניינים נזכרו בס' מגלת המגלה שחבר הנשיא ר'
אברהם צאחב אל שורטא ז"ל.

Et plus loin (fol. 103ᵛ):

ועתה נזכור מדברי הנשיא מה שיקרה בדבוק הגדול שהחל בשנת ד' אלפים תתק"פו
... ונלקט מהם מה שצריך לפי כונתינו ולפי העת שאנחנו בו שהוא שנת ה'
אלפים ונ"ט לעולם.

La date donnée ici, 1299, indique à ne pas en douter que, quatre ans après avoir terminé, en 1295, la deuxième rédaction de *LḤ*[12], Lévi b. Abraham a voulu communiquer au public quelques idées encore et qu'il a ajouté à son chapitre astrologique un passage supplémentaire (tiré, du reste, de מגלת המגלה de Abraham bar Ḥiyya).

De fait, ce passage manque dans la plupart des manuscrits consultés, et il semble se trouver à sa place uniquement dans le manuscrit de Rome, Vittorio Emanuele 12. Ainsi, le collationneur du manuscrit de Cambridge

12. C. Sirat, *op. cit.* (note 3), p. 168.

a cru qu'il avait affaire à une septième «omission», tandis qu'en vérité il s'agissait d'un addendum.

Notons par ailleurs qu'un texte anonyme dans le manuscrit de Paris, BN héb. 1066, fol. 107ʳ-109ʳ, qui porte le titre שנה ראשונה על דבוק הרב שהעיד על לידת משה רבינו ע"ה בסוף ג' שנים peut, maintenant, être attribué avec certitude à Lévi b. Abraham, ce texte étant identique à la septième «omission» du manuscrit de Cambridge.

ANNEXES

I. Les manuscrits du troisième livre (astronomique) du *Liwyat Ḥen*[13].

Note: Pour chaque manuscrit, nous indiquons sa cote à la bibliothèque où il est conservé ainsi que la cote de son microfilm à l'Institut des microfilms de manuscrits hébreux à l'Université hébraïque de Jérusalem; cette dernière cote est précédée par les lettres «IM».

1. Cambridge Add. 1563 (IM: 17475).
   a) Fol. 1ʳ-81ᵛ: 40ᵉᵐᵉ chapitre (partie astrologique).
   b) Fol. 92ʳ-104ᵛ: sept passages complétant le précédent.
2. Vatican, Vat. ebr. 383 (IM: 464).
   Le seul manuscrit complet contenant les 39 chapitres astronomiques et le 40ᵉᵐᵉ chapitre (astrologique).
3. Paris, Bibliothèque nationale, héb. 1047 (IM: 14650).
   Fol. 174ᵛ-220ᵛ: extraits des 39 chapitres astronomiques.
4. Paris, Bibliothèque nationale, héb. 1066 (IM: 33999).
   a) Fol. 1ʳ-106ʳ: 40ᵉᵐᵉ chapitre (partie astrologique).
   b) Fol. 107ʳ-109ʳ: texte astrologique de Lévi ben Abraham, complétant le précédent (il devait être inséré fol. 105ᵛ; voir ci-dessus).
5. Londres, Jews College, Montefiori Collection, 484 (IM: 6113).
   a) Fol. 30ʳ-36ʳ: extraits des chapitres 31 et 40.
   b) Fol. 44ᵛ-50ᵛ: extraits des chapitres 1-6, suivis (fol. 49ʳ sqq.) par des extraits de la fin du 40ᵉᵐᵉ chapitre (astrologique).
6. Rome, Vittorio Emanuele 12 (IM: 404).
   Fol. 37ʳ-139ᵛ: 40ᵉᵐᵉ chapitre (astrologique); la fin manque.

---

13. Rappelons que les manuscrits des autres parties de *LH* ont été signalés par Mᵐᵉ Sirat (*op. cit.* note 3, p. 169). Mᵐᵉ Sirat mentionne le manuscrit du Vatican, Vat. ebr. 198, mais cela semble être une erreur.

7. New York, Jewish Theological Seminary of America, Mic. 2559 (IM: 28812).

Fol. 1ʳ-107ᵛ: commence abruptement avec la fin du 4ᵉᵐᵉ chapitre et se termine abruptement au milieu du 28ᵉᵐᵉ chapitre.

8. Oxford, Bodleian Library, Michael 39 (= catalogue Neubauer, nᵒ 2023) (IM: 19308).

Fol. 83ʳ-93ʳ: extraits des 14ᵉᵐᵉ, 15ᵉᵐᵉ et 6ᵉᵐᵉ chapitres.

9. Varsovie, Zydowski Instytut Historyczny, 255 (IM: 10122).

Fol. 68ʳ-71ʳ: extraits des 14ᵉᵐᵉ, 15ᵉᵐᵉ et 6ᵉᵐᵉ chapitres.

Remarquons que le texte intitulé ספר הכולל, dans le manuscrit Oxford, Bodleian Library, Reggio 13 (= catalogue Neubauer nᵒ 2028) (IM: 19313), fol. 1ʳ-76ᵛ, qui dans le catalogue de cette bibliothèque (Additions, tome I, p. 1160) est attribué à Lévi b. Abraham, se distingue des chapitres correspondants de *LH*; le même texte se trouve dans le manuscrit de Vienne, Nationalbibliothek, Hebr. 57 (IM: 1334), fol. 3ᵛ-93ʳ. (Le catalogue de Krafft [nᵒ 184] l'attribue à Abraham Ibn Ezra; celui de Schwarz [nᵒ 187] postule qu'il appartient à *LH*.) Ce texte se trouve également dans les manuscrits de Varsovie, Zydowski Instytut Historyczny, 253 (IM: 10120), pp. 44-167 et New York, Jewish Theological Seminary, Mic. 2553 (IM: 28806), fol. 49ᵛ-189ᵛ et Adler 1743 (IM: 28854), fol. 90-147.

II. Liste des chapitres du troisième livre du *Liwyat Ḥen,* d'après le manuscrit du Vatican, Vat. ebr. 383, fol. 1ʳ-5ʳ.

Note: Entre parenthèses nous indiquons les numéros des feuillets où se trouve le chapitre lui-même, donnant ainsi une idée sur la longueur relative des chapitres. Le texte donné en tête du manuscrit a parfois été corrigé d'après le texte donné au début de chaque chapitre. Notons que la numérotation des feuillets n'est pas continue: elle passe de 75 à 82, de 127 à 134, de 160 à 165, et de 213 à 219 bien qu'il ne manque apparemment rien dans le manuscrit.

השער הראשון: בכונת חכמת התכונה ובקיום התנועה המערבית ובמספר הכדורים הגדולים הכוללים המקיפים את הארץ וסדרם ובקיום מציאות הגלגל היומי. וכי תנועתו המזרחית אחת פשוטה שוה. וכי יש לו ימין ושמאל. וכי השמים פשוטים ונצחיים ותנועתם נצחיית ולא ישתנו רק במצב. (5ᵃ-9ᵃ)

השער השני: לבאר שאין השמים מארבע יסודות ולא מאחד מהם ואינם בעלי איכות ומראה ואין להם קולות כי הם יסוד חמשי ולזה תנועתם סביב המרכז. (9ᵃ-11ᵃ)

השער השלישי: בהיות השמים וכוכביהם בדמות כדור מתנועעים בסבוב על שני קוטבים קיימים. (11ᵃ-14ᵇ)

השער הרביעי: לבאר שהארץ בתמונה כדורית מכל פאותיה. (16ᵃ–14ᵇ)

שער חמישי: באשר אין תנועה לארץ וכי היא באמצע הכל ואינה נוטה לשום צד ושהיא אצל גלגל המזלות כנקודה באמצע העגולה. (19ᵃ–16ᵇ)

שער ששי: בעגולים הרשומים בכדור והם החמשה הנכוחיים המחלקים הארץ לחמש רצועות. ושני עגולים מהלך קטבי המזלות. ועגול המזלות. ושני עגולים אחרים העוברים על קטבי העולם. ועגול חצי היום וששת עגולים העוברים על שני הקטבים ועל ראשי המזלות. (22ᵃ–19ᵃ)

שער שביעי: בעגולת האפק הישר והנוטה ובמדתו. (25ᵃ–22ᵃ)

שער שמיני: במה שיקרה באופקים הנוטים מעגול משוה היום עד המקום שיתנשא הקטב הצפוני על האופק ס"ו מעלות ויעבור על ראש קטב המזלות והוא סוף הישוב. (26ᵇ–25ᵃ)

שער תשיעי: במה שיקרה במקומות שיקביל נקודת הראש מעגול המזלות עד הקטב הצפוני. (28ᵇ–26ᵇ)

שער עשירי: בעובי כל כדור וכי אין רקות בין הכדורים אבל כל חלקי העולם הפשוטים האחד מקיף את חברו כגלדי בצלים וכל אחד מן כדורי השמים ב׳ שטחים מרכזם מרכז העולם ונמצא העולם בכללו גשם כדורי מקשיי ועצם השמים ספיריי ובהיר והכוכבים מאירים לא ספיריים. (29ᵃ–28ᵇ)

שער י"א: בהיות הכוכבים מטבע הגלגל וחלקים ממנו. וגלגל חוזר ומזלות קבועים. ולמה השמש והכוכבים יחמו יותר משאר חלקי השמים. וכי אור הירח קנוי ונאצל מן השמש. ומה ענין האופל הנראה בתוך עצם הירח ומהו סבת המראים המתחלפים אשר תראה בירח. ואם אור שאר הכוכבים פרש מאור השמש כענין בירח אם לאו. (32ᵃ–29ᵃ)

שער י"ב: בעלת חֻמם השמש והכוכבים. ומה המקומות המיושבות ושאינן מיושבים מרצועות הארץ וכמה מדת המיושב בצפון ברוחב ובאורך וערכו מהארץ איך יודע. וכמה מדת הארץ מכל האקלים. ולמה הונח גובה רום השמש בצפון. והיכן תשוב בצפון. ולמה קראו צד דרום ראש. ולמה יתחלף פועל השמש בעתים אחדים מן השנים. ונזכיר בו צורות הכוכבים. (42ᵇ–32ᵃ)

שער י"ג: בהתחלק הקף הארץ כפי התחלק השמים לש"ס מעלות. וכמה מדת שעור מעלה מן הארץ. וכמה מדת קטרה ותשבורת כל גוף ושבר כל. שטחה. ומדת הישוב ומדת כל אקלים מז׳ אקלימיה. ומדת היום הארוך ויוסיף בכל אחד ולמה כל אשר יעמיק בצפון ימעט מדת רוחב האקלים ויוסיף היום ארך במעט מדה ונזכור בכל האקלימים הצל בחצי היום. ונזכור איך יודע צל הרגלים מפני הצל הישר וההפוך. ונזכור בכל אקלים מדינות נגביל ארכן ורחבן. (48ᵇ–42ᵇ)

שער י"ד: בשני מיני הצל הישר וההפוך. או אמוד השוחה ואיך יודע האחד
מן השני. ואיך יודע כל אחד מהם מגובה השמש או איך יודע גובה השמש
מהם. ואיך יודע מרחב המדינה מפני גובה השמש בחצי היום בכל עת והפך
זה. ואיך יודע מהגובה השעות שעברו בכל עת. ואיך יודע הגובה מהשעות
שעברו. ואיך יודע השעה והמעלה הצומחת בלילה מגובה כוכב אחד תכירהו.
ואיך יודע מספר השעות המעוותות בכלי השעות. ואיך יודע גובה הקוטב
השוה למרחב. ואיך יודע אורך המדינה. (48ᵇ‑56ᵇ)

שער ט"ו: איך יודע מהצל גובה כל דבר נצב. או עומק הבור והבקעה והמרחק
שבין שני מקומות וכל התלוי בזה. (56ᵇ‑61ᵃ)

שער י"ו: לידע נטיית השמש מהקו השוה בכל מעלה שתבקש מגלגל המזלות ואיך
תודע. (61ᵃ‑62ᵇ)

שער י"ז: בשעורי הכוכבים אצל מדת הארץ קוטרם ושטחם וגופם ומחלקותם.
ומספר כל חלק הכל נשער במדת הארץ אשר הקדמנו זכרה ואם היה יותר
ראוי לחבר זה השער אצל שער ל"ב. (62ᵇ‑68ᵃ)

שער י"ח: בתכונת גלגל השמש ותנועתו ודרך תקונו אצל גלגל המזלות ובראשית
השנה וציירתי צורת הארץ וחלוקה לשבעה האקלימים. וצורת המיושב והאופק
הנוטה במרחב. ותכונת הכדורים ועגוליהם ומרכזיהם וגלגל ההקפות. (68ᵃ‑75ᵃ)

שער י"ט: בכוון הרוחות ובקדימת הזריחה והשקיעה במקומות השונים וכפי מקום
החמה במזלות הצפון והדרום ולמצוא מהר תקופת האמת מן המהלך השוה
ובמדת שנת החמה לפי החכמים. (82ᵃ‑86ᵇ)

שער כ': במנין האומות וראשית שנותם. ולדעת באיזה יום יעשה פסח הנצרי
בכל שנה. ומדבר במולדות ובעבור ולדעת סדר השנה. ולדעת מתי יכנס
חדש כל אומה. ולהשיב מנין כל אומה אל אומה. ומהלך הלבנה בשנת הלבנה
ובשנת החמה והתלוי בזה. (86ᵇ‑106ᵃ)

שער כ"א: בתכונת גלגל הירח והרכבת עגולו ודרכי תנועתו. (106ᵃ‑110ᵇ)

שער כ"ב: במצעדי המזלות באופן הישר ובאופן הנוטה ונגביל עלייתן באקלים
הרביעי ויתרון יומו הארוך על היום השוה המתקבץ מתוספת העליות.
(110ᵇ‑118ᵇ)

שער כ"ג: לדעת קשת היום וקשת הלילה מפני מצעדי המזלות ודרך ידיעת
תוספת היום הארוך או יתרון יום מהמעלה שהחמה חונה בה באי זה מקום
שתרצה. ולדעת באי זה יום שתבקש מספר שעותיו הישרות או אמד הבינוניות.
וכמה מעלות השעה הזמנית ושברי המעלה מפני קשת היום ואיך נשיב המעלות
והשברים לשעות וחלקי שעה והפך. ולדעת בכל יום שתרצה מעלות כל
שעה זמנית שבו. ולדעת מפני השעה המעלה הצומחת ותחלת כל בית משנים עשר

בתים. והמעלה הצומחת היא תחלת הבית הראשון. ונניח לוח מצעדי הארץ
הזאת. (118ב−127א)

שער כ״ד: בחלוף מעלת מהלך השמש בכל יום וסבת החלוף ומדתו בכל יום.
וסבת הניחנו רגע ראשית היום מחצי היום ונניח שרש למולד בניסן ל״א
לפרט בזמן האמצעי ומעמד שני המאורות האמצעי וחק הלבנה וראש התלי.
ונבאר כמה נוסיף לדעת עת הנגוד האמצעי וכמה נוסיף לשנה פשוטה ולמעוברת
ולמחזור אחד. (134א−139א)

שער כ״ה: בעלּיית הכוכבים וירידתם ובמחנות הירח ונעתיק בסוף שתי לוחות
לדעת תכלית מרחק נגה וככב מזרח ומערב בראש כל מזל ומזל. (139א−141א)

שער כ״ו: בסבות הסתר הכוכבים תחת ניצוץ השמש וזמן הסתר כל אחד מחמשת
כוכבי הנבוכה. (141ב−146א)

שער כ״ז: בתקון מקום הלבנה וידיעת מרחבה מן האופן בכל עת ובידיעת המולד
האמיתי או הנוכח האמיתי ואתן לפי דרכי קצת טעמי התקונין בקצרה.
(146א−151א)

שער כ״ח: בסבות קדרות הירח ודרך חשבונו ודקדוק כמותו ועתו ואורך זמנו וצד
התחלתו ותכליתו והוראתו. ונניח לוח לדעת קוטר הלבנה והחמה בכל עת.
(151א−165ב)

שער כ״ט: בהסתר הירח והגלותה פעם בארוכה ופעם בקצרה ונתן סבת זה ואיך
נדע ליל הראיה אם בזמנו או בליל עבורו. ואפרש לך קצת הלשונות שזכר
הר״ם בהלכות קדוש החדש. ואבאר קצת טעמי דבריו בקצרה. ואפרש קצת
מדברי רבותי ז״ל האמורים בענין זה במסכת ראש השנה בענין אמרם נולד
קודם חצות. (165ב−171ב)

שער ל׳: לבאר מה שיקרה בחלוף ההבטה באורך וברוחב. (171ב−176א)

שער ל״א: בסבת קדרות השמש. ובמדת הזמן שיתכן להיות בין שני קדריות השמש או
בין שני קדריות הירח. ודרך חשבון קדרות השמש וכל הצריך לבאר בו.
(176א−186ב)

שער ל״ב: לבאר מדת גובה הכוכבים ומרחקם מן הארץ ויביא מדת יציאת מרכז גלגל
הנושא מכל אחד וחצי קוטר גלגל הקפתו מכל אחד מהששה שהם שצ״ם נכ״ל.
(186ב−190א).

שער ל״ג: נתאר עגולי הכוכבים הנבוכים ותבניתם ומיני תנועותם באורך וזמני הקפתם
וחבורם עם השמש ובתנועתם בגלגל המזלות. (190א−199א)

שער ל״ד: במרחבי חמשת הכוכבים מאזור המזלות ומקומות חתוכיהם וצדי תנועות
גלגלי הקפותם ובפתילות גלגלי נוגה וכוכב. (199א−203א)

שער ל״ה: איך נכנס בלוחות [הנשיא, ר׳ דף 204א] לדעת מקום כל כוכב האמצעי
ונניח שרש למקום כל כוכב ומקום גובה רומו וראש תנינו. ונניח קצת חבורים
ונגודים אשר בהם גדרי לקות להיות לשרש למי שיבא לחשב מכאן ולהבא.
(203א–206א)

שער ל״ו: בתקון מקומות כוכבי הנבוכה וקצת טעמי התקון ולדעת אם הכוכב ישר
במהלכו או נזור וכמה מדת נזורותו ומתי יחל ויכלה ולהקל הדרך בחשבון
מקום כל כוכב בחצי היום ולדעת עת התחבר שני כוכבים או ג׳. ולדעת
מרחב ה׳ הכוכבים הנבוכים. (206א–211ב)

שער ל״ז: בסבת נזורות הכוכבים החמשה בגלגל המזלות ולהגביל מדת הנזורות
בכל אחד משני צדי גלגל ההקפה אצל מרחק הקרוב ונביא לוח לדעת
כמה מעלות ישוב כל כוכב מן החמשה לאחוריו וכמה ימי נזורו לפי מקומו.
(211ב–213ב)

שער ל״ח: בהתחלפות שיקרה בין כוכב למעלתו הרשומה באזור המזלות בעברו
בחצי השמים ובזריחתו ובשקיעתו. ולדעת מעלות מעבר הכוכב ולדעת עם
אי זו מעלה הכוכב זורח ועם אי זו מעלה ישקע ובניהוג מקום הלבנה
או זולתה. (219א–220א)

שער ל״ט: בידיעת מקומות כוכבי שבת מגלגל תשיעי ולידע מקומות כוכבי לכת
מגלגל השמיני שהוא גלגל המזלות ובטעם שמות המזלות והבדלי הכוכבים
במהירות ובמראה ובהירות שבין המשרתים ובין כוכבי שבת ולהכיר הכוכב
המסופק בכלי הנחשת ולידע מעלתו וכל הדומה לזה. ולדעת זמן תקופתו אחת
בכל שנה. ולידע המעלה הצומחת בעת ההיא. (220א–223א)

שער הארבעים: בכחות הכוכבים ומפעלם ומשפטיהם על צד הכלל ובחתימתו יבאר
שאין האישים חוזרים חלילה. (223א–315ב)

VIII

Distinguishing Two
R. Joseph b. Joseph Naḥmias:
The Commentator and the Astrologer[*]

Historians of fourteenth-century Jewish thought are familiar with R. Joseph b. Joseph Naḥmias, known to have written commentaries on the Pentateuch, Jeremiah, Psalms, Proverbs, Ecclesiastes, Esther, Avot, tractate Nedarim, and the *'avodah "'Atta konantanu"*. Most of these works are extant and were published by Moses Loeb (Aryeh) Bamberger (1869–1924) at the end of the nineteenth and in the first decade of the twentieth century. These commentaries were reprinted under the title, *Perushey Rabbi Joseph ben Naḥmias 'al Mishley, Megillat Esther, Yirmiyahu, Pirqey 'avot we-Seder 'avodat Yom ha-Kippurim* (Jerusalem, no publisher and no date indicated [5742/1982?]).

In the introductions to his editions, Bamberger demonstrated conclusively that Joseph Ibn Naḥmias was a member of the circle of R. Asher ben Yeḥi'el (the Rosh) in Spain. At least some of his commentaries were written in the Rosh's lifetime. In his commentary on Esther, for example, he adds after the Rosh's name the letters *n.r.w.* (= may God preserve and bless him). In his commentary on Proverbs he a few times mentions R. Israel b. Joseph Israeli, brother of Isaac Israeli, the author of *Yesod 'olam*, another member of the Rosh's circle.

Scholars have ascribed to Joseph Ibn Naḥmias yet another work, of a quite different sort. I refer to an Arabic treatise entitled *Nūr al-'ālam* (The light of the world; MS Rome, The Vatican, Ebr. 392, fols 51a–87; Institute for Microfilmed Hebrew Manuscripts, Jerusalem: No. 473). A Hebrew version of this work, *Or 'olam,* is also known (MS Oxford, Bodleian Library, Canon. Misc. 334 [= Neubauer 2778], fols 127b–101a [the manuscript is numbered from the end to the beginning]; Institute for

---

[*] First published in Hebrew in *Qiryat sefer* 62 (3–4) (5748–49) (1988–89), pp. 917–19.

Microfilmed Hebrew Manuscripts, Jerusalem: No. 22731). The name of the translator is unknown.

Neubauer believed that Joseph Ibn Naḥmias the commentator was identical with Joseph Ibn Naḥmias the author of *Nūr al-ʿālam*;[1] his opinion was accepted by Steinschneider[2] and Bamberger.[3] In what follows, I will show that this identification is mistaken and that there were two different men of the same name, who lived about three generations apart.

Or ʿolam (I used the Hebrew version rather than the Arabic original, whose microfilm is difficult to read) is an astronomical treatise. It belongs to the tradition that sought to reconcile mathematical astronomy — based essentially on Ptolemy's *Almagest* — with physical theory, namely Aristotelian physics. In Spain this tradition goes back to Ibn Ṭufail (d. 1185) and culminated in al-Biṭrūjī and Averroes. In the Hebrew tradition, it is echoed in Maimonides' *Guide for the Perplexed;* Gersonides is its most brilliant representative.[4] Joseph Ibn Naḥmias explains his goal right at the outset:

> Let me state that the masters of mathematical [= astronomical] science, when they discussed the principles of celestial motions, agreed to ground theory on two principles, both of which have been rejected by some of the masters of natural science. However, the latter have not themselves put forward a satisfactory theory accounting for the observable motions and changes in the heaven. Indeed, their intention has been merely to make evident, on the basis of purely natural proofs, that it is impossible that these two principles be the rules [governing] the celestial bodies. The first of these [mathematical] principles is the [astronomers'] positing of eccentric and epicyclic orbs, and the second is their positing of contrary celestial motions. (fol. 127b)

Joseph thinks that al-Biṭrūjī has failed in his attempt to reconcile physics and mathematical astronomy; although, he adds, "he should be

---

[1] A. Neubauer, "Joseph ben Joseph (Jose) Nahmias," *Jewish Quarterly Review* 5 (1893), pp. 709–13.

[2] M. Steinschneider, *Die arabische Literatur der Juden* (Frankfurt, 1902), pp. 166–67; see also idem, *Die hebraeischen Übersetzungen des Mittelalters und die Juden als Dolmetscher* (Berlin, 1893; repr. Graz, 1956), p. 597.

[3] See his *"Toledot ha-meḥabber u-sefaraw,"* in M. L. Bamberger, ed., *Perush Pirqey avot le-ha-Rav R. Yosef b. R. Yosef Naḥmias z.ṣ.l.* (1907) (not paginated).

[4] See my article, "Human Felicity and Astronomy: Gersonides' Revolt Against Ptolemy" (Hebrew), *Daʿat* 22 (1989), pp. 55–72.

much praised for having been the first to become alert [to the issue] and for having composed a treatise on it" (fol. 126b). Like Gersonides, whom he does not mention, Joseph advances a mathematical theory grounded in principles compatible with physical theory. Joseph's ambitions were however apparently beyond his capabilities. Efodi, who read his treatise and commented on it (see below), remarks with unveiled sarcasm: "This being so, if the learned R. Joseph can indicate the cause of this ... then I, too, will accept [his authority] and will call him the head and the chief of the mathematicians. Yet I wonder how, given his subtlety, he erred in this [matter]" (ibid., fol. 100b).

In the Bodleian manuscript, *Or 'olam* is immediately followed by a short treatise carrying the title: "The Rejoinder by E.f.d. [Profiat Duran][5] to the book by R. Joseph Ibn Naḥmias" (fol. 101a ff.). Its *incipit* reads as follows: "I have received the New Heavens [cf. Is. 66:22] set up by the excellent scholar R. Joseph Ibn Naḥmias, who should be lauded for his industry, diligence, and effort [to acquire] what the souls of the choicest ones long to know." The absence of the letters *z.ṣl* (= of blessed memory) after the author's name (occurring also in the sequel, e.g., fols 100a–b) shows that Ibn Naḥmias was still alive when Efodi wrote his rejoinder. The fact that Ibn Naḥmias is referred to as "the excellent scholar" shows that he was no longer young when these words were written; hence he may be supposed to have belonged to Efodi's generation. It follows that *Or 'olam* was written about the year 1400.

This conclusion that the author of *Or 'olam* is not the author of the various commentaries is borne out by a perusal of the latter.

The commentator Joseph Ibn Naḥmias often refers to his other writings: every commentary refers the reader to discussions in other commentaries. But none of his commentaries mention *Or 'olam*; nor does the latter refer to any other works by the author.

Moreover, in his commentary on Esther, Joseph Ibn Naḥmias explains the words "from Hodu to Kush" (1:1) as follows: "The learned R. Israel *z.ṣl* has explained that the earth is like a globe, and Hodu and Kush are situated near one another at the extremity of the globe. Kush is the extremity of the world if you set out from Hodu to walk around the globe until you reach

---

[5] "E.f.d." is a Hebrew acronym standing for *'ani Profiat Duran*, i.e. "I [am] Profiat Duran." Efodi, as he came to be known, was a noted philosopher and grammarian in the circle of Hasdai Crescas and died ca. 1414. See *Encyclopedia Judaica* (Jerusalem, 1972), vol. 6, col. 299–301.

Kush; and Hodu is the extremity of the world if you set out from Kush to walk around the world until you reach Hodu."[6] R. Israel is, as already noted, the brother of the astronomer Isaac Israeli, and he was indeed interested in cosmological matters.[7] It stands to reason that a writer who cannot explain the sphericity of the earth himself and must quote someone else on the subject is not the author of a treatise on astronomy.

Thus we may conclude that the author of the commentaries and student of the Rosh is not the author of *Or ʿolam*, who lived three quarters of a century later. This raises the question of what we know of an astronomer called Joseph Ibn Naḥmias.

The only information available seems to be that he dealt with astrology. He composed a prediction for the year [5]238 (=1478/9), which is preserved in an epistle by R. Abraham b. Eliezer ha-Levi (born in Spain ca. 1460, died in Jerusalem after 1528). In this epistle, R. Abraham copied astrological-prophetical notices by his brother-in law, the well-known astronomer Abraham Zakut (1452–1515), including the following passage:

> Let me tell you what a great astrologer in Spain, whose name is R. Joseph Ibn Naḥmias, may his memory be a blessing, wrote in an [astrological] judgment that he cast for the sun's eclipse of the year [5]238. He said: "Being unprejudiced and not seeking to flatter any religion or law, I say that a man will arise, a great disputant, a subjugator, a powerful man, despising depravity [perhaps: conspiracy], who will assemble great armies, establish a new religion, and destroy houses of worship and their priests. In his days, Jerusalem will be rebuilt with black stones, undressed. I do not want to make this matter explicit." So far his words. He further said that he did not wish to announce the time of his coming, for the reason mentioned there.[8]

---

[6] R. Joseph ben R. Joseph Naḥmias, *Perush Megillat Esther*, ed. Moses Aryeh Bamberger (Krakow, 1899), p. 6.

[7] His *Maʾamar Gan Eden* is extant. It has been published in *Jubelschrift zum neunzigsten Geburtstag des Dr. L. Zunz* (Berlin, 1884), pp. 21–42. See Y. Tzvi Langermann, "'The Making of the Firmament': R. Hayyim Israeli, R. Isaac Israeli and Maimonides" (Hebrew), in *Shlomo Pines Jubilee Volume on the Occasion of His Eightieth Birthday*, Part I (*Jerusalem Studies in Jewish Thought*, vol. VII, 1988), 461–76.

[8] The best text of this epistle is in I. Robinson, "Two Letters of Abraham ben Eliezer Halevi," *Studies in Medieval Jewish History and Literature*, ed. I. Twersky (Cambridge, Mass., 1984), pp. 403–22. The text had been published with minor variations in H. H. Ben-Sasson, "The Jews Facing the Reformation" (Heb.), *Proceedings of the Israel Academy of Sciences and Humanities* 4 (5729–31 [1969–71]), p. 76; Ben-Sasson shows how R. Abraham ha-Levi applied the forecast made at the beginning of the fifteenth century to

In sum, it can be said that the Ibn Naḥmias family produced two scholars who bore the name Joseph ben Joseph. The first was of the younger contemporary of Asher ben Yeḥi'el and belonged to his circle. Like most of the Rosh's students, he devoted himself to traditional subjects of study. He is the author of the commentaries to several biblical books, to Avot, and to tractate Nedarim. The second was "a great astrologer" who lived some three generations later, also in Spain. He did not limit himself to the practice of astrology but was interested in astronomical and physical theory as well— although, as it appears, not on a very high level.

Two Notes on *Sefer Meyasher 'aqov* by Alfonso, alias Abner of Burgos[*]

In an article published in 1933, S. Luria (1891–1964) called attention to a Hebrew mathematical-philosophical treatise, entitled *Meyasher 'aqov* (Straightening the curvature), found in MS London, British Museum, Add 26984, whose author is a certain "Alfonso". Luria intended to publish the treatise, but for obvious reasons this did not happen.[1] Luria devoted some space to this treatise in a book published in Russian in 1935, commenting that "the scholar who makes this text available to the scientific world would thereby render a great service to the study of the history of mathematics."[2]

In 1960, Luria obtained a photocopy of the manuscript and suggested that Dr. G. M. Gluskina prepare a scientific edition and translation of the treatise. Gluskina undertook the task. Her book was published in Moscow in 1983, after she overcame many difficulties. It includes a scientific edition of the text, a translation into Russian, many explanatory notes, geometrical diagrams (missing in the manuscript), and a facsimile of the manuscript.[3]

---

[*] First published in Hebrew in *Qiryat sefer* vol. 63 (3) (5750–5751) (1990–91), pp. 984–86. Research for this paper was done during the academic year 1987–88, which I spent as a Visiting Scholar at the Sidney M. Edelstein Center for the History and Philosophy of Science, Technology and Medicine, The Hebrew University of Jerusalem. I am grateful to the Edelstein Center for having allowed me to pursue research during one year in Jerusalem.

[1] S. Luria, "Die Infinitisimaltheorie der Antiken Atomisten", *Quellen und Studien zur Geschichte der Mathematik, Astronomie und Physik*, Abt. B: *Studien*, vol. 2 (1933), pp. 106–185.

[2] G. M. Gluskina, *Alfonso, Meyasher 'aqov* (Russian) (Moscow, 1983), p. 9.

[3] Ibid.

2   *Two Notes on Sefer Meyasher 'aqov*

A. Who is Alfonso, the Author of *Meyasher 'aqov*?

Even M. Steinschneider, who believed that *Meyasher 'aqov* was a translation, could not identify this Alfonso.[4] In two studies (in Russian)
published in 1974 and 1978, Gluskina proposed that he was none other than
the well-known convert Abner of Burgos, whose name as a Christian was
Alfonso of Valladolid.[5] Gluskina repeated this claim in her 1983 book. She
based herself on several considerations, and notably the similarity between
Abner's and Alfonso's philosophical views: a preference for Plato's
philosophy over that of Aristotle; the view that matter is composed of
primary particles and that the vacuum exists; the juxtaposition of the
finitude of all created beings with God's infinitude; and the view that a
"medium" (*'emṣa'i*) is situated between existence in potentiality and
existence in actuality. This identification was supported by stylistic
considerations.

I was not entirely convinced by Gluskina's arguments and tended to
reject her identification. I turned to Abner's Hebrew writings, both
published and still in manuscript. The latter are found in MS De Rossi 533
(2440), in the collection of the Biblioteca Palatina in Parma (Institute for
Microfilmed Hebrew Manuscripts, Jerusalem: film no. 13444). Although I
set out to refute Gluskina's thesis, I found myself corroborating it; a
comparison of Abner's writings with *Meyasher 'aqov* lends further — and
in my view decisive — support for the claim that he was its author.

[1] In *Meyasher 'aqov* we find the following impressive methodological
statement:

> I wandered through the alleys of inquiry, [namely] the books of Aristotle and
> his followers. I saw the various rebukes and criticisms that he addressed to
> Plato and Empedocles and the other Ancients and they did not seem right to
> me. This followed from one premise, which is both intellectually and morally
> right, that the [Holy One] Blessed be He implanted in my heart. ... This
> premise is that in my view it is inappropriate to ascribe to the scholars, who
> spent all their days inquiring into the sciences, such errors as are committed
> by persons who have never occupied themselves with science. *A fortiori* one

---

[4] Moritz Steinschneider, *Die hebraeischen Übersetzungen des Mittelalters und die
Juden als Dolmetscher* (Berlin, 1893; repr. Graz, 1956), p. 626.

[5] For a full bibliography of studies of *Meyasher 'aqov,* see Gluskina, *Alfonso,* pp.
128–29.

should not ascribe to scholars who spent all their days with science and inquiry and composing books about them errors that are not committed even by irrational animals. Rather, when we find [in a scholar's writings] anything that is seemingly erroneous, it is without doubt appropriate to search for a way to do justice to it, so as to render [his affirmations] true and [show why] they seem acceptable to his intellect.[6]

Now Abner of Burgos upholds precisely this principle of charitable interpretation:

> It is inappropriate to ascribe to the scholars, especially not to King Solomon and to David his father, such a silly error as one ascribes to other men.[7]

The striking similarity between these two passages, in both content and formulation, strongly suggests that they were written by one and the same person.

[2] Abner of Burgos takes care to write God's name with three *yod*s and connects this usage with the Trinity.[8] In the prooemium to *Meyasher 'aqov*, in which the author thanks God for having allowed him to discover the "measurement of the circle," God's name is also systematically written with three *yod*s.

[3] Abner's Hebrew shows a marked tendency to use the ending -*in* for the plural of the present tense: *yoshevin, soverin, ḥoleqin*, etc.[9] The same tendency is conspicuous in *Meyasher 'aqov*: *ṣerikhin* (98a20), *ne'emarin* (99b10), *yoṣe'in, shawin* (100b22) etc.

Taken together, and in conjunction with Gluskina's arguments, the similarities of both substance and form between *Meyasher 'aqov* and Abner's writings seem to establish beyond doubt that he was indeed the author of *Meyasher 'aqov*.

    Last but not least, I should mention that this identification was accepted by the late Professor Shlomo Pines. He intended to devote a paper to the

---

[6] *Meyasher 'aqov*, ed. Gluskina, fol. 97a:10ff. (the numbering is according to the manuscript, indicated in the edition). See also fols 97b:18 ff.; 100a:20ff.

[7] Parma, Biblioteca Palatina, MS De Rossi 533 (2440), fol. 22a.

[8] See I. Baer, "The Kabbalah in Abner of Burgos' Christological Doctrine" (Hebrew), *Tarbiz* 27 (1958), 279–280, and pp. 278–79 (n. 2, *in fine*).

[9] Ibid.

4   *Two Notes on Sefer Meyasher 'aqov*

importance of *Meyasher 'aqov* for the history of Hebrew scientific thought but unfortunately did not live to carry out this project.[10]

B. Who is R. Moses ha-Levi Mentioned in *Meyasher 'aqov*?

A certain "R. Moses ha-Levi from the town of Seville" is mentioned three times in *Meyasher 'aqov* as someone who advanced a proof for Euclid's fifth postulate (the so-called parallel postulate). Gluskina failed to identify him,[11] nor does his name appear in Steinschneider's *Mathematik bei den Juden*. I would suggest that he is the thirteenth-century physician-philosopher Moses b. Joseph ha-Levi, known in Arabic as Abū 'Imrān Musa' al-Lawī al-Ishbilī, of whom a short treatise on the First Mover, in the Avicennian spirit, is extant.[12]

---

[10] S. Pines, "Some Views Put Forward by the Fourteenth-Century Jewish Philosopher Isaac Pulgar, and Some Parallel Views Expressed by Spinoza" (Hebrew), in J. Dan and J. Hacker, eds, *Studies in Jewish Mystics, Philosophy and Ethical Literature, Presented to Isaiah Tishby on his Seventy-fifth Birthday* (Jerusalem: The Magnes Press, 1986), 395–457, on pp. 445–46. I am grateful to Dr. Dov Schwartz of Bar-Ilan University for calling my attention to this passage in Pines' article.

[11] Gluskina, *Alfonso*, index of names, p. 130.

[12] See Steinschneider, *Die hebraeischen Übersetzungen*, p. 410; G. Scholem, "Joseph Ibn Waqār's Arabic Treatise on Kabbalah and Philosophy" (Hebrew), *Kiryat sefer* 20 (1943–44), p. 157. Moses ha-Levi is mentioned by Hasdai Crescas, Joseph Albo, Isaac Abravanel, and Abraham Ibn Waqār; see *Encyclopedia Judaica* (Jerusalem, 1972), vol. 12, col. 421–22. On his philosophy, see: H. A. Wolfson, "Crescas on the Problem of Divine Attributes," *Jewish Quarterly Review* 7 (1916–17), pp. 40–43; idem, "Averroes' Lost Treatise on the Prime Mover," *Hebrew Union College Annual* 23 (1950–51), pp. 683–710; G. Vajda, "Un champion de l'avicennisme: Le problème de l'identité de Dieu et du premier moteur d'après un opuscule judéo-arabe du XIIIe siècle," *Revue thomiste* 48 (1948), pp. 480–508; idem, *Recherches sur la philosophie et la kabbale dans la pensée juive du Moyen Âge* (Paris, 1962), pp. 133, 213–15; idem, "R. Moses ha-Levi's View on Divine Providence" (Hebrew), *Melila* 5 (1955), 163–68. A passage from a work on music theory by Moses ha-Levi is preserved in *Sefer 'eyn kol*, Shem-Tov Ibn Shaprut b. Isaac's commentary on Avicenna's *Canon*: see Steinschneider, *Die hebraeischen Übersetzungen*, p. 689; idem, in *Hebraeische Bibliographie* 19 (1879), pp. 43–44; I. Adler, *Hebrew Writings Concerning Music in Manuscripts and Printed Books from the Geonic Times up to 1800* (= *International Inventory of Musical Sources / Répertoire international des sources musicales*, vol. B IX^2) (Munich, 1975), p. 239, no. 540. Adler republishes the passage in the notice devoted to Ibn Shaprut. (ibid., pp. 175–79, no. 370). This passage shows that Moses ha-Levi was indeed interested in mathematics.

Al-Fārābī on the Foundations of Geometry

In this talk I wish to sketch, in bare outline, a short medieval treatise devoted to the philosophy of mathematics, or rather of geometry. This is the *Commentary on the Beginning of the First Book of Euclid's "Elements"* by Abū Naṣr al-Fārābī (870-950). The text is preserved in a single Arabic manuscript and in four manuscripts of the Hebrew medieval translation by Moshé ibn Tibbon. My edition, French translation, and study of the *Commentary* are forthcoming in vol. 11 (1989) of *Jerusalem Studies in Arabic and Islam* (where also the appropriate references are to be found).

1. Al-Fārābī's *Commentary* addresses itself to what, in his times as in ours, is the fundamental problem of the epistemology of mathematics: how do we acquire the mathematical, *ideal*, notions and what is their epistemological grounding. Specifically, what is the relationship between the entities which are the subject of the geometrical definitions, axioms, and theorems - *viz.* a surface with no depth, a line with neither depth nor breadth, a point devoid of all three dimensions - and the corresponding physical, *real*, objects? In antiquity, and indeed today too, there have been essentially two approaches to the problem: the *a priorist*, Platonic, one, and the *a posteriorist*, Aristotelian, one. Here we will be concerned only with the latter, which, instead of postulating the existence of transcendent ideal mathematical entities, purports to obtain the mathematical notions on the basis of tangible objects. This indeed is Aristotle's stance: take a snub nose, abstract away the matter, and you get the notion of a curved line.

Obviously Aristotle's empiricist philosophy of geometry hinges entirely on the notion of *abstraction*. And the philosophical significance of al-Fārābī's *Commentary* lies precisely in his reflections on, and criticism of, this notion.

2. From the outset, al-Fārābī raises the crucial question: All the objects referred to in the Euclidian definitions - the point, the line, and the surface - are present in the physical bodies and are, therefore, perceived by the senses. But perception necessarily involves also the *qualities* of these objects - hot and cold, dry and moist, hard and soft, etc. In geometry, however, these objects are to be apprehended by the *intellect*, and such an intellection pertains only to their *essences*, not to their accidental qualities. How, then, do we pass from the sensible object to its intelligible counterpart?

Al-Fārābī's point of departure, we see, is primarily Aristotelian: All knowledge derives from sense-perception, and this holds for geometry too. Hence al-Fārābī's task is to explicate the epistemological foundations of geometrical knowledge within the framework of Aristotelian empiricist theory of knowledge.

The first problem - how to rid the perceived point, line, surface, and body from their sensible qualities - is fairly easily disposed of. For al-Fārābī sees this type of abstraction, where the sensible qualities are thought away, as the working of the human intellect, or more specifically of the *actual* intellect. As is well known, al-Fārābī holds that the human intellect is at the outset all potential, hylic. Through the senses, and with the indispensable and unfailing assistance of the intellect agent, the potential intellect gradually acquires certain *forms* of substances in the physical world: having apprehended these forms, the hylic intellect passes to the state of an actual intellect, in which the apprehended forms are so to say imprinted. The actual intellect thus possesses abstract forms, i.e. forms to which no qualities any longer adhere. Therefore, within al-Fārābī's theory of the intellect, to strip the point, line, surface, and body of their sensible qualities is a trifling matter. It is a trifle, of course, because al-Fārābī has introduced into his epistemology the essentially Neoplatonic notion of the intellect agent, a notion which in a way brings back the Platonic transcendent Ideas through the back door. Therefore, although al-Fārābī formulates his problem in Aristotelian terms, in truth one part of his theory of abstraction - namely that concerning the thinking away of the qualities - is founded on Neoplatonic premises. It is this Neoplatonic element in al-Fārābī's thinking which also allows him to spot the decisive weakness in the other, and more essential part of Aristotle's theory of abstraction. That is what we now come to.

3. The fundamental point of al-Fārābī's criticism is conveyed in the following sentence: "Just as the [geometrical objects] are bound up with sensible [qualities] ... so also they are bound up with one another: in its existence, the point is inseparate from the line, the line inseparate from the surface, the surface inseparate from the body." But, al-Fārābī reasons, geometrical science considers each of these separately. How then, do we obtain these separate notions of the geometrical objects? The problem is that a mere abstraction - a simple thinking away of matter - does *not* yield these required *separate* geometrical objects. We shed more light on the meaning of this idea, perhaps also on its origin, by taking note of a similar remark made by Proclus in his own *Commentary on the First Book of Euclid's "Elements"*. Unlike al-Fārābī, Proclus *constructs* the extended geometrical objects through the *motion* of their boundaries: a line is produced through the motion of a point, etc. But, Proclus says, by thus producing a geometrical object of a higher dimensionality, the boundary becomes a part of it and thereby loses its independent existence. "The partless partakes of divisible existence and the breadthless of breadth; and the limiting elements are no longer able to preserve their simplicity and purity" (trans. G.R. Morrow). Here is the significant insight shared by Proclus and al-Fārābī: in as much as the point, the line, and the surface are "in" the three-dimensional body, they partake of the latter's three dimensions. From al-Fārābī's Aristotelian vantage point, this means that abstracting from matter does not yield a dimensionless point, nor a line without breadth, nor a surface without depth. Since these *ideal* notions simply do not exist in *reality*, they *cannot* have their origin in sense perception and, *ipso facto*, cannot be obtained by abstraction. Although al-Fārābī does not say it in so many words, it is clear that he is criticizing and rejecting Aristotle's simplistic theory of abstraction. This critique, we can now see, again takes its cue from the Neoplatonic premises in al-Fārābī's philosophy: the geometrical objects, it will turn out, are intellectual *constructs*, not mere reflections of physical existents. Yet al-Fārābī is too much an Aristotelian to construct them in Neoplatonic fashion through the motion of the boundary. He must therefore construct them differently: "since it is in the manner of the intellect to separate everything whose substance is intelligible from the substance of everything else, it [the intellect] seeks to define these [geometrical objects] so that they be separate from one another." But how?

4. One may think that al-Fārābī is wasting his time: are not the geometrical objects defined - separately - in the *Elements*? This argument misses its target: it confuses two different approaches to the construction of geometry. Indeed, al-Fārābī tells us that geometry can be constructed in two distinct ways. You can begin with what is better known - *viz.* the body - or with what is more intelligible - *viz.* the point. Al-Fārābī is here following Aristotle who, in the *Topics* (6.4, 141a24-142a9) distinguishes two senses of what is "better known," namely what is better known "to us" and what is better known absolutely. (Aristotle even invokes the geometrical objects as an example: the body is better known to us, although absolutely it is the point which is best known.) Al-Fārābī goes somewhat beyond Aristotle however: in a section of his *Iḥsā' al-ʿUlūm* (a section which, incidentally, is preserved only in the Hebrew translation) he calls these two procedures "analysis" and "synthesis," respectively (terms which he probably took over from Galen's *Small Art*). These terms he defines in his *Kitāb al-Mūsīqī al-Kabīr* as follows: "analysis is the inverse of synthesis. In analysis we must put the elements in the order in which we know them; in synthesis, by contrast, we put them in the order in which they exist: what exists primarily is placed before the others." Thus, al-Fārābī follows Aristotle and sets in parallel the principles of inquiry with those of existence, epistemology with ontology.

Al-Fārābī takes these two approaches to be *complementary*: "When we seek the study [of geometry] we will first apply the order according to the sensible; the art [i.e. geometry] itself, however, will apply the order according to the intelligible." The student will set out from the sensible body and pursue analysis until he reaches the geometrical point. The synthetic method being the one followed by Euclid in the *Elements*, al-Fārābī's purpose is to supply the analysis, the procedure complementary to the already existing synthesis. This analysis he intends as a propedeutic to geometry itself.

5. We are to begin with the body, then. But what *is* a body? It is, to be sure, that which extends into three dimensions. But is extension an essential attribute of body or an accidental one? In a fairly long digression al-Fārābī describes both stances which, as is well known, were under debate in late antiquity and in the Middle Ages. Al-Fārābī concludes that the geometer anyway need not take a position on the issue, for he

considers the three dimensions *abstracted* from matter and supposes that they exist separately. Al-Fārābī is obviously intent upon making his discussion acceptable to the widest possible circles.

6. Suppose, then, that we have at our disposal the notion of the matter-less, geometrical, three-dimensional, body, i.e. of pure spatial extension. How do we proceed in the analysis to obtain the geometrical surface, line, and point, so that they be separate?

Al-Fārābī now makes the following, quite startling statement: "The geometer calls extension 'length' and posits it as a universal common to the body, the surface, and the line." He emphasizes that "length" here has a special, technical, meaning: "The mass of people apply the term 'length' to the largest extension of a body extended into all directions, calling the shorter extension 'breadth'. ... The geometer, however, does not mean this notion. Rather he has in mind absolute extension. Thus, when the geometer speaks of the body, the surface, and the line as being length, he means [their] extension." Clearly, al-Fārābī's meaning is that the body, the surface, and the line *are* length in the sense that they are extended. Indeed, in his *Paraphrase of Aristotle's "Categories"* al-Fārābī says just the same thing: "That which has parts having positions is called by the mathematicians 'length'. They divide it [into three kinds]: there is [length] without any breadth, namely the line; there is [length] which is only length and breadth, namely the surface; and there is [length] which is length, breadth, and depth, namely the body."

To be sure, this leaves us not a little puzzled. What is the idea behind this quite curious counter-intuitive stipulation? Al-Fārābī does not tell us. My suggestion is the following: al-Fārābī introduces his idiosyncratic notion of length in order to make the Euclidian definitions of the line, the surface, and the body into definitions that conform to the Aristotelian normative notion of definition. Aristotle, as is well known, postulated that a scientific definition should specify the essence of the *definiendum* and do this by indicating the genus and the differences. Now, the Euclidian definitions apparently come nowhere near to conforming to this notion of definition; they do not indicate the *essence* of the line, the surface, the body. Should we conclude that the basic definitions of the paradigmatic science of geometry are not scientific? Not if we follow al-Fārābī. For by investing the term "length" with the meaning "absolute extension," one can consider "length" as indicating the genus, with the expressions "with/without breadth/depth" specifying the differences. Thus, the body, the surface, and the line become *species* of the genus "length." On this reading, Euclid's definitions become "good" definitions, for one can reply to the question: "What is a line, a surface, or a body," by saying: "a length."

7. Al-Fārābī now draws on his notion of length as absolute extension also to show how analysis yields the ideal, and therefore separate geometrical objects. He first points out that the extensions in the three dimensions are *independent* of one another: "The three directions can be construed each separately, or [all] together, or again any two of them can be construed together, excluding the third." On this basis he now *constructs* one by one the three extended geometrical objects: "When we say 'length, breadth, and depth' ... we thereby indicate extension into three directions which can be united and apprehended all together, the intelligible [object] then being the geometrical body." We next make the first step of the analysis: "If we delete one of the dimensions and apprehend what has resulted - namely length and breadth only - the intelligible [object] will be the surface." Repeat the same operation once more, and you obtain the geometrical line as an intelligible object.

Al-Fārābī has thus accomplished an important part of the task he set himself: he showed how the analysis of the geometrical body yields the notions of the geometrical surface and line. His procedure is *constructive*: on the assumption that the dimensions are independent, he twice deletes one dimension, obtaining the ideal geometrical objects. These objects are now indeed separate of one another - the surface does not partake of depth, the line does not partake of depth or breadth.

8. We now proceed to the construction of the point, which to be sure, cannot be obtained by a reiteration of the same procedure as before. Al-Fārābī takes his cue from the ontological distinction between a geometrical object and its boundary. Thus, the body can be construed either with or without its boundary, roughly as we distinguish today between the closed and the open segment. Now if we can apprehend the *same* body with or without its boundary, clearly the boundary itself is not a body. And al-Fārābī reasons: "it follows that the boundary of the body is not a body; rather it is the surface which is the boundary of the body." The surface in turn is extended in breadth and length only. Al-Fārābī

specifies that it is precisely in that dimension where the surface is no longer divisible - namely that of depth - that it constitutes a boundary. The conclusion, to be generalized later, is this: "From that side where [the surface] constitutes a boundary, it is indivisible." The same procedure is repeated with the surface as an object: the boundary of the surface is the line, which is divisible from one side only, which is not the bounding one, namely breadth. Hence the general conclusion: "the side constituting a boundary is indivisible, whereas the divisible side is not a boundary."

Al-Fārābī's procedure is to construct a series of geometrical objects with decreasing dimensionality: we made two transitions from a given geometrical object to its boundary. In each case, the boundary has one dimension less than the bounded object, the "missing" dimension being precisely the one along which the bounding is effectuated. In general terms, al-Fārābī's idea is as follows: for each geometrical object, there exists a distinct geometrical object, which constitutes the boundary of the former and which is divisible in one direction less than it; the additional indivisibility belongs precisely to the side in which the bounding is effectuated. Al-Fārābī does not formulate this principle generally, but it clearly underlies his extrapolation of the series body-surface-line to culminate in the point: "since line is extended in one direction only, the boundary of the line will be obtained from the absence of this extension too, so that [the boundary] will have no extension in any direction whatsoever. ... The boundary of the line is called by the geometers 'point'."

Al-Fārābī has thus achieved his aim: having defined the point as the boundary of the line, the principle he has established allows him to construe the point as something - namely a boundary - devoid of any extension and, therefore, indivisible. His analysis has thus yielded the point precisely as defined in the *Elements*. Al-Fārābī has also succeeded in constructing the point so as to make it *separate*, and he is very perspicacious about this: "the term 'boundary' refers to [its object] in as much as it is contiguous to something, whereas the term 'point' refers to [its object] in as much as it can be apprehended separately, without the line." Al-Fārābī notes, however, that the definition of the point is incomplete because it does not specify the essence of the *definiendum*; he is satisfied, however, that it is "complete enough" for the needs of geometry.

*

9. Let me now say some words on the contexts in which we have to set al-Fārābī's *Commentary*.

Al-Fārābī was one of the heads of tenth century Bagdadi school of logic. Indeed, for him the study of logic, and of philosophy generally, had an ethical, eudomonic, value. Specifically, in a period of great theological and philosophical discussions, with political implications, logic was supposed to bring *certitude*, to show how to distinguish the true from the false. Al-Fārābī indeed believed that the (Aristotelian) philosophers who use logic as an instrument of investigation are "absolutely the elite" of society.

Yet logic can guarantee only the *validity* of arguments, not the *truth* of the conclusion, which, of course, depends also on the truth of the premises. This is why, in various writings, al-Fārābī attached particular importance to the epistemological status of the premises of an argument. From this derives also his interest in the premises of geometry, which he considers as a model of certain, immutable, knowledge. Therefore, the *Commentary* originated within the context of Bagdadi school of logic, specifically its reflection on the first principles on which knowledge depends; it is, so to say, a contribution to an ongoing discussion within this philosophical school.

But the *Commentary* simultaneously conveys other messages as well, addressed to critics of philosophy, and to these we now turn. Of the discussions about Greek philosophy at about the time the *Commentary* was written we have a particularly good picture through the account of the famous debate which took place in Bagdad in 942 between the logician Abū Bishr Mattā ibn Yūnus, al-Fārābī's teacher, and the mutakallim Abū Saʿīd al-Sīrāfī. I give a very brief summary of al-Sīrāfī's criticism of the study of philosophy; we will then see how our *Commentary* can be understood as a response to it.

Al-Sīrāfī argues that the sensible world is so complex that it cannot possibly be apprehended by intellection; rather, it can be apprehended only directly, so to say by immediate intuition. The philosophers who recommend long studies as the only way to knowledge and to felicity do this only out of their corrupt interests, namely because they charge high tuition fees. Moreover, these philosophers deny that philosophical knowledge can be acquired in everyday, natural language, arguing that only their own technical vocabulary is adequate to seize universal intelligibles. This is a fundamental error, however: the philosophical technical vocabulary derives from the Greek and can therefore convey only the intelligibles of that language. Indeed, all intelligibles are relative to a given nation and can be apprehended only with its natural language. Logic too is conventional and depends on the specificities of the Greek language just as grammar depends on Arabic. The logicians' belief that their science is universal and complete is an illusion, arising from their shutting themselves up within the bounds of an artificial language. Only natural language can express differences of opinion and thus contribute to the progress of knowledge.

The great modernity, indeed philosophical and political actuality of these arguments, cannot detain us here, where we will only look at the counter-claims our *Commentary* can now be seen as advancing implicitly.

To the claim that there is no systematic, intellectual, way leading from the world of sensibles to that of intelligibles, the *Commentary* answers that geometry is a decisive counter-example: the intelligible objects of geometry are obtained from the sensible body through analysis. Consequently, the technical vocabulary of geometry is within reach of everyone; the claims that philosophers draw on a vocabulary the access to which is restricted are false.

Moreover, the intelligible objects of geometry are not apprehended on the basis of any *language*, but rather directly from the sensible body. Therefore, although geometry is originally a Greek science, its truths are universal: geometry is not conventional and is not specific to any nation. Again, the homonymy of terms occurring both in natural language and in science (e.g. "length") is a possible source of confusion for the student, and demonstrates that natural language is not the best, and certainly not the only basis on which science can be erected.

10. Al-Fārābī's *Commentary*, therefore, should be read on a number of levels. To the logicians, it showed that Euclid's geometrical science conforms to the norms of Aristotle's metascience. For the Aristotelian philosophers preoccupied by questions of foundational epistemology, it describes a new theory of abstraction which takes into account Neoplatonic criticisms of Aristotle's simplistic notion. To the teachers of geometry it shows how to guide the student from the sensible objects to the knowledge of the intelligible, ideal, geometrical objects. Finally, the

epistemological account of Euclidian geometry refutes the arguments of the adversaries of Greek science, by showing that their claims on behalf of natural language, common sense, and relativism, are false.

XI

The Theory of the Opposites and an Ordered Universe: Physics and Metaphysics in Anaximander

In a bold and suggestive paper published in 1947, Professor Gregory Vlastos has given an incisive and intriguing account of Anaximander's cosmogony and cosmology. The "hard core" of Vlastos' reconstruction amounts to the statement that Anaximander held, indeed invented, a "philosophical concept of nature as a self-regulative equilibrium, whose order was strictly immanent, guaranteed through the fixed proportions of its main constituents".[1] This thesis, I will argue, is false. Regrettably so, because, as Vlastos justly notes, such a view of nature is of greater significance and import than Anaximander's strictly physical hypotheses. Much is at stake here: our analysis will suggest that Anaximander's physical theory – the theory of the opposites – which (in a modified form) was to be embraced by Aristotle and the Peripatetic school, *could* not found the notion of a self-regulative, immanent, natural order. Consequently, in Anaximander as well as in Aristotelian philosophy of nature, natural order had to be sustained and upheld by external factors: in particular, what was to become the sublunary world was not a closed system. As long as the theory of the opposites prevailed, physics could not do without metaphysics.

The theory of the opposites will be at the center of our study, but we will have occasion to reflect also on other issues in Anaximander's doctrines. In particular, scholars have long been puzzled by the following apparent contradiction: Anaximander, we know, held that the opposite constituents of the world are in a state of permanent equality; yet, it seems, he also maintained that the world is gradually drying up and that it will ultimately be reabsorbed into the Boundless from which it issued. How can these two ideas be reconciled? Our inquiry into Anaximander's theory of the op-

---

* I express my gratitude to M.D. Grmek and to D. O'Brien for criticism and advice on an earlier draft.
[1] G. Vlastos, "Equality and Justice in Early Greek Cosmologies", *CP* 42, 1947, 156-78; reprinted in: D.J. Furley and R.E. Allen (eds.), *Studies in Presocratic Philosophy* (London, 1970), I, 56-91 to which reference is made. The quotation is from p. 82.

posites will provide a precise answer to this question and show that Anaximander's teaching was perfectly consistent.

A word on method. This paper suggests what I take to be simple, straightforward solutions to long-standing central problems in Anaximander. These solutions do not depend on hitherto unknown evidence, nor do flashes of new insights emerge from meticulous philological analysis. Indeed, although philology has undeniably been indispensable for clarifying both the evidence and the problems, I believe that in studying Anaximander scholars have all too often been absorbed, indeed distracted, by textual questions. The result has been that neither the problems which animated Anaximander's thinking, nor, consequently, his solutions to them, have been appreciated adequately. As will be seen, some commonly held views on Anaximander simply make no concrete physical sense: they may do justice to the words, but not to the ideas. This paper, then, is an attempt to take Anaximander seriously as a first-rate thinker who, in the context of his times, asked himself important questions concerning nature and who tried to provide them with fairly rational answers. Rather than save a collection of texts, I will try to save Anaximander the theoretician.[2]

I. Vlastos' strong thesis on Anaximander

Professor Vlastos in fact adduces three distinct theses:

(i) The basic constituents of Anaximander's world are equal opposite powers balanced against one another in a dynamic equilibrium.

(ii) This equilibrium is strictly *inner*-worldly: it is maintained by the sole equality of the opposites and, in particular, does not involve an intervention of the Boundless.

(iii) Anaximander's notion of the world as a self-regulative equilibrium in which cosmic justice prevails, depends on contemporary societal order subsumed under the political notion of *isonomia*.[3]

---

[2] Since I do not engage in philological analysis, I will refer to texts and translations easily available in the secondary literature. Of these, often cited works are referred to as follows: G.S. Kirk and J.E. Raven, *The Presocratic Philosophers* (Cambridge, 1957): KR (the chapter on Anaximander is practically unaltered in the 2nd edition with M. Schofield (Cambridge, 1983)); W.K.C. Guthrie, *A History of Greek Philosophy: Vol. I: The earlier Presocratics and the Pythagoreans* (Cambridge, 1962): HGP, I; Charles H. Kahn, *Anaximander and the Origins of Greek Cosmology* (New York, 1960): Kahn, *Anax.* Where appropriate I add also the DK reference: H. Diels and W. Kranz, *Fragmente der Vorsokratiker*, 5th edn (Berlin, 1934-7).

[3] In "Equality" the type of relationship (causal or other) between political order and the notion of nature is not specified. In a later paper: "Isonomia", *AJP* 74, 1953, 337-66,

Of these, *i* is uncontroversial; *ii* I take to be Vlastos' strong thesis which I shall undertake to refute; *iii* is a distinctively sociological thesis which will not concern us here (although my criticism of *ii* will have obvious damaging consequences for *iii*).

Vlastos' reconstruction takes off from Anaximander's sole surviving fragment:

> And into those things from which existing things take their rise, they pass away once more, "according to just necessity (χρεών); for they render justice and reparation to one another for their injustice according to the ordering of time."[4]

What emerged from the Boundless, Vlastos holds, are certain opposites, namely (following Simplicius) "The hot, cold, dry, moist, and the rest."[5] The "justice" consists in their being permanently balanced against one another: when encroachment occurs, it is duly compensated by reparation. Indeed, the state of equilibrium is perpetual: the opposites are balanced in the Boundless itself (where they have no self-identical existence)[6] and issue from it "*together* in balanced proportions", so that they continue to be balanced in the constituted world too. The equibalance of the opposite powers is an essential feature of the world, without which it would be "not cosmos but chaos": at no time will one of the opposites "be strong enough to dominate another. When encroachment occurs, it will be compensated by 'reparation', as e.g. in the seasonal cycle the hot prevails in the summer, only to suffer commensurate subjection to its rival in the winter." In sum: "overall justice is preserved throughout the life-process in the world despite the occurrence of injustice; and this by the equation of reparation to encroachment."

So far Vlastos' thesis *i*. We now come to what I take to be the crucial question concerning Anaximander's notion of nature: what is it that

Vlastos is more explicit: the idea of a natural self-regulative order between equals is there said to be a "generalization from politics", to have been "projected from politics to cosmology", or "modelled on a notion of political justice" (pp. 347, 361, 362). But the sociology of knowledge is a hazardous enterprise: T.J. Tracy, *Physiological Theory and the Doctrine of the Mean in Plato and Aristotle* (The Hague/Paris, 1969) believes that it was the medical theory which had an "influence" on the political thinkers (e.g. p. 75, n. 62): by itself, the fact that analogous ideas are concomitant does not allow us to determine which, if any, influenced the other.

4 DK, A9 and B1: Vlastos' translation ("Equality", 73), which I quote for the sake of consistency although e.g. Kahn's (*Anax.*, 166) is closer to the Greek. All the quotations in the following two paragraphs are from "Equality", pp. 73-83.

5 On the identification of the opposites cf. below notes 8 and 45.

6 This particular idea has been rejected by most scholars but it does not bear on our discussion.

guarantees and preserves the postulated justice and equilibrium? Vlastos' answer – his thesis *ii* – is clear and unambiguous: for Anaximander, he affirms, order is *immanent* in nature; it depends solely on – and in fact is identical with – the self-maintaining, self-regulative equilibrium of the opposites. The overall cosmic justice is "assured through the invariant equality of the opposites." This notion obviously leaves the Boundless with a remarkably undistinguished function. True, it is said to "govern" the world: yet this, Vlastos argues, cannot refer to any "direct action by the Boundless upon the inner structure of the world," but merely to a "safeguarding [of] the original equality of the opposites with one another." On this view, then, the Boundless' only role is to ensure that the world remains a *closed system*: "since the world is 'encompassed' by the Boundless, nothing can enter or depart to upset the balance fixed upon the opposites in the process of generation." In sum: it is not the Boundless which guarantees the unfailing natural order, but rather the "self-regulative equilibrium" of the opposites, "the equal power of the basic constituents of nature to hold each other in check."[7]

Consider the two theses. Thesis *i* attributes to Anaximander the view *that* justice prevails, *that* the opposites are in equilibrium. As far as it goes, this is a pure statement of facts; but there is nothing in it by way of an account or an explanation of these facts, for it does not specify *why* the equilibrium is maintained, nor, in particular, whether the world-order is self-regulative or dependent on something external (*viz.* the Boundless). Thesis *ii* indeed fills in precisely this gap: it attributes to Anaximander a theoretical explanation according to which the natural order is *immanent* and depends entirely upon a self-maintaining equilibrium of the opposites, with the Boundless' role limited to that of an impermeable external boundary. Now most scholars (whatever their views on the precise nature of Anaximander's opposites[8]) accept something like thesis *i*, i.e. the idea that, in Anaximander's view, the world is constituted of certain opposites in a state of permanent equilibrium. Unfortunately they have mostly remained vague and uncommitted on the *cause* of that equilibrium, i.e. they have not specified their position on thesis *ii* (or something like it).[9] In this sense,

---

[7] Vlastos, "Isonomia", 362.

[8] For a recent balanced view on this, cf. G.E.R. Lloyd, "Hot and Cold, Dry and Wet in Early Greek Thought", *JHS* 84, 1964, 92-106, quoted after Furley and Allen, *Studies*, I, 254-280, on pp. 259-270.

[9] E.g. G.S. Kirk in "Some Problems in Anaximander", *CQ* 5, 1955, 21-38, reprinted in Furley and Allen, *Studies*, I, 323-49, p.342, affirms: "cosmological events are maintained by a fluctuating balance of power between opposed masses"; but what guarantees that

Vlastos' thesis *ii*, which unequivocally states that the equilibrium of the opposites, justice and order, is *inner*-worldly and self-caused, is a *strong thesis*: whether true or (as I shall do my best to show) false, it spells things out and thus makes the underlying crucial issue explicit and subject to critical discussion.

In what follows I will endeavour to show that, appealing though it is, thesis *ii* cannot be maintained. We must, I submit, distinguish two different kinds of equilibrium – static and dynamic – both of which, I believe, play important roles in Anaximander's system. Once this is admitted, it will transpire that the equality of the opposites guarantees only a static equilibrium; by contrast, the alleged dynamic equilibrium is not self-maintaining, but rather depends upon the Boundless.

II. Static Equilibrium: The notion of health in the Hippocratic corpus

The notion of equilibrium is paramount in the Hippocratic corpus, where it underlies the notion of health as an equilibrium of the body's constituents. Although most Hippocratics wrote some two centuries after Anaximander, we may, I believe, take our cue from their theories in order to try to understand Anaximander's. Indeed, it is generally agreed that Anaximander is the fountainhead of the theory accounting for the world-order in terms of a balanced interaction of opposite constituents and the claim seems warranted that the medical versions of this theory, such as the one in *On the Nature of Man*, go "right back to Anaximander".[10] It will be seen that, the heterogeneity of the Hippocratic treatises notwithstanding, the notion of

"fluctuating balance"? How does he construe the "self-perpetuating machine" to which he enigmatically refers? It seems equally insufficient to invoke the "legalistic metaphor" which allegedly "motivated" Anaximander to uphold "both the continuity and the stability of natural change": KR 119. Kahn, *Anax.* 187f. is vague, but seems to attribute to Anaximander (and to Aristotle) the idea that an equilibrium of the opposites "guarantees that the fundamental order of the universe will persist unchanged . . .''; but this comes near to being a tautology, for the equilibrium *is* the order, and we should like to be told what guarantees the equilibrium in the first place. Guthrie, *HGP*, I, 83 applauds the Milesian notion of φύσις "which is something essentially internal and intrinsic to the world"; he remains unclear, however, on whether "the world" includes the Boundless or not, and thus whether he subscribes to thesis *ii*. G.E.R. Lloyd, "Hot and Cold" 269, refers to "a continuous self-regulating interaction" of Anaximander's "unnamed subjects" but does not make this idea explicit.

[10] Lloyd, "Hot and Cold", 270; cf. also Kahn, *Anax.*, 191.

equilibrium was the subject of a large consensus.[11] It is in the light of this received notion, I will suggest, that we should interpret what little evidence we have of Anaximander's conceptions of equilibrium.

Let us then look briefly into the Hippocratic notions of health and disease. Medical writers – whatever their particular views on the constituents of the human body (humors, powers or qualities, elements, etc.) – generally construed health as an harmonious, equilibrated blending of these constituents. Correlatively, disease was taken to be a state in which one constituent was in excess or in defect. The classic formulation of this view is that of Alcmaeon of Croton: health, he said, is the "equality (ἰσονομία) of the powers (δύναμις)," or "a proportionate blending (κρᾶσις) of the qualities (τῶν ποιῶν)"; disease ensues when one of these powers gains supremacy (μοναρχία) over the rest.[12] Similarly, according to *On Ancient Medicine*, man is healthiest when a perfect blending prevents any particular power from manifesting itself.[13]. How deeply this notion was entrenched becomes particularly clear from *On the Nature of Man*, where it is used as the *premise* of an argument. To refute the monist view, on which the human body has a single constituent, the author (usually identified with Polybus, Hippocrates' son-in-law) argues that this theory cannot explain disease and pain: these states, he assumes, ensue when the body's equilibrium is disrupted, with one of the contrary qualities (hot, cold, moist, dry) gaining predominance; but the postulate of a single constituent of the body obviously undermines the notion of opposite qualities and with it that of equilibrium and so does not allow us to account for disease. The author, himself adhering to the humoral theory, then states that the body "enjoys the most perfect health when [the humors] are duly proportioned to one another . . . and when they are perfectly mingled. Pain is felt when

[11] The idea that the Hippocratic corpus is "systematically connected", i.e. that its various treatises share some basic notions despite their doctrinal heterogeneity, has been put forward in Owsei Temkin, "Der systematische Zusammenhang im Corpus Hippocraticum", *Kyklos* 1, 1928, 9-43. The "undue emphasis placed there on the humoral theory" (Owsei Temkin, *The Double Face of Janus* (Baltimore, 1977), 14, n. 24) does not undermine its central theses. What follows owes more to Temkin's doctoral dissertation than is visible through the footnotes.

[12] DK, 24 B4.

[13] Hippocrates, *On Ancient Medicine*, 19; cf. Hippocrate, *L'Ancienne médecine*, introduction, traduction et comentaire par A.-J. Festugière (Paris, 1948), p. 17: "L'homme se trouve dans la condition la plus excellente quand tout est en coction et en repos, sans manifester aucune force particulière".

one of these elements [sc. humors] is in defect or excess or is isolated in the body without being compounded with all the others."[14]

Health, then, was generally construed as a physiologically-defined state of equilibrium. Yet – and this must be particularly stressed – it is a basic tenet of Hippocratic physiology that there are indefinitely many states of equilibrium: in different persons the constituents are equilibrated differently. This idea implies that "health" is multiform – there is perfect health and relative health. Traditionally, the Greeks thought of health as an ideal associated both with a perfect bodily harmony and beauty and with intellectual and moral perfection.[15] But realities of medical practice, together with a more general, socially-founded trend toward relativism, fostered the notion of "relative" health: after all, not all men, not even all Greeks, are Apollos, and a cripple may well be in good general health.[16] The idea thus emerged that every individual has a personal normal, equilibrated, blending of the constituents, defining *his* particular state of health: this is that person's constitution (φύσις), with respect to which his health and disease are defined.[17] It is indeed a central Hippocratic doctrine that different environmental conditions give rise to different constitutions. The treatise *Airs, Waters, Places*, in particular, is primarily concerned to determine how, under the influence of various external factors – wind, heat, dryness, differences of nutriment – different states of equilibrium are established: each of these states defines the constitution – the norm of health – in the corresponding environment. Moreover, even in one and the same person, the constitution changes over time: the equilibrium defining the health of the young differs from that of the old (cf. below, § V).

We must now ask why a state of equilibrium does not last – how, that is, disease sets in. For the Hippocratics, mainly in the Coan school, non-

[14] Hippocrates, *On the Nature of Man*, 2, 4; quoted after W.H.S. Jones (Loeb).
[15] Cf. Henry E. Siegerist, *A History of Medicine* (New York, 1961), vol. 2, 299: "Health was not only a possession but an ideal. Jaeger, very justly, even goes so far as to say 'that the Greek ideal of culture was the ideal of health' [*Paideia* (New York, 1944), vol. 3, 45]."
[16] Cf. Fridolf Kudlien, "The Old Greek Concept of Relative Health", *Journal of the History of the Behavioral Sciences* 9, 1973, 53-9, where the origin and the social context of that notion are discussed.
[17] On the meanings of φύσις in the Hippocratic corpus, cf. Hans Leisegang, "Physis", in Pauly-Wissowa, *Realencyclopädie*, 39 (1941), 1130-1164, at coll. 1139-42; Felix Heinimann, *Nomos und Physis* (Basel, 1945), 103-4. August Bier, "Hippokratismus: Die Physis", *Münchener Medizinische Wochenschrift* 78 (1), 1931, 355-9, 408-11 contains interesting insights despite its explicitly a-historical perspective.

traumatic disease, just like health, was a physiologically-defined state of the entire body, namely one of imbalance.[18] In particular, "disease" was not an ontological or nosological concept: there was more concern with the notion of ill-being of a person than with that of disease.[19] How, then, is the state of disequilibrium brought about? There is general agreement that a person's equilibrium is upset by external imbalanced influences – e.g. intemperate conditions of weather, intake of food in which some quality predominated, etc. "It is changes that are chiefly responsible for diseases, especially the great changes, the virulent alterations both in the seasons and in other things."[20] Similarly, diseases may arise from excess or deficiency in nutriment or exercise.[21] This doctrine is the theoretical foundation of the basic Hippocratic therapeutics – dietetics.[22] In fact, the idea that disease is a perturbed equilibrium brought about by unbalanced external influences assigns the physician two roles. First and perhaps foremost, he has to advise the healthy person how to live "in conformity to nature" (κατὰ φύσιν): the objective of the hygienic mode of living is to neutralize, as far as possible, the unwholesome, destabilizing, influences. For instance, the extreme conditions of the seasons "must be counter-balanced by way of living": during (say) the hot summer, one should eat food contributing toward making the body cold and soft.[23] Since each individual has a different constitution – i.e. a particular equilibrium of the constituents is the norm for his health – the prescribed way of living must be adjusted both to the individual's constitution and to the prevailing external conditions at a given time and place. Ideally, the physician could determine a diet that would completely offset all external influences: the equilibrium would then not be upset, and health would be lasting.[24] The physician's second role, of course,

---

[18] Fridolf Kudlien, *Der Beginn des medizinischen Denkens bei den Griechen. Von Homer bis Hippokrates* (Zürich/Stuttgart, 1967), 70-71.

[19] "[Der Grieche] kennt überhaupt weniger Krankheit als Krank-Sein", Temkin, "Der systematische Zusammenhang," 14; cf. also O. Temkin, "Die Krankheitsauffassung von Hippokrates und Sydenham in ihren 'Epidemien' ", *Archiv für Geschichte der Medizin* 20, 1928, 327-52. On the emergence of this notion of disease see Gert Preiser, *Allgemeine Krankheitsbezeichnungen im Corpus Hippocraticum* (Berlin/New York, 1976), particularly 89-113.

[20] [Hippocrates], *Humors* 15, quoted after Jones; cf. also *Aphorisms* 3.1.

[21] *Regimen* 1; *De Victu* 2.

[22] For what follows, cf. Siegerist, *History*, vol. 2, 237; Ludwig Edelstein, "Hippocratic Prognosis" (1931), in his *Ancient Medicine* (Baltimore, 1967), 65-85, on p. 70, and "The Dietetics of Antiquity" (1931), *ibid*, 303-16.

[23] E.g. *Regimen in Health* 1.2.

[24] "If indeed . . . it were possible to discover for the constitution of each individual a due

is to endeavour to reestablish a patient's health if the equilibrium has already been knocked off balance. In this case, he strives to force back the component which has become dominant, namely, according to the prevailing view, by prescribing a diet containing its contrary – "contraries are cured by contraries".[25]

Consider now the notion of equilibrium. Two postulates, I wish to suggest, underlie it. The first is this: A person is healthy only when the balance of his body's constituents conforms to his φύσις; any deviation from it means disease. In a perfectly equilibrated or neutralized environment, the balance prevailing in a given body will therefore persevere – the equilibrium is disrupted if and only if the body is affected by uncompensated external influences. Thus, the Hippocratic concept of health is that of a *static, albeit labile and unstable equilibrium.*

The second postulate concerns the "dynamics" of disequilibrium. Just as an equilibrium is disrupted only under the influence of an external imbalanced impulse, so also it will need an external counterbalancing action to redress the already disequilibrated body. More precisely: just as in the absence of external imbalancing influences the body's constituents remain equal, so, once out of balance and under the same conditions, they will remain unequal. This means that once a constituent has come to overpower the others, then, left only to the interplay of its constituents, the body will never return to a state of equilibrium. Worse: the body necessarily slides ever farther away from equilibrium. For the Hippocratic doctrine excludes the notion of a static (chronic) disease, i.e. the idea of one constituent permanently dominating the others: unlike health and equilibrium, disease and disequilibrium cannot persevere. Rather, disease is generally construed as "a process in time",[26] whence the central role of prognosis: the patient's state must evolve – either back to health or toward death. The

proportion of food to exercise, with no inaccuracy, either of excess or of defect, an exact discovery of health for men would be made" (*Regimen* 1.2).

[25] *Aphorisms* 2.22; similarly *Breaths* 1. Cf. also *On Ancient Medicine* 13: "according to the theory of the new school, if the injury was caused by one of the opposites, the other opposite [i.e. the remedy] ought to be specific." See also Kahn, *Anax.* 130-31. The same idea is found in the chirurgical treatises. In *De fracturis – De articulis* traumas are construed as being due to forces acting "against nature"; healing requires that the physician exert a stronger, opposite force to bring the body back to a state "in accordance with nature". Cf. Markwart Michler, "Die praktische Bedeutung des normativen Physis-Begriffes in der hippokratischen Schrift 'De fracturis – de articulis' ", *Hermes* 90, 1962, 385-401, on p. 392.

[26] Oswei Temkin, "Greek Medicine as Science and Craft" (1953), in *The Double Face of Janus*, 137-153, on p. 149; cf. also p. 445.

physiological notion underlying this view is that once a constituent has gained dominance, it will continue to move the system away from equilibrium: the feebler constituents will be unable to outbalance the stronger one which presumably gets ever stronger as the process advances.

To reverse this inherent movement of the diseased body toward death, an *external* intervention is necessary: this is, according to a prevailing view, precisely the role of the physician.[27] What in and for itself is the physiological necessity (ἀνάγκη) governing an illness, appears to the sick person as a fate (τύχη). To counter this fate, the physician's τέχνη is needed: without it, the author of *On Ancient Medicine* says, the patient's condition would be left solely to the law of chance. This is indeed why, according to the same author, medicine emerged in the first place: "The fact is that sheer necessity has caused men to seek and to find medicine, because sick men did not, and do not, profit by the same regimen as men in health."[28] In other words: if a diseased person is subjected to the equilibrated regimen appropriate to a healthy person, his state worsens; only if the physician deliberately counters the constituent which has gained supremacy with a suitable regimen, can the person's health be reestablished. To be sure, patients occasionally recover without the physician's intervention: but all these cases of apparent spontaneous recovery, the author of *The Art* argues, are in fact due to the counteractive power of some food the patient has unwittingly taken in.[29] In sum, therefore, the causes of disease and the physician's τέχνη are strictly symmetrical: both are powers external to the body and acting on its φύσις at a given moment. Moreover, the causes of disease are divine in the sense that they have their origin outside man, in environmental factors (e.g. the weather); in parallel, the physician's τέχνη is also construed as a divine gift, and although its successes are obtained naturally, they cannot (as an inscription says) be realized "without a divine power."[30]

---

[27] For what follows cf. W. Nestle, "Hippokratika", *Hermes* 73, 1938, 1-38.

[28] *On Ancient Medicine* 1, 3.

[29] *The Art* 6.

[30] Nestle, "Hippokratika", 3-5. My interpretation is at variance with that of a number of scholars who ascribe to some Hippocratics, notably to the Coans, the view that each body's φύσις is a healing φύσις by which (at least sometimes) it can reestablish its disrupted equilibrium with no need for an external (the physician's) intervention. This interpretation seems to me to run against the principles of the Hippocratic physiological doctrine, however. As Charles Kahn (*Anax.*, 130) aptly put it, this doctrine implies that "In the conflict of opposites [. . .] the weaker power will require succor or support if it is not to meet destruction at the hands of the stronger." Similarly, Temkin pointedly writes that the idea of a healing φύσις can hardly be reconciled with the notion of the physician's

Back now to Anaximander. The above analysis has shown what "equilibrium" means in early Greek medical thought: a system's static equilibrium perseveres unless disturbed from outside; but once disequilibrated, the system will never be balanced again unless redressed through external intervention. Bringing (as I think we should) these notions to bear on Anaximander, we readily see that to credit him with the idea of a self-maintaining dynamic equilibrium is fairly implausible. For imagine that at a given moment, e.g. immediately after the initial separation, the op-posites are equal. They would then hold one another in check, thereby creating a complete and lasting standstill: no encroachment and no repara-tion – i.e. no movement or change – would ever come to pass. For the Boundless is perfectly uniform and balanced and thus cannot cause destabilization. Thus just as the health of a man who follows the ideal diet perseveres unaltered, so also Anaximander's unprocessed world would endure in its initial static equilibrium. Again, assume that the static balance of the opposites was (somehow) disrupted, be it ever so slightly: this will never result in a "dynamic equilibrium". For if (say) the hot momentarily overpowers the cold, then, if the system is left to itself, the feebler cold will never be able to outbalance, let alone dominate, the stronger hot. Thus, in as much as the Hippocratic notions of equilibrium can be assumed to throw light on Anaximander's, we must conclude that the idea of a *self-regulating* process of alternately dominating opposites is unwarranted: Vlastos' thesis *ii* is false.

"quite drastic interventions" invoked in the concluding aphorism (O. Temkin, "Der systematische Zusammenhang", 33, n. 1). Unfortunately, discussions of this issue all too often reflect opposing views in *contemporary* biology and medicine: those who advocate a holistic idea of the human body detect in the Hippocratic corpus their own notion of a healing φύσις, whereas those who favor a reductionist view of biological phenomena deny that such a notion can be found there. Representatives of these two views are, respectively, A. Bier, "Hippokratismus: Die Physis", and Ch. Daremberg, *Histoire des sciences médicales* (Paris, 1870), I, 116-120. More modern authors are less prejudiced, but have not studied the question with sufficient detail and conceptual clarity. In particular, they have tended to assume tacitly that the specific φύσις of each individual is necessarily also a healing φύσις. L. Bourgey, for example, writes that "the organism which heals itself only continues the movement through which the living body creates itself" (Louis Bourgey, "Hippocrate et Aristote: L'origine, chez le Philosophe, de la doctrine concernant la nature", in: M.D. Grmek (ed.), *Hippocratica. Actes du Colloque hippocratique de Paris (1978)* (Paris, 1980), 59-64, on p. 64): he overlooks the crucial fact that growth – as a process throughout which the opposites are always balanced – is something fundamentally different from healing – the reestablishment of a disrupted equilibrium.

III. A role for the Boundless

Anaximander's fragment postulates alternating "encroachments" and "reparations" and this, obviously, is not a static equilibrium: the opposites are cyclically advancing and retreating. Scholars are in general agreement that Anaximander had *meteorological* phenomena in mind: the regular variation of the seasons, perhaps also the variation of day and night.[31] We must now ask how these regular processes come to pass. The system of opposites producing the meteorological phenomena is one which, like a pendulum, oscillates between two opposite extremes: but what is it that reverses the movement at the extremes, counteracting the immanent forces postulated by the theory of the opposites? What, for instance, forces the heat, at the height of its power in midsummer, to a gradual retreat leading to a (temporary) victory of the cold?[32]

The answer is not far to seek: Anaximander affirmed the Boundless to have controlling powers, to "direct" or to "steer all".[33] This statement scholars have understandably found difficult to reconcile with the notion of the self-regulative natural order they ascribed to Anaximander.[34] We now see, however, that the minimal directive action which must be attributed to the Boundless is to swing the opposites to and fro, to guarantee that upon encroachment reparation will follow. The Boundless must alternately counteract the opposites, just like the physician: for a time, one of the

[31] Cf. *HGP*, I, 80, 101; Vlastos, "Equality", 82; KR 118; U. Hölscher, "Anaximander and the Beginning of Greek Philosophy", in Furley and Allen, *Studies*, I, 281-322, p. 298 (= U. Hölscher, *Anfängliches Fragen* (Göttingen, 1968), 28-9, to which I refer because the English translation at times renders Hölscher's already opaque text entirely incomprehensible); Kahn, *Anax.*, 184.

[32] Guthrie in fact comes close to recognizing this problem: " . . it is difficult to see how the Hot, having once been allowed to gain the supreme victory – or commit the supreme injustice – could ever be forced to give up its ill-gotten gains" (*HGP*, I, 101). He argues here against ascribing to Anaximander the idea that a "great winter" necessarily succeeds a "great summer", but precisely the same argument also applies to ordinary summer and winter. Neither Guthrie nor, as far as I can see, other scholars have noticed that the principles of the theory of the opposites do not allow us to account for cyclical changes; attributing the latter to a "self-regulating equilibrium" (cf. above n. 9) is false and, worse, sweeps the problem under the carpet.

[33] Aristotle, *Physics* 203b7ff. (DK, A15). Cf. *HGP*, I, 88; KR 114-6; Gigon, *Ursprung*, 65; Kahn, *Anax.*, 44.

[34] E.g. M.C. Stokes, "Anaximander's Argument", *Canadian J. of Philosophy*, 1976, suppl. vol. 2, 1-22, on p. 13, writes that "It is far from clear that there is any need for a sustaining eternal body in such a balanced universe." Less explicitly: KR 114-6; Kirk, "Some Problems", 344, n. 25; cf. below n. 40.

opposites is allowed to gather force, with its opposite diminishing correspondingly; then, following the Boundless' intervention, the feebler
opposite will strengthen at the expense of the stronger, something it could
never do on its own, until it will equal, then surpass it. The process will then
be reversed again, and so on and on.

All this the Boundless does following "the ordering of time": what
determines the point of reversal is not how powerful/weak the opposites
have become, but rather something independent – *time*. This is presumably
how Anaximander interpreted the *periodical* regularity of day and night
and of the seasons. Indeed, days differ in brightness, summers in heat, yet
night and winter follow at regular temporal intervals.[35] That the Boundless
steers all things according to the ordering of time probably also accounts for
movement and change as such: why, that is, the equal opposites do not, in
the first place, hold one another in check, producing a static, frozen and
immobile world. Thus, we are fortunate to have regular periodic
phenomena because the Boundless steers them according to time. This the
Boundless certainly does *not* do by merely "safeguarding the original
equality of the opposites with one another", nor by just "having initiated
the world in such a way as to provide a continuing rule or law of change".[36]
Unfortunately, it is impossible to say anything more positive. From our
vantage point, however, the crucial point is *that* the world-order depends on
the Boundless, *that* left to itself, the world would either remain in a
stalemate or slide into disorder. Indeed, what little we can know of how
Anaximander construed the Boundless' control over the world confirms
this conclusion: apparently because it "encompasses" the world, the
Boundless holds it fast and prevents its disintegration. For in the actual
world the outer circumference is made of fire which, due to its greatest
purity, has gone farthest from the center: were it not for the Boundless, we
gather, this fire would have continued its centrifugal motion indefinitely.[37]
The Boundless thus counteracts the inherent centrifugal motion of the fire
and only thereby are the opposites held together in a single coherent world.
Our world of oppositions would necessarily disintegrate into chaos if order

[35] Kirk ("Some Problems", 346) perhaps meant something similar when writing that "In
any case Time does not control the *amount*, but rather the *period* in which the fixed
proportion must be paid." (Cf. also KR 120). Similarly, but elusively, Kahn, *Anax.* 179:
"In the fragment, the conditions of payment are fixed by the arbiter Time, and his law is a
periodic pendulum of give and take."
[36] Vlastos, "Equality", 81; KR 115-6.
[37] Cf. Kirk, "Some Problems", 344, n. 25; Vlastos, "Equality", 81, n. 128; Kahn, 90-91
(with n. 2 on p. 91).

were not imposed upon it by the Boundless. Disappointing though it is, there is no escape from the conclusion that Anaximander's physics did not allow him to frame the idea of an immanent order of nature.

No wonder, then, that Anaximander held the Boundless to be divine.[38] For the Presocratics construed as divine whatever accounted for *order*: "Their theme," Vlastos pointedly writes, "is nature, and their object is to explain the how and why of its unfailing order. When they find in this a moral meaning . . . they may express the trust and reverence they feel for it by calling it 'god'. . . . [They] dared transpose the name and function of divinity into the realm conceived as a rigorously natural order and, therefore, completely purged of miracle and magic."[39] This is precisely the Boundless' role: it imposes structure, law and order on the conflicting powers making up the world and thus is an absolute, supreme power. It is not a capricious ruler however, for it controls the world in accordance *only* with "the ordering of time", i.e. *regularly*. Although Anaximander's physical world is *not* a closed system with an *immanent* natural order depending on the sole interplay of the opposites, the rigorous justice, the periodic *lex talionis* established by the divine Boundless, still makes it a *lawful*, ordered, regular world.[40]

---

[38] *HGP*, I, 87-9; KR 116; DK, A15.

[39] G. Vlastos, "Theology and Philosophy in Early Greek Thought", *Philosophical Quarterly* 2, 1952, 97-123; the reference is to Furley and Allen, *Studies*, I, 92-129, on pp. 114, 119.

[40] Vlastos, who concedes to the Boundless no role in ordering the world, is understandably at a loss when it comes to accounting for the Boundless' divinity. In "Equality" (p. 81, n. 128) he suggests that it suffices for the Boundless to "encompass" the world to earn the title "divine". But if so, does not encompassing imply "governing" in the sense of controlling and ordering, as Vlastos' own (though slightly later) analysis, quoted above, implies? Of this crucial question Vlastos says that it "need not be laboured". In "Theology and Philosophy" (p. 114f.) he chooses to deny altogether that Anaximander called the Boundless "divine". T.G. Sinnige, *Matter and Infinity in the Presocratic Schools and Plato* (Assen, 1968), 1-4, suggestively argues that Anaximander's Boundless is a successor-concept to the ancient mythical notion Time (χρόνος), construed as an "omnipotent and active Ruler, embracing the universe". On this interpretation, "encompassing" indeed implies direct ruling, a notion which gives credence to Guthrie's surmise (*HGP*, I, 88) that the Boundless might have been ascribed "some form of consciousness". Similar consequences follow from the suggestion that the origin of Anaximander's Boundless is to be sought in an ancient Iranian theological tradition (Walter Burkert, "Iranisches bei Anaximandros", *Rheinisches Museum* 106, 1963, 97-134). All this obviously takes us still further away from the notion of an *immanent* natural order.

IV. Static equilibrium in Anaximander: the stability of the earth

Anaximander, I now want to suggest, in fact did reason in terms of equilibrium – but this is *static* equilibrium, accounting for the stability of the earth. According to Hippolytus, Anaximander held that "the earth is aloft, not dominated by anything; it remains in place because of the similar distance from all points [of the circumference]." Aristotle slightly enlarges upon this report: "There are some who say that the earth remains in place because of similarity [or symmetry], as did Anaximander among the ancients; for a thing established in the middle, with a similar relationship to the extremes, has no reason to move up rather than down or laterally; but since it cannot proceed in opposite directions at the same time, it will necessarily remain where it is."[41] Unlike the other Milesians, who postulated a single and, therefore, from an Anaximandrian point of view, "dominating" elemental body on which the earth could then be assumed to repose, Anaximander accounts for the stability of the earth in terms of equilibrium. This clearly is a *static* equilibrium: like health, the earth remains stationary if not disturbed.

Anaximander's argument obviously involves considerations of symmetry. Scholars have therefore interpreted it as distinctively *formal*, i.e. as independent of any particular *physical* system: "a brilliant leap into the realm of the *a priori*", indeed a foreshadowing of Leibniz's principle of sufficient reason, having a "specifically mathematical character". Anaximander's earth, Jonathan Barnes spiritedly comments, was "mathematically suspended by abstract reason".[42] But this is misconstrued. Behind the formal symmetry, I will now show, lurk physical forces. It is the opposite actions of these forces which result in the static equilibrium: Anaximander's earth was suspended physically.

To extract Anaximander's physics, we begin with his cosmogony. According to a well-known report:

> He says that at the birth of this cosmos a germ of hot and cold was separated off from the eternal substance, and out of this a sphere of flame grew about the vapour surrounding the earth like the bark round a tree. When this was torn away and shut off in certain rings, the sun, the moon and stars came into existence.[43]

In general outline, the process here described is generally agreed to be as

[41] Quoted after Kahn, *Anax.*, 76 (DK, A11; A26).
[42] KR 135; Kahn, *Anax.*, 77; G.E.R. Lloyd, *Magic, Reason and Experience* (Cambridge, 1975), 68; J. Barnes, *The Presocratic Philosophers* (London, 1979), I, 26-7.
[43] Quoted after *HGP*, I, 90 (DK, A10).

follows:[44] first, a germ or seed was detached from the Boundless. It then developed – separated off – into a ring (or, though less probably, sphere) of fire on the one hand, and a cold moist mass at the center, on the other.[45] Here ends the truly cosmo*gonic* stage, for the subsequent development follows strictly physical laws which have continued to prevail in the world ever after. Through the action of the fiery, hot and dry circumference on the cold and moist center, the dry land becomes separated from the sea. In fact, according to Aëtius (whose report is confirmed by Alexander), "Anaximander says that the sea is a relic of the primal moisture, the greater part of which has been dried up by the fire."[46] Thus, not only is the heat endowed with the tendency to move toward the periphery (this is why the sun is farthest from the earth and why its fire is purest), but it can also endow the formerly cold moisture with a similar movement; indeed, the fiery heavenly bodies nourish themselves on moisture driven upwards.[47]

Let us now notice that, contrary to what most scholars tacitly assume, the roles here attributed to the opposites are strictly *asymmetrical*: one of the opposites moves out, the other hangs on immobile at the center; the one heats and sets in motion, the other gets heated and moved. Moreover, in nourishing itself on the earthy moisture which it raises, the heavenly fire even assimilates the moisture to its own substance: this is obviously a (partly) irreversible destructive process, for the moisture is steadily diminishing. Again, the heavenly fire has the power to inform and vivify the inert moist stuff at the center: for according to Anaximander's zoogony, animal life emerges through the action of the sun's heat on mud and slime.[48] In short: the centrifugal, fiery stuff is *active*, its inert moist opposite is *passive*.[49]

Anaximander's *physical* account of the earth's stability can now be spelt out. The cosmogonic process results in a system of concentric fiery rings (or

---

[44] For what follows cf. *HGP* I, 92; Kahn, *Anax.*, 86-7, 102; KR 129-133, 139.

[45] With this formulation I avoid committing myself on the precise nature of the opposites which originally separated off. Important as the question may be for the history of the theory of the opposites, it does not directly bear on the question I am studying here (cf. n. 8 above).

[46] Quoted after *HGP*, I, 92 (DK, A27). Cf. also KR 139-40; Kahn, *Anax.*, 65ff., 102ff. Most interpreters believe that Aristotle, *Meteorologica* 353b5ff. and 355a21ff., has Anaximander in mind.

[47] Kahn, *Anax.*, 86, 89-90, 103 (DK, A27). The idea that fire nourishes itself on moisture is widespread in the Hippocratic corpus; Cf. e.g. *De victu* 3; *De Flat.* 3.

[48] Cf. *HGP*, I, 101-4; Kahn, *Anax.*, 109-13; KR 141-2 (DK, A11, A30).

[49] Cf. below, n. 66.

spheres) with a passive core at their center. The passive stuff is inherently *inertial*: only by and by does the heating endow its moist part with centrifugal movement, thus effecting evaporation. The earth (including the remaining moisture in and on it) is therefore stable in "mid-air" primarily because, in Anaximander's physics, *its stuff has neither an inherent nor an acquired motion*.[50] That the earth is (roughly) symmetrically situated at the center of the fiery circumference – this fact itself is already a result of its inertness – presumably guarantees that it will not be destabilised by imbalanced evaporation (of the moisture) or scorching (of the dry matter) on one of its sides. (In fact, according to Anaximander, evaporation produces wind, i.e. motion, and, similarly, excessive dryness or moisture inside the earth may have been the cause Anaximander invoked to account for earthquakes.[51]) The earth thus remains stable for much the same reason as a man remains healthy in a perfectly equilibrated environment. It is Anaximander's physics, in sum, and not the purported geometrical symmetry of the world by itself, which accounts for the stability of the earth: the notion of an immobile, unsupported earth rests, first, on a "law of inertia" applying to the passive stuff at the center and, second, on the notion that, as a consequence of the world's symmetry, the potentially destabilizing heat acting on the passive center is equibalanced. The stability of the earth thus results from a physical static equilibrium about an inert center.

We may now round off our account and suggest a solution to a riddle that has puzzled more than one. Anaximander, we know, held the form of the earth to be cylindrical, shaped like a column-drum whose "depth is one third of its width".[52] This is truly perplexing, for judging from the general features of his thinking, we would expect Anaximander to opt for a spherical earth. In his cosmogony, the circumference heats a primal moist and inert mass suspended at the center: it would be natural to assume this mass (like a drop of water) as well as the emerging earth to be spherical. Again, the notion of a cylindrical earth is at odds both with Anaximander's "sense of aesthetic symmetry" which informs all the details of his cosmology and

---

[50] John Robinson ("Anaximander and the Problem of the Earth's Immobility", in John P. Anton and George L. Kustas (eds.), *Essays in Ancient Greek Philosophy* (Albany, 1971), 111-8, proceeds on the correct premise that Anaximander's "theory of the earth's immobility is founded upon the supposition that the earth has no absolute weight, that is, that it has no natural tendency to fall" (p. 113). Yet, since he cannot conceive that Anaximander in fact held this view, he concludes that "the view imputed to Anaximander by Aristotle not only *was* not but *could* not have been held by him" (p. 116).

[51] Kahn, *Anax.*, 63, 68, 100-4 (DK, A11, A24, A28).

[52] Kahn, *Anax.*, 55-6, 81-2; KR 134 (DK, A10).

with "the geometric spirit which seems to dominate [his] thought" in which circles and spheres loom large.[53] Gigon has spelt out what other scholars have subtly implied: Anaximander's argument for the stability of the earth, he perspicaciously remarks, is not valid for a cylindrical earth whose surface is *not* equidistant from the circumference. "Anaximander has not noticed this error. Once it was noted, the conclusion became inevitable that the earth must be a sphere."[54]

Instead of scolding Anaximander for his errors, we should, I suggest, ask how he arrived at the notion of a cylindrical earth and why he stuck to it. Not only, as noted, is this notion strikingly out of tune with his cosmogony and cosmology, but it seems simply implausible that one who has accomplished such an outstanding intellectual breakthrough as to conceive of the earth as freely suspended in space, who is regarded as the "founder of Greek astronomy" who may have invented the spherical astronomical model,[55] should not be aware that a cylinder included in a ring or in a sphere is not equidistant from it: a glance at a potter's wheel would reveal as much. Finally, Anaximander may have postulated the existence of antipodes and may have believed that the surfaces of the earth are not flat but convex:[56] two potential obstacles on the route toward construing the earth as spherical may thus have already been removed.[57] Let us, then, be generous with this "intellect of truly amazing grasp and audacity"[58] and adopt the working hypothesis that he might have had good reasons to believe the earth was cylindrical.

Consider the first stage of the cosmogony. We postulate a seed, we have to account for a world. More specifically: we assume a mass of strictly homogeneous, indistinguishable matter[59] in which opposites as such are

[53] Vlastos, "Equality", 75f.; Kahn, *Anax.*, 89-98.

[54] Olof Gigon, *Der Ursprung der griechischen Philosophie* (Basel, 1945), 86f.; similarly, though less severely, *HGP*, I, 100. Kahn, *Anax.*, 79-81, 116, repeatedly remarks that the doctrine of the sphericity of the earth was but a "perfection" of Anaximander's view. He does not pause to ask why Anaximander has not accomplished this perfection himself.

[55] T.L. Heath, *Aristarchus of Samos* (New York, 1981), 36; Charles H. Kahn, "On Early Greek Astronomy", *JHS* 90, 1970, 99-116.

[56] Kahn, *Anax.*, 55-8, 84-5; *HGP*, I, 98-9; (DK, A11).

[57] Cf. the contradictory appraisals of the bearing of observation on the question of the earth's flatness in Kahn, *Anax.*, 117 and *HGP*, I, 99. Cf. also Gigon, *Ursprung*, 88-9.

[58] N. Rescher, quoted in Barnes, *Presocratic Phil.*, I, 20.

[59] That the Boundless is a kind of material matrix from which the world (or worlds) emerges is an almost generally accepted view; for numerous references cf. Elisabeth Asmis, "What is Anaximander's *Apeiron*?" *Journal of the History of Philosophy* 19, 1981, 279-297, on p. 279, n. 1.

inexistent; we have to show how the initial separation came about, how stuffs of opposite characteristics (warm, dry, bright, etc. vs cold, moist, dark, etc.[60]) moved out into opposite directions. (The passive stuff, although not moving *away* from the center, must still be given some motion, for presumably the earth *grew* out of the original, smaller, seed.) The model most obviously suggesting itself here is the growth of a plant, particularly of a tree, out of a seed.[61] Put a seed in a glass of water: you hardly have a matter which is more homogeneous, insipid and indistinguishable than water. Yet the seed, drawing in the water around it, apparently *separates* it into two bulks: a relatively bright one moving upward (notably the stem) and a dark one, extending downward (the root).[62] We may recall that Empedocles indeed associated the two halves of the plant with specific elements and motions: he took fire to be a component of the green part of the plant, earth of its root, the two elements accounting for the respective motions upward and downward.[63] (In relation to Anaximander, Empedocles of course inverted explicandum and explanans.) This, I suggest, was Anaximander's model for the primal separation of the opposites out of the Boundless.[64] The seed produces two opposite bulks of matter: the dark one, the root, simply hangs on where it is – this is the earth; by contrast, the light one, the trunk, supplies the celestial matter and will duly be expanded. The question why Anaximander held the earth to be cylindrical now receives a straightforward plausible answer: its

[60] Kahn, *Anax.*, 161.
[61] The term in the doxographical report (DK, A 10), γόνιμον, may be Anaximander's own and "in any case suits the image of the growing tree": Hölscher, "Anaximander", 292 (= *Anfängliches Fragen* 21-2). Cf. also *HGP*, I, 90.
[62] I am indebted to Emmanuel J.A. Freudenthal for an enlightening discussion of this point.
[63] Aristotle, *De Anima* 415b28ff.
[64] Our interpretation suggests that it is not impossible that Anaximander's doctrine grew out of a wish to resolve a problem in Thales' account. Anaximander's first inspiration may have been to postulate a seed in order to explain how the homogeneous *water* is separated into two bulks of matter. When it came to the theory of the opposites, however, Anaximander would have been led to substitute the indeterminate Boundless for water. From this vantage point, the water and the Boundless are analogous in that they are both indistinguishable and neutral. (The idea that Thales chose water as his "element" because of its neutrality or formlessness was suggested by G.W.F. Hegel, *Vorlesungen zur Geschichte der Philosophie* (Leipzig, 1971), Vol. 1, 297; this suggestion does not seem to have found the favor of modern students of Thales.)

form is that of a *root*.[65] Our interpretation can be pursued further. So far the tree model accounts for the vertical separation of matter into an "up" and a "down", but not yet for the formation of the heavens as a surface surrounding the earth. This, we see, is precisely the role of the notorious bark on which so much ink has been spilt. The bark is a surface enveloping the stem of a tree and it gets broader and wider as the tree gets higher. This is the gist of the model: with the tree as the cosmogonical model, the separation of the opposites (the vertical growth of the tree) *concomitantly* brings about, and thus explains, the formation of an envelope which in due course becomes the heavens. On this interpretation, Anaximander's model is admirably elegant and simple: it accounts for the formation not only of the (cylindrical) earth, but also of the heavens by postulating only a vertical separation through a seed. We may now proceed to the last step of the cosmogony: the bark surrounds the *upper* part of the tree and it is thus constituted of the active, centrifugal stuff. It will now – as Anaximander indeed says that it is – be "torn away": through this Big Bang it is separated into several rings on which are fixed the heavenly bodies.[66] One surmises that the entire stem, consisting of concentric rings, is thus expanded into the heavenly circumference surrounding the former root, the earth, at the center.[67]

[65] Anaximander, we are told, compared the earth to the drum of a stone column whose breadth is three times its depth (DK, A10; *HGP*, I, 98; Kahn, *Anax.*, 55f. 81f.). The root-model admittedly does not account for this specific shape (roots are oblong cylinders). However, scholars have failed to make sense of Anaximander's figures, so we must, I believe, admit that we do not know whether and how Anaximander reconciled his cosmology with his cosmogony.

[66] As far as I can see, no scholar has noticed that (already) in Anaximander one set of the opposites is active, the other passive. Accounts of Anaximander tacitly assume that the roles of the opposites are symmetrical so that the centrifugal movement of fire is in need of explanation. Guthrie, for instance, suggests that the reason why the fiery spherical "bark" parted, is doubtless "the increasing pressure from the mist or steam caused by its own action in evaporating the watery centre" (*HGP*, I, 94). But this obviously involves a *petitio principii*: if heat can bring about evaporation and thereby pressure, this is surely because it is capable of endowing the (inert) water with centrifugal motion, something it can do only because it has itself an inherent centrifugal motion. The notion that fire or heat has such an inherent motion was, of course, widespread in early Greek philosophy (cf., most characteristically, the Hippocratic *On Fleshes*). This notion, it seems, is connected with the idea, shared by Anaximander, that the action of heat on slime can vivify it, i.e. endow it not only with motion, but even with the capacity for self-motion. Anaximander's Big Bang, in short, depends on the postulated inherent centrifugal motion attributed to heat.

[67] My interpretation differs from the prevalent view according to which "Anaximander conceived his cosmogony on the analogy of early views concerning the seed of animals and the development of the embryo" (*HGP*, I, 90f.; cf. also Kahn, *Anax.*, 86-7; KR 131-

Anaximander's belief in the cylindricality of the earth, we conclude, was not gratuitous: rather, he derived it from a simple and powerful cosmogony according to which the earth owes its form to the fact that it grew out of a seed as a root. Nor is his tenet of the earth's stability in space grounded in mere abstract symmetry: the earth persists motionless ever since its growth came to a halt (presumably when the stem was torn off and transformed into the celestial rings) because the *physical* principles underlying Anaximander's cosmogony and cosmology implied that its matter was inert and that, due to the symmetrical arrangement of the world, its motionlessness is not impaired by the action of imbalanced forces.

V. Shifting equilibrium: dying in good health

Anaximander is reported to have held two views which modern scholars find difficult to reconcile with his basic tenet concerning the equality of encroachment and reparation. The first of these is that ever since the moist earth came into existence, it has continuously been dried up by the sun so that "some time it will end up by all being dry".[68] The second is contained in Aëtius' report: "Anaximander of Miletus . . . says that the first principle of existing things is the Boundless; for from this all come into being and into it all perish. Wherefore innumerable worlds are both brought to birth and again dissolved into that out of which they came."[69] Now these two views appear to complement one another: although we have no explicit evidence, it seems natural to think that Anaximander believed that the world would

3). This interpretation, which goes back to H.C. Baldry ("Embryological Analogies in Pre-Socratic Cosmogony", *CQ* 26, 1932, 27-34) appears attractive in that it directly produces a spherical world out of a spherical egg (a motif already present in the mythical world-egg); the interpretation suggested above shows, I believe, that this is not necessary. On the other hand, the embryological interpretation has to explain away the bark simile and it obviously cannot reconcile the cosmogony it attributes to Anaximander with his view that the earth is cylindrical. G.E.R. Lloyd, *Polarity and Analogy* (Cambridge, 1971), 309-12, comes close to the interpretation I suggest. He does not, however, connect the separation of the opposites with the tree model and consequently errs (in my view) in assuming that the bark surrounds the *earth*. Lloyd (311-2) makes interesting suggestions on how Anaximander may have construed the "breaking off" of the heavenly circumference.

[68] Aristotle, *Meteor.* 353b6f., quoted after KR 139; the view is attributed to Anaximander by Alexander of Aphrodisias, and this attribution is confirmed by Aëtius (DK, A27), cf. *HGP*, I, 92.

[69] Quoted after *HGP*, I, 100 (DK, A14).

perish into the Boundless through a sort of dehydrated death.[70] Now the notion of an end of the world seems incompatible with the basic postulate of Anaximander's physics: "How does the world pass away, if it forms a self-perpetuating system?" scholars ask perplexed.[71] The question remains disturbing even if we replace the notion of a self-perpetuating system with the more modest and better warranted one of equality of encroachment and reparation. In what follows I shall argue that the solutions hitherto proposed to the problem are unacceptable and I shall suggest a new way of looking at it.

The most popular strategy for escaping from the apparent contradiction has been to postulate that Anaximander regarded the ongoing drying up of the earth as belonging to an extended period of a "great summer" upon which in due course a "great winter", a flood, will follow.[72] This suggestion, which goes back to Cornford,[73] obviously preserves the balance. It involves serious difficulties, however, and I believe that it is wrong. To begin with, it is entirely *ad hoc*, for there is no evidence whatsoever to connect Anaximander with this view which, by contrast, is explicitly attributed to Xenophanes.[74] Further, this interpretation cannot make sense of the statement that the worlds are finally reabsorbed into the Boundless.[75] Lastly and

[70] Cf. Gigon, *Ursprung*, 94; Hölscher, "Anaximander", 300 (= *Anfängliches Fragen*, 29); more hesitatingly *HGP*, I, 100.

[71] Kirk, "Some Problems", 342. Similarly: KR 140 ("this would be a serious betrayal of the principle enunciated in the extant fragment"); *HGP*, I, 101 ("The permanent victory of the hot and dry would obviously disorganize the whole world-order").

[72] Other strategies need not detain us. Vlastos sweeps the problem under the carpet: he never addresses the question why, if "overall justice is preserved throughout the life-process of the world," a reabsorption of the world into the Boundless should be not only possible, but even necessary to produce a "complete and absolute end of all injustice" ("Equality", 82). By contrast, it is the evidence which Kahn, *Anax.*, 185 (with n. 3), prefers to sweep under the carpet: he declares ample evidence to be unreliable and then postulates, now without any evidence, that Anaximander believed in a Magnus Annus. Carl Joachim Classen, "Anaximandros", in Pauly-Wissowa, *Realencyclopädie*, Supplementband 12 (Stuttgart, 1970), cols. 30-69, at col. 59, ll. 57-59, mistakes an admission of ignorance for an explanation: "Da auch das Werden ohne jeden Grund von A. angenommen wird, warum nicht auch ein entsprechendes Vergehen?"

[73] F.M. Cornford, *Principium Sapientiae: The Origins of Greek Philosophical Thought* (Cambridge, 1952), 183-5. This view has been accepted e.g. in: Kirk, "Some Problems", 336f.; KR, 139-40; Hölscher, "Anaximander", 299 (= *Anfängliches Fragen*, 29); Kahn, *Anax.* (see preceding note).

[74] KR 139-40, 177-8.

[75] Cf. *HGP*, I, 101. Guthrie goes on to argue that the notion of a Great Year is incompatible with Anaximander's doctrine because it would be impossible to force back a victorious opposite. This argument does not hold: in Anaximander's cosmology there is no

most important, the view with which Anaximander is generously credited in fact runs counter to the fundamentals of his philosophy. For at the very heart of Anaximander's physics and metaphysics is the idea of an opposition between the Boundless and the world(s) to which it gives rise, between the limitless and the limited.[76] This fundamental opposition dissolves if we attribute to Anaximander the notion of a Great Year which implies that, once it came into existence, the world too, and not only the Boundless, will exist eternally,[77] a view which was in fact held by those who accepted the idea of a cyclical return.[78] Moreover, for a Presocratic the belief that the world is everlasting would immediately imply that it is divine.[79] Yet Anaximander definitely held only the Boundless, and not the world, to be divine.[80] This, in my view, conclusively rules out the possibility that Anaximander believed in a Great Year. Here indeed is the crux of the matter: we have to find a way to reconcile the idea that the opposites are constantly paying justice to each other with the idea of a progressive drying up and ultimate perishing of the world. Or else we must join in Guthrie's candid and resigned conclusion and admit that "we have indeed failed in our interpretation of [Anaximander] and there is little chance of success".[81]

The way out of the dilemma, I suggest, is to hold fast to both its horns and to realise that the problem is conceptual, not textual: the difficulty is

difference of principle between a yearly and a "great" summer or winter, and it is precisely this consideration that led us (n. 32 above) to conclude that *all* periodical changes must be due to the Boundless.

[76] Cf. Gigon, *Ursprung*, 62-8; Hölscher, "Anaximander", 300 (= *Anfängliches Fragen*, 31).

[77] Kirk, without much argument, in fact ascribes this view to Anaximander; cf. KR 122-3. Guthrie (*HGP*, I, 112) rejects this interpretation and emphasizes the contrast between the Boundless and the (or rather: any) world-order. Similarly, Paul Seligman, *The Apeiron of Anaximander* (London, 1962) fittingly characterizes Anaximander as a "dualist" in the sense that "there is a decisive distinction between "principle" and nature, between the everlasting *apeiron* and the things that come-to-be and must pass away" (p. 54f.). Against those who make Anaximander's world eternal he raises the question: "why did Anaximander insist that the ἄπειρον, not the world, was ἀΐδιος if the continued existence of the world was to be the essential import of his doctrine? . . . In Anaximander's scheme, indestructibility is reserved to the divine ἀρχή . . . " (p. 77).

[78] For an overview cf. B.L. Van der Waerden, "Das Grosse Jahr und die ewige Wiederkehr", *Hermes* 80, 1952, 129-55.

[79] Cf. Aristotle, *Physics* 203b14; W. Jaeger, *The Theology of Early Greek Philosophy* (Oxford, 1947), 28-32.

[80] When Anaximander speaks of ἄπειροι οὐρανοί as gods he means the stars, traditionally conceived as divine; cf. *HGP*, I, 112.

[81] *HGP*, I, 112. Such admissions are rare. Guthrie's refusal to espouse pseudo-solutions to real problems cannot but be admired.

inherent in the theory of the opposites itself and does not arise from faulty or missing textual evidence. This can easily be perceived if we reflect that the medical writers faced much the same problem: they had to reconcile their view of health as an equilibrium of constituents with the stubborn fact that people get older and ultimately die; although someone may be healthy throughout his life, he is still continuously sliding toward old age and death. To cope with this problem, Hippocratic writers drew on their notion of the constitution (φύσις) of a person: in different (healthy) persons, the constituents are *differently balanced* and, moreover, in one and the same person the center of the (static) equilibrium is shifting as time passes. We thus hear that "constitutions of men differ: dry constitutions, for instance, are more or less dry as compared with themselves or as compared with one another. Similarly with moist constitutions . . . "[82] This means that, although a person may be in perfect health – i.e. the constituents in perfect equilibrium –, he may still be dry or moist. Further, independently of the idiosyncratic variations of constitutions among persons, the constitution of each individual gradually moves from one extreme to its opposite. The author of *On the Nature of Man* thus holds "that a man is warmest on the first day of his existence and coldest on the last". A more elaborate version, involving a further pair of opposites, is found in *Regimen I* where the child is said to be moist and warm, a young man warm and dry, a man dry and cold and finally old men to be cold and moist.[83] Medical theorists, we see, held both the view that health is a stable balance of the constituents *and* the view that different periods of life are characterized by distinct constitutions. In each person, the equilibrium of the opposites is thus continually *shifting*, without necessarily ever getting imbalanced: people may die (almost) healthy.

That Anaximander should have held an analogous theory in respect to successive ages of the world is what one would expect. For the idea that the earth is irreversibly moving toward death is implicit in his biological-botanical cosmogonical model when applied to the entire life-history of the world: like a living being, the model implies, the world comes into existence from a seed, then gradually passes from one "constitution" (moist)[84] to its opposite (dry), and finally perishes. And yet, at any moment throughout this process the world remains "healthy": i.e. the opposites are always subject to the law of justice. If Anaximander perceived a difficulty in reconciling the idea of a continually changing "constitution" of the world

---

[82] [Hippocrates], *Regimen* 2.67.
[83] *On the Nature of Man* 12 (similarly: *Aphorisms* 1.14); *Regimen* 1.33.
[84] "The seeds of all things have a moist nature": Aristotle, *Metaphysics*, 983b26.

with the postulate that among the opposites justice always prevails, he could console himself with the thought that the very same anomaly beset (or was to beset) medical theories too. Would that he had foreseen that some twenty-five centuries later Thomas S. Kuhn was to proclaim unsolved puzzles to be perennial in science!

A further point may now be stated. Although the *structure* of Anaximander's theory – how it construes change in stability – is analogous to that found in certain Hippocratic writings, there is a difference of *content*: Anaximander associates the perishing of the world with the dry, but the author of *On the Nature of Man* links death rather with the moist. There were indeed two conflicting traditions on this subject. As G.E.R. Lloyd has noted, a number of archaic usages (Homer and Hesiod, notably) suggest that death was (also) conceived as dry, life as moist.[85] The same view underlies an old medical theory, going back at least to Hippon of Samos, which associated life with a *natural moisture*, death with its drying up.[86] This notion was later combined with that of a natural or innate vital heat, a synthesis that is reflected, for example, by Aristotle when he surmises that Thales chose water as his "element" because, among other things, "heat itself [i.e. the principle of life according to Aristotle's own belief] is generated from the moist and kept alive by it".[87] One rationale of the theory of natural moisture (a theory which was to prevail until the overthrow of Galenic-Avicennian medicine and physiology[88]) was precisely that, in contradistinction to the doctrine of health as equilibrium, it offered a plausible account of gradual ageing and of death. When the natural moisture becomes dried up or otherwise corrupted, life is extinguished:[89] the theory thereby implied the notion of a "natural span of life", the maximum age an individual can attain if he leads a hygienic life.[90] It immediately transpires

[85] Lloyd, "Hot and Cold", 271-2.

[86] Thomas S. Hall, "Life, Death and the Radical Moisture", *Clio Medica* 6, 1971, 3-23, at p. 6.

[87] *Metaphysics* 983b23. The identity of the biological theory involved here with Anaximander's meteorological theory has been perceived by W. Jaeger in his *Aristotle* (Oxford, 2nd ed. 1948), 150, n. 3.

[88] Hall, "Life, Death . . . ".

[89] In the later versions of the theory, life was taken to depend on natural moisture *via* the vital heat; cf. the following note.

[90] Cf. e.g. Aristotle, *On Respiration* 479a15f. where we are told that in old age only little vital heat remains in the body "for most of it has been breathed away in the long period of life preceding". Hence: "natural death is the exhaustion of the heat owing to lapse of time, and occurring at the end of life" (*ibid.*, 479a33f.). Natural death is therefore "involved from the beginning in the constitution of the organ, and not an affection derived from a foreign source" (*ibid.*, 478b24f.). Quotations are from the Oxford

that Anaximander's belief that life has its beginning in moisture[91] and his idea of a dehydrated death of the world are in close agreement with the theory of natural moisture. This interpretation is corroborated by a small, albeit significant, point of detail. Augustine reports that Anaximander held that the worlds are dissolved "according to the age to which each is able to survive":[92] this idea, it seems, unmistakably echoes the notion of "natural span of life". It is also noteworthy that the *irregularity* which is suggested by Augustine[93] is indeed implied by the biological notion, but does not fit into Anaximander's general scheme of regular periodic phenomena established by the "ordering of time".

In sum, then, it appears that the much debated question: "how does the idea that the world perishes square with the tenet that justice prevails?" receives a surprisingly simple answer. In all probability Anaximander held that the one *complements*, rather than contradicts, the other: justice continually characterizes the relationship among the opposites as long as the world exists, and is not at all violated by a *natural* death, which, in Aristotle's words, is "not an affection derived from a foreign source".[94] The world perishes once it has reached the end of its natural life span (depending on the quantity of primeval moisture). Unlike cyclical (daily and yearly) meteorological phenomena, this is a distinctively irreversible process: this, I believe, is also the idea behind Alcmaeon's saying that men die because "they cannot join the beginning to the end".[95] *Death therefore ensues although the opposites have continuously followed the law of justice.* In Anaximander's view, I submit, *the death of the world is the end of any encroachment and reparation, a final standstill and end of change* within that

translation by G.R.T. Ross (Oxford, 1908). For some interesting remarks concerning death in Aristotle cf. Stephen R.L. Clark, *Aristotle's Man* (Oxford, 1975), 164-73. In Islamic and, to a lesser extent, Jewish mediaeval theology the notion of a "natural life span" became involved in the debate over *ajal*, viz. whether the duration of life of a person is determined by God at birth, or depends on behavior, chance factors, etc. On this, as well as on Galen's and Avicenna's views of death, cf. Gotthold Weil, *Maimonides über die Lebensdauer* (Basel, 1953); I have used the Hebrew translation by M. Schwarz (Tel-Aviv, 1979) to which the translator has added many supplementary notes.

[91] Cf. n. 48 above.
[92] Quoted after KR 125 (DK, A17).
[93] Kirk, who draws attention to this note of irregularity, maintains that it is "foreign to the idea of a sequence of single worlds, but . . . is essential to the atomistic conception" (KR 125). I hope to have shown why this irregularity is equally compatible with Anaximander's medical-physiological notion of death.
[94] Cf. above n. 90.
[95] DK, 24B2; quoted after KR 235.

world. Contrary to what most scholars hold, then, death is not a "supreme victory" of one of the opposites and consequently constitutes no encroachment. The tenet of justice as formulated in the fragment, it follows, does not at all bear on the question of the end of the world.

How should we now interpret Aëtius' report according to which Anaximander held that the worlds are "dissolved into that out of which they came"? Perhaps we may surmise that it was again biological ideas that guided Anaximander. At death, living beings disintegrate into that from which they came – dust. This commonsense notion certainly went into the natural moisture theory which indeed involves the tenet that a complete drying up implies death – the loss of form, disintegration. "Earth has no power of cohesion," Aristotle was to say; "the moist is what holds it together. For it would fall to pieces if the moist were eliminated from it completely."[96] Anaximander may thus have thought of the world's dry death as implying a total disintegration into dust. Therefore, just as a corpse is reabsorbed by the earth, so the disintegrated substance of the dead world, retransformed into its initial, utterly amorphous "indeterminate" state, fades again into the Boundless.

Let me lastly hazard a remark concerning the issue of "innumerable worlds". The argument against ascribing to Anaximander the notion of innumerable worlds coexisting in space depends on two premises: (i) "It is clear that after Epicurus had popularized the atomic doctrine of innumerable worlds in infinite space there was a tendency to read this view back into all earlier physical theory . . . Statements which refer unambiguously to innumerable coexistent worlds . . . arose from a confusion between οὐρανοί and κοσμοί. This confusion is not due to Theophrastus, but had its origin in assumptions natural enough once the Epicureans had made generally familiar the atomists' doctrine of innumerable worlds arising haphazard at different points in infinite space."[97] (ii) Anaximander himself *could* not have held the same view: "the belief of the atomists in innumerable worlds was closely reasoned from their ideas about the nature of body and of space. Not only is there no evidence for these ideas in Anaximander, but one may say with confidence that a clear philosophic distinction between body and empty space was not made before the fifth century . . . [Anaximander's] ἄπειρον is not empty space but body . . . "[98] Now *i* seems fairly plausible, i.e. it certainly may have been the case that atomistic

---

[96] Aristotle, *De gen. et corr.* 335a2f. (Harold H. Joachim's translation).
[97] *HGP*, I, 108, 111.
[98] *Ibid.*, 113-4.

ideas colored reports on earlier thinkers. By contrast, *ii* comes alarmingly close to affirming the consequent.[99] To be sure, underlying the atomists' idea of infinite world-orders arising haphazardly was the notion of infinite empty space. Yet, and here is the crucial point, *Anaximander's own, quite different premises naturally implied an identical view.* For Anaximander's Boundless is devoid of any particularizing marks – spatial, temporal, or others: in it, different spatio-temporal points are indistinguishable (as are the opposites). Now we know that according to Anaximander the world was – or rather: any world is – generated through a seed which issues from the Boundless. But from the Boundless' vantage point, we should now realize, there is no reason why a world-seed should be produced at one spatio-temporal point rather than at any other.[100] On Anaximander's premises, the generation of a world is an essentially indeterminate event. This implies that just as our world was produced, other worlds may – indeed: must – have been produced and will be produced elsewhere in the Boundless.[101] Anaximander's Boundless, I suggest, may be likened to a glass of soda water throughout which world-bubbles ceaselessly form, hover and then perish. Or rather to a huge, seasonless (three-dimensional) matrix in which seeds are sown stochastically so that at every moment plants at all ages and stages of development coexist. This image fits strikingly well Simplicius' report that Anaximander (just like the atomists) believed worlds "to be coming-to-be and passing away for an infinite time, with some of them *always* coming to be and others passing away".[102] The assumption therefore seems plausible to me that Anaximander's biological-botanical cosmogony guided him to a view according to which worlds eternally arise haphazard throughout the Boundless:[103] at any moment a multitude (or even an

---

[99] Also: "If we find evidence that Theophrastus treated Anaximander's worlds as both coexistent and successive, this will suggest strongly that he was applying atomistic reasoning to Anaximander" KR 124. Cf. also the entirely *a priori* arguments against the possibility of ascribing to Anaximander the idea of successive worlds in Kirk, "Some Problems", 335; similarly Kahn, *Anax.*, 50-51.

[100] The situation was strikingly analogous for mediaeval philosophers who had a hard time reconciling the idea of an eternal and unchanging God (who cannot be supposed to have fits of will) with the tenet of a unique, temporal creation of the world.

[101] Aristotle, *Physics* 203b23ff. (cf. KR 123-4) uses an analogous argument, except that he proceeds on the assumption of an infinite void. Whether or not Anaximander held the Boundless to be spatially infinite does not bear on the question we are discussing now (cf. below n. 104).

[102] Similarly Augustine's report: "those worlds, he thought, are now dissolved, now born again, according to the age to which each is able to survive". Both quotations from KR 124 (DK, A17); cf. above n. 93.

infinity) of worlds coexist and the process of coming-to-be and passing away of the worlds continues eternally. Anaximander and the atomists arrived at essentially identical views,[104] although their routes to them were rather different.

VI. Conclusion

In the preceding pages I have suggested new interpretations of some of Anaximander's doctrines: the cosmogony; the tenet of the earth's immovability; the idea of a "just" world gradually moving toward a dry death. From these interpretations Anaximander emerges for the first time as a *consistent* thinker in both his cosmogony and his cosmology. We may

[103] Gigon, *Ursprung*, 66-7, in a minority view, ascribes to Anaximander the belief in innumerable coexisting and successive worlds. Apparently little impressed by Cornford's British empiricism (there is "nothing in the appearance of the natural world" to suggest innumerable coexistent worlds": *Principium Sapientiae*, 177), he takes Anaximander to have made a "purely speculative inference": "Sein Ausgangspunkt ist . . . ausschließlich der Gedanke gewesen, dass nur dann die Dinge im Gleichgewicht seien, wenn dem Einen Unbegrenzten grenzenlos viel Begrenztes gegenüberstünde."

[104] There may be one difference, however: it is not quite clear whether or not Anaximander took the Boundless to be infinitely extended, not only in time but in space too. (Cf. *HGP*, I, 85; KR, 109; Kahn, *Anax*. 233; Barnes, *Presocratic Phil.*, I, 28-37.) Therefore we cannot be certain that Anaximander believed in an infinity, and not just in a multitude, of coexisting worlds. This point, reports on which could indeed easily have become atomistically colored, is of relatively minor importance however. What matters is that at any moment numerous worlds coexist and that there is no reason to ascribe to Anaximander the notion of a succession of single worlds (cf. *HGP*, I, 111). Indeed this notion has little to recommend it: what, precisely, does it mean that the worlds *succeed* one another? Why and how should the perishing of the world be related to the generation of the next one by the Boundless? Recently Asmis ("What is Anaximander's *Apeiron?*") has gone as far as to *identify* the infinite succession of single worlds with the Boundless itself. This is, strictly, meaningless. (What *is* the Boundless during the "long (on this interpretation: temporal) intervals" which, according to Cicero, separate the worlds?) It looks as if those who ascribe to Anaximander a belief in a succession of single worlds somehow unwittingly assume that there is a *causal* link between the perishing of one world and the generation of the next. On Anaximander's premises, such an assumption is, however, entirely unwarranted. One wonders whether those who interpret Anaximander along these lines, attributing to him the relatively uninspiring notion of a single world at a time, suspended at the center of a limited, presumably spherical Boundless, have not themselves projected later – namely Aristotelian – ideas into him. That the study of the Presocratics is prone to such systematic biassing is cogently argued in Denis O'Brien, *Theories of Weight in the Ancient World*. Vol. I: *Democritus Weight and Size* (Paris/Leiden, 1981), xiii-xx.

therefore wholly subscribe to the conclusion reached by Nicholas Rescher (albeit on the basis of his own, entirely misguided interpretation): "Anaximander in the sixth century B.C. possessed a conception of the development of the universe in the sequence of natural occurrences and the working-out of natural processes that is more detailed, more scientific, and partly more sophisticated than has usually been credited", and he should therefore be considered as "the founder of scientific cosmology". In particular, we owe to Anaximander the idea (and first example) of an explanation founded on a *model*.[105] My principal target, however, has been the theory of the opposites: *that* justice prevails, I argued, is one thing; *how* it is sustained is another. We have in fact seen that in terms of the theory of the opposites, only a static, but not a dynamic, equilibrium can be immanent in nature. In other words: where movement and change occur, the observed equilibrium must be sustained by something external: for Anaximander the guarantor of order and permanence was the Boundless.

The theory of the opposites occupied a central place in Peripatetic and medical thought for no less than two millennia. Aristotle provided the opposites warm-cold and moist-dry with a substratum, transforming them from "powers" into "qualities" of the elements and the mixed bodies, and his physics became the framework for the mediaeval theory of matter. It may therefore prove interesting to raise the following question: what is it that has now assumed the function which Anaximander had attributed to the Boundless? What, in other words, sustains the equilibrium between the Aristotelian opposite qualities which, left to themselves, we surmise, would produce "not cosmos but chaos"?

Mediaeval philosophy had a ready answer for this question, one founded on the enigmatic Aristotelian notion of *Active Intellect* which, in an interpretation ultimately deriving from Alexander of Aphrodisias, Plotinus, and Themistius, was supposed to endow sublunary substances with *forms*: it was generally held that the Active Intellect both informs matter and maintains the forms, thus preventing the substances from disintegrating under the opposite action of their constituents.[106] Consider, as one among many,

---

[105] N. Rescher, "Cosmic Evolution in Anaximander" (1959), in his *Essays in Philosophical Analysis* (Pittsburgh, 1969), 3-32, on pp. 28-30.

[106] For a most thorough treatment of this neglected subject cf. Herbert A. Davidson, "Alfarabi and Avicenna on the Active Intellect", *Viator* 3, 1972, 109-178. Cf. also Richard Walzer, "Aristotle's Active Intellect: Nous Poietikos in Greek and Early Islamic Philosophy", *Atti del convegno internazionale sul tema: Plotino e il Neoplatonismo in Oriente e in Occidente* (Rome, 1974) (= *Problemi attuali di scienza e di cultura* 198), 423-36; Charles Touati, "Les problèmes de la génération et le rôle de l'Intellect agent chez

the following version of the theory, from the pen of the fourteenth century Provence Jewish philosopher, Levi ben Gershom (Gersonides):

> Down here there are opposite elements, and it is in the nature of opposites to destroy one another, thus destroying whatever is combined out of them. It necessarily follows [sc. since such destruction does not occur] that down here there are some causes which bring about and preserve the existence of the elements and of whatever comes to be from them . . . In fact, the preservation of the sublunary realm depends on the equilibrium of the constituting elements. The cause of this equilibrium is the action reaching the elements from the celestial bodies . . . The celestial bodies are found to be of the utmost possible perfection in view of the providence they exercise and the excellence they confer upon things down here. Indeed it is through them that the sublunary realm preserves whatever it has of the good and the perfect . . . For it is they which preserve, as perfectly as possible, the opposite elements on which depends the permanence of all composite bodies. It is they, moreover, which preserve, as long as possible, the elemental heat in the living beings. Indeed, if the action of the celestial bodies reaching the sublunary things were to cease for even a tiny moment, then the good and perfect in them would be wanting; similarly, no life would remain to any living being.[107]

The message is clear: the relative stability of animate and inanimate sublunary substances cannot be accounted for on the premises of Peripatetic physics alone. That substances exist and last rather depends on something external, namely (as we learn elsewhere) on the separate intellects whose action is mediated by the celestial bodies. (Here the mediaeval philosopher would introduce an argument from design.[108]) The continuity is striking: the Active Intellect does for the mediaeval philosopher a job comparable to the one the Boundless had done for Anaximander no less than twenty centuries earlier. In fact, the very premises of the theory of the opposites implied that stability and permanence depend on an extrinsic cause, that the sublunary world cannot be construed as a closed system. To take cognizance of order, to account for the world's being cosmos and not chaos, the theory of the opposites *had* to be backed up by metaphysics. Anaximander's Boundless and the Active Intellect are in the same – divine – boat. Anaximander and

Averroès", in *Multiple Averroès. Actes du colloque international . . . Paris 20-23 septembre 1976* (Paris, 1978), 157-164.

[107] Levi ben Gershom (Gersonides), *Sefer Milḥamot ha-Shem* Bk IV, Chs. 3, 6 (Riva di Trento, 1560): p. 27ᵛ, col. a, ll. 2-5, 9-12; p. 28ᵛ, col. b. ll. 43-51; (Leipzig, 1866): pp. 161, 170 respectively; my translation.

[108] On Levi ben Gershom's particular version of this physico-theological theory cf. Gad Freudenthal, "Cosmogonie et physique chez Gersonide", *Revue des études juives*, 145, 1986.

the Peripatetics alike did not frame the idea of an immanent physical order which need not be imposed and sustained from outside by an eternal unchanging divinity.

(AL-)CHEMICAL FOUNDATIONS FOR COSMOLOGICAL IDEAS: IBN SÎNÂ ON THE GEOLOGY OF AN ETERNAL WORLD

I. INTRODUCTION

Historically, the foremost problem of cosmology is arguably that of cosmo*gony*. Immanuel Kant still considers the question whether the world had a beginning in time to provide one of the four antinomies of pure reason.[1] The problem came to a head with Aristotle. In the ancient Mediterranean societies, the idea that the universe came to be after it had not been was never questioned, of course.[2] Also the Milesian philosophers, followed by Plato, took the coming-to-be of the universe to call for an account. The radically new view according to which the actually existing world may not at all have had a temporal beginning was framed by Aristotle. Not only the underlying matter of the world, he argued, but the world as it *is*, i.e., with its very structure, has existed since all times: the heavens and all *forms* in the sublunary world, notably the species of plants and animals, are eternal. This set going a heated debate that was to continue for more than two millennia.[3]

The upholders of eternity were notably Aristotelians, later joined by the Neoplatonists; the defenders of the thesis of the timely origin of the world were at first Stoic and Atomist philosophers, later those who upheld creationist cosmogonies, which they wanted to be in conformity with both the principles of natural philosophy and the Scriptures. The pros and cons were derived from almost every domain of thought — from the most abstract theological and metaphysical speculations and analyses (e.g. of the nature of God, time, motion, infinity) down to physical considerations drawing on empirical evidence. This paper will be concerned with the latter. We will see how, beginning with Plato and the Stoics, adversaries of the eternity thesis argued that certain observable geological phenomena — namely those processes we today call 'erosion' — cannot be reconciled with the view that the world has existed since ever: erosion, they reasoned, is a unidirectional process and if it had been at work since an infinite time, all accretions on the surface of the earth would have been planed down long ago. In other terms: the observable existence of mountains empirically refutes the eternity postulate. But theories, we know, can all too easily be accommodated even to the apparently most adverse evidence. In fact, the upholders of eternity, beginning with Aristotle and Theophrastus, were not slow to retort: by postulating *generative* geological processes compensating for erosion, they sought to defuse their opponents' criticism.

For many centuries, however, Aristotelians — these are the empirically-minded proponents of the eternity thesis with whom we will mainly be concerned here — were unable to say what these theoretically-postulated generative geological processes were: their theory of matter could not provide an account of how the four sublunary elements *cohere* to form stones and mountains. The difficulty was this. According to the received Aristotelian theoretical notions, cohesion was brought about by the moist component of substances: a clump of matter from which all moisture was eliminated, crumbled into dust. But stones, in as much as they are solid, were construed as 'dry.' How, then, do they hold together? Indeed, a simple empirical fact such as the hardening of clay through heating could not straightforwardly be integrated within the Aristotelian theory of matter. *A fortiori*, Aristotelians were at a loss to explain the formation of stones or mountains. Thus, within Peripatetic philosophy, an account of the formation of mountains remained a desideratum with crucially important metaphysical and theological implications.

This is the context, I will suggest, in which we have to place the petrological and orogenic theory put forward by Ibn Sînâ (Avicenna) at the beginning of the eleventh century. Ibn Sînâ adduced an innovative geology, founded, I will show, on notions deriving from chemistry and alchemy. Specifically, Ibn Sînâ draws on the theoretical notion of *unctuous moisture* — a non-evaporable moisture —, a notion whose origin goes back to the fifth century B.C., but which came to prominence only in the wake of the widespread use of fractional distillation by Arab (al-)chemists. From chemistry, the notion passed to natural philosophy and by the tenth century it had become a well-entrenched theoretical concept with great explanatory import. By definition, unctuous moisture is a moisture capable of conferring cohesion and Ibn Sînâ founds on this notion his account of how stones and mountains can be formed through desiccation.

II. GEOLOGY AND THE ETERNITY OF THE WORLD[4]

The existence of continents and islands separated by masses of water posed a double challenge to Aristotle's postulate that the past of the world extends to infinity: (i) the first has its point of departure in the Aristotelian notion of natural motion, i.e., motion toward the natural place. Aristotle himself tells us that 'some people' — this seems to refer to Plato[5] — have argued that if the world were eternal, each of the four elements should long ago have reached its natural place; the sublunary world should then have consisted of four perfect, immobile concentric spheres of earth, water, air, and fire. The existence of elevated land, reaching out into the spheres of water and air seems to disprove this conclusion. (ii) The second challenge is that, if one accepts, with Aristotle, the idea of exhalations raised by the sun's heat,[6] one would expect, with certain Presocratics, the sea to dry up constantly;[7] hence, if it had existed eternally, it would already have dried up completely.

Aristotle's answer to both arguments is essentially the same. Eternal movement, Aristotle argues, cannot but be *cyclical*. The paradigmatic instance, to be sure, is the movement of the heavenly bodies,[8] but the principle holds of change in the sublunary world too. Thus, in response to the first argument Aristotle adduces the idea that the sun, owing to its double movement (daily and seasonal), causes a constant transformation of the four elements into one another and, moreover, brings about regular generation and corruption of the sublunary substances.[9] Hence, generation and corruption can go on, and indeed has and will go on, eternally. The laws of nature do not imply that in an indefinite time the world must end up in a static state.

Against the second argument, Aristotle makes the point — in fact a corollary of the former — that evaporation is compensated by precipitation: this cycle is eternal too and thus evaporation does not lead up to a static final state either. Specifically, contrary to what the Presocratics had believed, the drying up of the sea is not a unidirectional process, so that the postulate of the eternity of the world does not conflict with the continued existence of seas.[10] Rather, this stretch of land may become submerged in the sea and that part of the sea may dry up, but the process will always follow an order and be *balanced*: land and sea constantly change their respective places, but both are eternal and maintain their original equilibrium.[11] Indeed, the relative quantities of the elements in the world are constant.[12] Now it was an easy task for

Aristotle to account for the submerging of land in water, since he accepted the widespread notion of occasional catastrophic deluges. (This idea, incidentally, also accounts for the fact that culture is of young age, although mankind, as all other species, must be eternal.[13]) The other side of the equation was more difficult to come by: Aristotle at one place[14] suggests that the drying of sea into land may be part of an ageing process (indeed, in Aristotelian terms ageing involves getting dryer[15]) — but does not go into any detail. He nowhere explains concretely how new land is formed in compensation for land that turns into sea.

Theophrastus, Aristotle's pupil and successor as the head of the Lyceum, was confronted with much the same problem. In fact, in the meanwhile the geological argument against eternity had been forcefully and systematically developed by Zenon of Citium, the founder of Stoicism. The Stocis, as is well known, believed that, although the underlying 'matter' of the world is eternal, the actually existing world is perishable: in a succession of 'world periods,' different worlds come into existence and are annihilated again in a conflagration (*ekpyrosis*).[16] This cosmogony supplied a conceptual framework within which unidirectional geological processes could be easily construed. One Stoic argument against the eternity of the world, as reported by Theophrastus (and preserved by Philo), is of particular interest to us:

If the earth had no beginning in which it came into being, no part of it would still be seen to be elevated above the rest. The mountains would now all be quite low, the hills all on a level with the plain, for with the great rains pouring down from everlasting each year, objects elevated to a height would naturally in some cases have been broken off by winter storms, in others would have subsided into a loose condition and would all of them have been completely planed down. As it is, the constant unevennesses and the great multitude of mountains with their vast heights soaring to heaven are indications that the earth is not from everlasting.[17]

The Stoics, we see, clearly recognized the class of phenomena we call today 'erosion' as constituting a unidirectional geological process. They invoked this process in order to prove that the world could not be eternal.[18] The upholders of the opposite view had therefore to show how the existence of continual destructive processes such as erosion can be reconciled with the postulate of the eternity of the world.

Theophrastus himself indeed sought to counter the Stoic arguments. His rebuttals are of particular interest to us, for they highlight the difficulties inherent in the Aristotelian position. In line with his general metaphysical conception of the world as an organic whole, Theophrastus claims that 'trees and mountains differ not in nature' and just as 'the trees shed their leaves at some seasons and then bloom again at others,' so also 'the mountains, too, have parts broken off but others come as accretions.'[19] In other words: erosion is compensated by the formation of mountains. But how are mountains formed? Theophrastus in fact goes beyond Aristotle and describes a sort of 'volcanic' theory, according to which the element fire, rising upward from the bottom of the earth, 'pulls up with it a large quantity of the earthy stuff,' thus giving rise to a mountain.

Theophrastus' theory, it seems, remained unsatisfactory even to its author. Indeed, although the theory was adduced as giving an account of how the formation of mountains counterbalances erosion, Theophrastus then takes it to *explain away* erosion altogether: fire and earth, he claims are so firmly held together that mountains are not at all destroyed.[20] Perhaps Theophrastus is here unwittingly revealing the soft point of his account: the premises of Aristotelian theory of matter (or 'chemistry') do not at all warrant the claim that fire and earth may hold together, nor, consequently, the idea that the rising fire can pull behind it earthy matter (except as an exhalation). We will in fact see in some detail below that in Aristotelian terms, natural substances require *moisture* in order to cohere. Earth and fire, therefore, cannot form a stone or a mountain. Theophrastus was perhaps not unaware of the difficulty: the subsuming of the formation of mountains under the biological analogy, or metaphor, which nothing links to the would-be physical account, may have been meant to secure the idea of a cyclical formation of land and sea on a conveniently abstract level, for want of something better. At any rate, Theophrastus' account did not find acceptance with subsequent natural philosophers.

A theory of the formation of mountains thus remained an important desideratum in the framework of Peripatetic physics and metaphysics: erosion was an easily observable process, but one rarely witnessed the formation of a mountain. The existence of constructive processes, counterbalancing the destructive ones, was a corollary of the metaphysics of the eternal world, but Peripatetics remained unable to specify their precise nature: why should mountains behave like trees and 'bloom' again after having been eroded? What — in Aristotelian terms — is the efficient cause of mountains? The question was crucial, for the postulate of the eternity of the world hinged on it. Indeed, as

52

will be seen, it came to the fore as soon as creationist philosophies began to attack Aristotelian metaphysics with the tools of its own physics.

The Stoics themselves, at least the later ones, tried to provide a positive account of the existence of the continents. Their main explanandum was this: Why is the surface of the earth uneven? In fact, were it not for the mountains, the sea would have covered the totality of the earth, making life impossible for man and most animals and plants. In answer to this, the Stoics maintained that existence of continents is due to Providence (*Pronoia*). Strabo, in a passage probably deriving from Posidonius, takes care to emphasize that if nature alone were at work, no continents would ever have existed:

The work of Nature is this, that all things converge to one thing, the centre of the whole, and form a sphere around this; and the densest and most central thing is the earth, and the thing that is less so and next in order after it is the water . . .

Therefore, Providence must be involved too:

But since water surrounds the earth, and man is not an aquatic animal, but a land animal that needs air and requires much light, Providence has made numerous elevations and hollows on the earth, so that the whole, or the most, of the water is received in the hollows, hiding the earth beneath it, and the earth projects in the elevations . . .[21]

As it stands, this account leaves however unexplained the changes taking place on the surface of the earth, changes which the Stoics were the first to underscore. If, in fact, erosion — i.e. nature — planes down rocks and mountains, may it not eventually nullify the work of Providence? Moreover, can the tenet that the continents are due to Providence be reconciled with the numerous observations — pertaining, e.g., to fossils — indicating that sea and land have occasionally changed their respective places? The Stoics sought to resolve the difficulty by the idea of continued *reciprocal* changes between sea and land, earth and water:

We must take it for granted, first, that the earth is not always so constant that it is always of this or that size, adding nothing to itself nor subtracting anything, and, secondly, that the water is not, and, thirdly, that neither of the two keeps the same fixed place, especially since the reciprocal change of one into the other is most natural and very near at hand; and also that much of the earth changes into water, and many of the waters become dry land . . . Why, then, is it marvellous if some parts of the earth which are at present inhabited were covered with sea in earlier times, and if what are now seas were inhabited in earlier times?[22]

Thus the Stoics were confronted with a problem not unlike the one with which they themselves confronted the Peripatetics: they disposed of the problem of the initial formation of the continents by attributing it to Providence; yet in order to account for the changes and the continued existence of the continents they too had to show how water can become land. Thus, an account of the efficient casue of mountains became a desideratum for both Peripatetic and Stoic philosophies.

The two Stoic geological arguments we have considered were eagerly seized upon by Arab upholders of a creationist cosmogony. The Ikhwân al-Ṣafâ', for example (joined by many others such as al-Bîrûnî and al-Ghazālî),[23] reproduced the Stoic argument for Providence: they argued that were it not for the mountains, water would have covered the entire surface of the earth; the world would have consisted of four concentric spheres of the elements, and man could not exist. The continents, therefore, were created by God, and they testify of His kindness.[24] Yet the Ikhwân recognized that both the land and the sea undergo perpetual changes. They maintained that these changes are cyclical and attributed them to the 36 000 years long revolution of the fixed stars, which modifies the conditions of heat and cold in the different parts of the world. The Ikhwân now had to show how changes in temperature transform land into sea and vice versa. The first half was relatively easy: when mountains are heated, they argued, their moisture evaporates and the remaining dry substance crumbles. The second half of the demonstration, however, remained a problem: the Ikhwân maintained that the rivers convey sand to the sea, at the bottom of which it is 'baked' into the hills and mountains. Yet the precise nature of this 'baking' was not specified, nor was it explained why the mountains should rise over the seas in which they were formed. The de-deficiency of the Ikhwân's account thus again highlights the difficulty of providing an adequate account of the formation of mountains.

Some more 'orthodox' authors adopted the position that no changes whatsoever in the respective positions of sea and land have taken place since the creation. Thus, the Arab author of the book known in Latin as *De elementis* argued on a double front. Against the Ikhwân al-Ṣafâ' he correctly pointed out that the supposed 36 000 years cycle is by far too short: on their hypothesis the sea would have had to shift by one degree in a century, yet 'history teaches us that a great number of cities have been located for many centuries at the same distance from the sea.'[25] He then directs his criticism against those who, on Aristotelian premises, tried to account for the existence of the continents on purely physical grounds. 'Some philosophers,' the author of *De elementis* reports,

claim that when the earth was formed it was perfectly round, without valleys or mountains. Its shape was then precisely spherical, like that of the heavenly bodies. Those valleys and mountains that we see on the surface of the earth are due to no other cause than the action of the waters. The waters hollowed out the less compact parts of the soil, and so the mountains were formed. These less compact regions, once hollowed out, became the places of the seas.

Aristotelian philosophers, we thus learn, sought to turn the argument from erosion in their favor. Yet is was an easy task to refute this argument: on the mentioned Aristotelian premises, the sublunary world initially consisted of four concentric spheres of the elements. Therefore:

Suppose that at the beginning the earth were a body perfectly spherical and smooth, without any valley or mountain. The terrestrial mass was then necessarily covered up entirely with a uniform layer of the mass of waters. Then, however, the rain falling from the upper regions of the air fell on the layer of water covering the earth.[26]

Evidently, then, this rain could produce no erosion, much less lead up to the formation of mountains.

Some Aristotelians, lastly, ignored the orogenic problem altogether by treating the cosmological issue on a conveniently abstract level. This is notably the case of Ibn Rushd. In the section of his *Epitome of [Aristotle's] De generatione et corruptione* corresponding to *De gen. et corr.* II, 10, Ibn Rushd reviews the range of sublunar phenomena in which the heavenly bodies play a part. The stars, Ibn Rushd, argues, are eternal and so are the processes of generation and corruption depending on them, notably the very fact that there is life on earth. Now a condition for the continued, eternal, existence of the living species is the existence of dry places. 'It is evident,' Ibn Rushd says, 'that it is the celestial bodies that will continue to preserve this [kind of dry] place in species. Otherwise water would prevail over it, for the natural thing for earth *qua* heavy is that it be submerged in all its parts under water, since it has already been proved that that is its appropriate limit. Thus it is apparent that this function of the stars and especially of the sun is an essential one.'[27] Ibn Rushd contents himself with the assurance that sea and dry land will continue 'in species' — he does not seem to be interested in the down-to-earth, concrete, geological processes some of his predecessors had considered.[28]

Everyone, in sum, except those who simply maintained that the world has subsisted changeless since creation, needed a theory of the formation of mountains: the Aristotelian *falâsifah* and their opponents who denied the eternity of the world agreed that the changes in the respective places of sea and land were reciprocal. While erosion and subsequent overflooding easily accounted for the transformation of land into sea, the converse change was generally postulated without its efficient cause being indicated.

III. IBN SÎNÂ: TOWARD A THEORY OF THE FORMATION OF STONES AND MOUNTAINS

A philosopher and a physician, Ibn Sînâ (980—1037) touched both upon the summits of metaphysical speculation and the most tangible particular natural phenomena. We will now see how Ibn Sînâ, drawing on contemporary chemical ideas, framed an innovative theory of the formation of stones and mountains in order to secure the Aristotelian tenet of the eternity of the world. We may observe in passing that within the context of Arab philosophy this tenet changed somewhat its significance. Indeed, Aristotelians writing in Arabic, foremost among them al-Fârâbî and Ibn Sînâ, incorporated into their metaphysics significant Neoplatonic elements, notably the idea that all that exists proceeds from the first Being through emanation. Thus, as is well known, Ibn Sînâ held that there is one Necessary Existent, God, who is the necessitating cause of all existents in the world. Since God is eternal, this stance implied that the entire universe — immaterial (intellects) and material (celestial and sublunar matter) — is necessarily eternal too. The Neoplatonically-colored Peripatetics thus went beyond Aristotle in making the coming-to-be of the world and its eternal existence into a necessary corollary of their metaphysics.

In the *Dânesh-namé*, his encyclopedic work written in Persian, Ibn Sînâ clearly indicates that the existence of land poses a serious problem to his general metaphysical scheme.[29] Following Aristotle, Ibn Sînâ defines the *place* of a body as the inner limit of the body containing it,[30] and moves on to elaborate a conception of natural place which slightly deviates from that of the Stagirite: the natural place of fire is the inner surface of the firmament; those of air, water and earth are the inner surfaces of the spheres of fire, air, and water, respectively. In fact, Ibn Sînâ explains, since all simple bodies necessarily have a single natural movement, it follows that each of them has only one natural place. A second consequence is that the spatial arrangement of the elements is spherical: it is impossible, Ibn Sînâ argues, that one and the same nature (form) should produce different figures. This reasoning implies that the entire sublunar world necessarily — and therefore *eternally* — forms one body constituted of four contiguous concentric spheres. This account, founded as it is on general metaphysical considerations involving necessity, raises with particular acuity the question why earth and water visibly do not form eternal static and concentric spheres. In answer, Ibn Sînâ says that 'the cause for the water not covering the entire surface of the earth is that water becomes earth, and vice versa. Where earth becomes something else than itself a cavity ensues; and where something else other than earth becomes earth, there is an accretion.' In the *Dânesh-namé* Ibn Sînâ does not elaborate these general assertions: he does not specify how something that is not earth can become earth. This missing link is supplied by Ibn Sînâ's geology, for which we must look elsewhere, namely in Ibn Sînâ's comprehensive *Kitâb al-Shifâ'*.

We have a succinct summary of Ibn Sînâ's views from the pen of Shmuel b. Yehudah Ibn Tibbon, the well-known 13th century Jewish translator and philosopher. Shmuel Ibn Tibbon is best known for his translation into Hebrew of Maimonides' *The Guide of the Perplexed*, but he was also interested in subjects directly relevant to us here, as can be seen from the fact that he translated and commented on Aristotle's *Meteorologica*.[31] His own work, *Ma'amar Yiqqawu ha-Mayim* (i.e., 'Let the Water be Gathered'; *cf. Genesis* 1:9), written not long after 1221,[32] takes as its point of departure precisely the problem we are concerned with: one of his erudite friends, the author tells us, has asked him to find out what the philosophers say of the fact that the element water does not surround the earth and does not cover it entirely.[33] After a brief survey of the opinions of 'the majority of the [Aristotelian] philosophers,' including Ibn Rushd, Ibn Tibbon gives a short account of what Ibn Sînâ — 'also a follower of Aristotle' — says of the problem 'in his great book called *Kitâb al-Shifâ'*, in its part devoted to natural science, in the section on meteorology.' Ibn Tibbon's expostion is so lucid and insightful (in addition to being faithful[34]) that it is worth translating *in extenso*:

Ibn Sînâ affirms the following: From what we have said previously it is clear that the nature of water and [that] of earth imply that the water should be above the earth, the earth in the midst of the water, so that the water would surround the earth from all sides. But existence [i.e., reality; *mezi'ût*] is not so. Rather, existence is according to what is necessitated by the global order.[35] Indeed, since it is the nature of the elements that parts of them change into one another, it is impossible for the earth to persevere in its natural state [*ha-ᶜinyan ha-tivᶜi la*], for it is in the nature of the earth that parts of it change into water or other elements; similarly, the other elements also change into earth. Now whatever earth chagnes into another element will be subtracted from the global body of the earth. Necessarily, therefore, there will be a defectiveness in the sphericity and the depth [of the earth], for earth is dry, and cannot aggregate so as to regain its natural [i.e., spherical] figure [*tekhûnah*; lit.: property]; rather, it will preserve its acquired figure, which is not natural to it. Again, whatever other elements will be transformed into earth and come down on it, will doubtless become an accretion and an eminence added upon it: indeed, unlike solidified [i.e., frozen] water, which, when poured [sic!] onto other [i.e., liquid] water, forms with it a single spherical body, it will not spread on [the globe of the earth]. It thus follows necessarily that on the globe of the earth crags, depths (?), and hills are formed.

The action of the stars is involved in making necessary this change [of earth into water — and thus of dry land into sea — and vice versa], in accordance with their position at the zenith of the thing which is changing and in accordance with their motion. This holds in particular of the planets which at times move northward, at times southward. ... It would seem that these are major causes [*sibot gedolot*] in bringing about an increase of water at one place [through] a displacement of water toward it, and a diminution of water at [another] place [through] a displacement of water away from it. Over a long span of time, each of these two [displacements] increases, until it achieves a notable effect in making the water pour to low places and in exposing the hills.

Other causes help along, however, for it is impossible [that the phenomenon in question comes about] unless clay is formed from water and earth, and unless the sun and the stars have an influence on this clay, causing it to become stone when it is exposed, thus forming the mountains. This being so, it is impossible but that there be dry land and sea. So far what has been said by the above-mentioned savant.[36]

The context of Ibn Sînâ's geology is thus defined by the very same ideas we have encountered in the *Dânesh-namé*. In addition, we find here the notion — so often invoked by the astrologers as a knock-out argument, but shared by the natural philosophers too — that the displacement from one place to another of masses of water on the surface of the globe is caused by the celestial bodies (notably by the moon). This sets the stage for the really crucial question: how is clay formed from earth and water, and how are stones and mountains formed from clay? For an answer to these queries we have to turn to Ibn Sînâ himself, namely to Treatise (*fann*) V of the Physics of his *Kitâb al-Shifâ'* (composed about 1022), where Ibn Sînâ expounds his petrological and orogenic theory. In 1927 Holmyard and Mandeville showed that the well-known medieval tractate appearing under the titles of *De mineralibus* and *De congelatione et conglutinatione lapidum*, is a translation of sections from this part of the *Shifâ'*.[37]

To establish the 'conditions of the formation of mountains,' Ibn Sînâ says, one has to examine successively the conditions of the formation of stones, rocks and, lastly, mountains: petrology is to provide the foundation for the theory of orogeny. How, then, are stones formed? According to Aristotelian mineralogy, the underlying matter of stones is earth if they are opaque, and water if they are transparent. Ibn Sînâ accordingly distinguishes two processes[38]: stones are formed either through the hardening (*tafkhîr, conglutinatio*) of clay, or through the 'congelation' (*jumûd*) of water. We can quickly dispose of the later: congelation, the transformation of liquid into solid, is brought about, Ibn Sînâ holds, by a 'petrifying virtue,' instanced by the alchemists' Virgin Milk, which is 'compounded of two waters which coagulate into a hard solid.' Yet Ibn Sînâ is prudent on the matter and holds that, for the most part, stones and mountains are formed from earth.

How, then, does earth become stone? Here we must understand the terms of the problem as it faced Ibn Sînâ. On Aristotelian theory, 'moist' is 'that which, being readily adaptable in shape, is not determinable by any limit of its own'; 'dry,' on the other hand, is 'that which is readily determinable by its own limit, but not readily adaptable in shape.'[39] Stones, then, are obviously dry; they must be formed through desiccation. But desiccation of what? One would expect the answer to be: of a mixture of earth and water. But this poses a serious difficulty. According to Aristotle, if drying completely eliminates the moisture from a clump of earth, nothing solid can possibly result: 'earth,' Aristotle says, 'has no power of cohesion without the moist. On the contrary, *the moist is what holds it together*; for it would fall to pieces if the moist were eliminated from it completely.'[40] Now this postulate is precisely Ibn Sînâ's theoretical starting point: 'pure earth does not petrify,' (this is the well-known formula: *terra pura lapis non fit*), he echoes Aristotle, 'because the predominance of dryness over the earth endows it not with coherence but rather with crumbliness.'[41] This means that stones have to be dry while still containing moisture, an idea which would have seemed nothing short of a straightforward *contradictio in adjecto* to most of Aristotle's Greek commentators. In Ibn Sînâ's time, however, this was no longer the case. The idea of a *non-evaporable moisture* was readily available in Arab alchemy and chemistry: this was the notion of an *unctuous moisture*, and it is on this theoretical notion that Ibn Sînâ draws to account for the formation of stones through desiccation.

Beginning with the *observation* that

Often clay dries and is changed first into something intermediate between stone and clay, *viz.* a soft stone, and afterwards is changed into a stone [proper],

Ibn Sînâ goes on to present the *theoretical* contention that

The clay which most readily lends itself to this is that which is unctuous (*lazij*), for if it is not unctuous, it usually crumbles before it petrifies.[42]

The key notion here, on which the entire theory depends, is *unctuous*. What does it mean and why does unctuous clay not crumble upon desiccation? The question has more to it than meets the eye. We will now in fact briefly see that the concept of an unctuous moisture was a well entrenched one within the tradition of Arab natural philosophy. Far from simply using a convenient metaphor to avoid confronting the problem head on, Ibn Sînâ in fact sought to ground his petrology on a received theory within another part of natural philosophy.

IV. CHEMICAL PREMISES: AQUEOUS VS. UNCTUOUS MOISTURE (AN HISTORICAL APERÇU)

On close inspection, we find that in his biological writings, Aristotle himself occasionally distinguishes two kinds of moisture, one *aqueous*, the other *unctuous* (or fat, greasy; *liparos*),[43] of which the latter is the less liable to decay. For instance, plants live longer than animals 'because they have an oiliness and a viscosity which makes them retain their moisture in a form not easily dried up' (notably, we infer from the context, through cold).[44] Similarly, 'Water animals have a shorter life than terrestrial creatures, not strictly because they are humid, but because they are watery, and watery moisture is easily destroyed, since it is cold and easily congealed' (i.e., dried up by cold).[45] Again, differences in the thickness of hair among animals are explained thus: 'if the moisture [of the skin] be watery it dries up quickly and the hairs do not gain in size, but if it be greasy the opposite happens for the greasy is not easily dried up.'[46] *Unctuous moisture, we may conclude, is resistant to desiccation by either heat or cold*; indeed, olive oil is dried up by neither heat nor cold.[47] And in a word: '*fat things are not liable to decay*'; indeed, 'a fat substance is incorruptible.'[48]

In the ps.-Aristotelian *Problems*, the notion is already used as a matter of course in a 'chemical' account:

For when the dough is kneaded and the lightest flour and the stickiest moisture are left, the bread, when it has been exposed to the fire, becomes glutinous and does not dry up; for that which is sticky cannot be separated.[49]

Aristotle's Greek commentators did not, as far as I am aware, make much of fats and of their perdurable properties. Indeed, the problem of cohesion gained in prominence only after a crucial development had taken place, namely when Aristotelian natural philosophy had to face — and to account for — a vast array of empirical findings, related to the disintegration of inanimate substances, which accumulated within alchemy, independently of the developments of philosophical reflection within the Schools. These findings emerged from the systematical use of distillation (a procedure unknown to Aristotle), first by Greek, but then, and mainly, by Arab alchemists. Arab natural philosophy sought to integrate the new facts (and indeed the alchemists' theoretical notions too[50]) within the theoretical framework it took over from Aristotle.

Arab alchemists distilled practically every mineral and animal matter.[51] The typical result of a fractional distillation was the following: first, a soft fire would raise a vapor which, upon condensation, became a clear liquid: this liquid was referred to as 'water.' Then, a second, stronger, heating would raise a further, colored and unctuous liquid, referred to as 'oil' (*duhn*). At the end of the process, a dry residue was left behind at the bottom of the alembic.[52] This kind of procedure and its typical results are described by the Renaissance chemist Conrad Gesner (1516—1565). Although his account is from a later period, Gesner describes the very same facts and is worth quoting for his clarity:

Of a plant or any other substance ordeined to be distilled, what parte of it is most meet to be extenuated and fyret (that is the purest parte, the lightest, the thinnest, the *moistest* and the most superficial parte . . .) being first of all fyret by the force of the heat, is lifted up; next suche other partes as in purenes cum nie to the first, and last such a *moisture* of the thinges as is more crosse *that held together the earthy partes*, a certain *fatness and oiliness*, by a stronger force of the fyre is separated, and taken up hooly; which once clean drawn forthe, the body remaineth dissolved and brought to ashes.[53]

Distillation thus *isolated* the two distinct kinds of moisture: the aqueous, which evaporates easily and which condenses to water; and the unctuous, which evaporates only with difficulty and upon the disappearance of which the body disintegrates. Only the unctuous moisture, then, brings about cohesion.

In the corpus of writings ascribed to Jâbir ibn Ḥayyân (for convenience, I will use the name 'Jâbir' to designate its authors), the number of fractions obtained was increased from three to four when it was observed that fractional distillation yields also an aeriform, volatile and inflammable substance (which we know today to be sal ammoniac[54]). This discovery allowed Jâbir to establish a one to one correspondence between the four fractions obtained in distillation and the four Aristotelian elements. Indeed, fractional distillation of organic matter became of capital importance to Jâbir because he took it to be a method by which the nature and composition of any given substance could be determined.[55] The first distilled, volatile and inflammable fraction, was naturally taken to derive from fire; the second was considered to be the element water; the third, oily, fraction was associated with the element air; lastly, the solid residue which remained behind was identified with the element earth.[56]

It is the notion of 'oil' that is of interest to us here. To Jâbir, just as to the other alchemists, oil was the second and last liquid to rise in distillation: its evaporation left behind only a powdery residue. This means that among the four components of a substance, the one responsible for cohesion is oil. Now, Jâbir believed that each of the four Aristotelian qualities could be isolated through distillation. Moisture, in particular, which — in conformity with Aristotle — Jâbir construed as the principle of liquidity and cohesion, he affirmed to be obtained through the distillation of oil: moisture, he claimed, is isolated 'when oil is distilled until a very glutinous and elastic substance is obtained . . . This substance never solidifies.'[57] For Jâbir, then, oil, or unctuous moisture, is the principle of permanent liquidity and hence of cohesion. Whence the idea that it is the *unctuous quality that brings about combination.*[58] Indeed, the standard account of calcination, as given for instance by the famous physician and physicist al-Râzî, is this: calcination is the 'destruction of bodies' (i.e. metals) through 'the burning of the sulfurs and oils they contain, [resulting in] their reduction to white lime whose parts cannot further be divided.'[59]

Ibn Sînâ himself, let us now note, draws on the concept of unctuous moisture in his great medical work, the *Canon*, namely precisely when he seeks to explain how stones are formed within the human body. The formation of calculi (*ḥaṣâh*), Ibn Sînâ says, involves matter — which is affected — and an active, efficient, cause. The latter, to be sure, is (vital) heat, or, more precisely, *imbalanced*, i.e., excessively strong, heat. As for the material cause, Ibn Sînâ says that 'the matter [of calculi] is a *thick unctuous moisture* [*ruṭûbah lazijah ghalîzah*],' which has its origin in thick nutriment: milk and cheese, agglutinative (*lazij*) bread, indigestible fruits giving rise to unctuous moisture, etc.'[60] Ibn Sînâ, we see, was familiar with the chemical notion of 'unctuous moisture' and drew on it in order to account for cohesion: heat — which usually eliminates moisture — can give rise to hard bodies when acting on matter containing unctuous, that is non-evaporable, moisture.

By the tenth century, we may conclude, 'unctuous moisture' has become a well-entrenched theoretical concept, accounting for the cohesion of living and non-living matter. This was an indispensable concept if one wished — and many natural philosophers in fact did wish — to account for cohesion of substances within a largely Aristotelian framework, in which cohesion was imputed, almost by definition, to 'moisture.' The Ikhwân al-Ṣafâ', Albertus Magnus and many other medieval philosophers drew on it. As late as the second half of the seventeenth century the notion is still alive and well: even corpuscularians use it, e.g., to explain electrical attraction as brought about by tiny elastic — because unctuous — threads issuing from the rubbed electrical body.[61] It is on this well-entrenched theoretical notion that Ibn Sînâ founds his petrology and his geology.

V. CONCLUSION: METAPHYSICAL INTENTIONS, CHEMICAL FOUNDATIONS

Back to Ibn Sînâ's geology. We are now in a postion to understand why Ibn Sînâ insisted that stones and mountains are formed from specifically unctuous clay. This is not an alleged observational statement, nor a mere metaphor. Rather, Ibn Sînâ was drawing on an entire chemical *theory*, a theory that allowed him to explain how clay — a mixture of earth and water — can petrify. On this theory, the action of heat on the unctuous clay evaporates the aqueous moisture, but the unctuous, non-evaporable one, remains behind. This unctuous moisture continues to inhere in the clay and thus endows it with cohesion, preventing it from crumbling upon desiccation.

Once this petrology was established, the account of orogeny followed easily. This process is construed by Ibn Sînâ as follows. At first, water or clay petrify, presumably (Ibn Sînâ does not specify this) at low places, where they are carried by rivers as silt. Then mountains can be formed by one or the other of two causes: (i) The essential cause of mountains is a 'wind' (*rîh*) which arises in the bowels of the earth; this is the same 'wind' which is also the cause of earthquakes, i.e., Aristotle's dry exhalation.[62] This 'wind,' Ibn Sînâ says, 'raises a part of the ground and a height is suddenly formed.' (ii) Mountains can also be formed by an accidental cause, namely when the petrified mass remains where it was formed and 'certain parts of the ground are hollowed out while others are not, by the erosive action of winds and floods which carry away one part of the earth but not another.'[63]

Ibn Sînâ was now in a position to construe the surface of the earth as eternally subject to the actions of two counterbalancing forces. One is destructive and easily observable: 'at present time,' Ibn Sînâ says, 'most mountains are in the stage of decay and disintegration.' Yet theory establishes that an opposite, generative, geological process, one which is less accessible to observation, is at work too: some mountains 'by God's wills ... increase through the petrification of waters upon them, or through the floods which bring them a large quantity of clay that petrifies on them.'[64]

Ibn Sînâ thus succeeded where Theophrastus had failed. Contrary to Theophrastus, he could explain how a *cohering* mass of earthy matter can be formed, to be then raised by a 'volcanic' action. (Theophrastus speaks of 'fire,' Ibn Sînâ of 'wind,' but lurking behind both notions is in fact Aristotle's dry exhalation.) He thus gave substance to Theophrastus' metaphor in which the earth was likened to a tree which throws out its leaves again after having shed them: generation and corruption of the surface of the earth can go on eternally without leading up to a static final state. Peripatetic philosophers were now, at long last, in a position to rebuke the argument from erosion: the postulate of the eternity of the world was 'saved.'

The gist of Ibn Sînâ's petrological account — his invoking specifically unctuous moisture — has been recognized and correctly appreciated by Albertus Magnus.[65] 'It is perfectly clear,' Albertus says, that earth alone 'does not cohere into solid stone.' The reason is that 'the cause of coherence and mixing is moisture, which is so subtle that it makes every part of the earth flow into every other part.' Hence, 'if the moisture were not soaked all through the earthy parts, holding them fast, but evaporated when the stone solidified, then there would be left only loose, earthy dust.' Thus, one sometimes finds, compressed within a stone, earth which appears to be solid, but which finally falls into dust. 'And the cause of this is simply that its moisture, which was not unctuous or viscous enough, evaporated when the [surrounding] stone solidified.' The conclusion is that the moisture is necessarily non-evaporable, i.e., unctuous: 'There must be something viscous and sticky so that its parts join with the earthy parts like the links of a chain.'[66] It is, in sum, 'the viscous and unctuous moisture which gives coherence to the material of stone.' Ibn Sînâ was right, Albertus Magnus concludes, to insisit that stones are formed from unctuous clay.

The metaphysical intentions of Ibn Sînâ's chemically-founded theory did not escape notice either. They have been very clearly perceived and pointed out by Shmuel Ibn Tibbon, who perceptively compares them to the Aristotelian *opinio communis*, as represented by Ibn Rushd. Since Shmuel Ibn Tibbon thereby highlights the weighty metaphysical stakes lurking behind Ibn Sînâ's geological endeavor, his *exposé* is again worth translating *in extenso*:

It is manifest that his [Ibn Sînâ's] opion is that the exposed [i.e., dry] land is something which is generated [*nithawa*] after it had not been, an opinion which differs from that of the other philosophers following the opinion of Aristotle. He also dissents from them concerning the preservation of that [kind of dry] place. This disagreement follows upon [Ibn Sînâ's] above-mentioned view concerning generation. For the other philosophers say: the preservation [of dry land] is necessary — *qua* species, [dry land] can neither change nor be annihilated; for it is impossible that the entire earth be covered with water, with the result that all beings living only on land be annihilated. Thus in his commentaries, Ibn Rushd made clear his view that the existence of exposed land is necessary, not possible. To prove this he says: if it were possible that the entire earth be submerged [in water], then this possibility would [already] have been actualized; for time is eternal according to their [the philosophers'] opinion. Now the actualization [of that possibility] would entail the annihilation of the plants and animals. But, [Ibn Rushd continues,] there are many animal species in which generation and existence require the support of an individual of the same species: for instance, man acquires existence only through another man — from a father and the mother. The same holds of many animal and plant species. [Ibn Rushd] therefore says that [if the above hypothesis were true, then] these species would not [anymore] exist today: the [proposition stating] their existence would be false. But they do exist, as we know from sense experience. Consequently, their annihilation, their non-existence, is of the class of impossible things, not of the class of possible ones. Hence, what exists today has existed ever since and will exist eternally, infinitely. Ibn Rushd's view is indeed that the existence [i.e. generation] of man not from man is of the class of impossible things; this is for him a first postulate, one that is verified conjointly by sense experience and by the intellect that had been created in man. Hence it is by supposing this [the generation of man not from man] to be impossible, that he demonstrates the impossiblity of the earth being submerged in water.

[As against this,] Ibn Sînâ has written in the above-mentioned book that it is not impossible that a flood will occur and cover all the inhabited earth, or a part of it, and annihilate all or some animals, and that afterwards [these animals] come to be [again] from the *mixis* of the elements, helped along either by the stars only, or by one of the separate intellects [as well]. To explain his view: according to him, it is not impossible that, say, the species of man be annihilated and that subsequently, during the eternal time (as they believe), a *mixis* will come to be in the earth, which is suitable to receive the human form. For in his view, man's generation from man is not necessary, but only what is most frequent; it is the most appropriate and the easiest [mode of generation], just as most frequently a mouse is generated from a mouse, a frog from a frog, although occasionally a mouse comes to be from earth, a frog from rain water. The same holds of other species too. Also the generation of man from earth is possible, according to his opinion: the difference between the mouse, the frog and others which are born and give birth, and man is only a quantitative difference: this kind of [spontaneous] generation is very rare in man, indeed most infrequent, whereas in the mouse and the frog it is not all that rare, although it still is infrequent. This is what Ibn Sînâ's words amount to.[67]

Shmuel Ibn Tibbon thus makes clear that although in devising his novel geology, Ibn Sînâ responded to a need Peripatetic philosophers had felt at least since Theophrastus, he in fact pursued intentions that went far beyond what the other Aristotelians had in mind. For most philosophers assumed that *qua* species, dry land and sea had always existed, that, like all other species, they were eternal: this they interpreted as the work of a particularizer, and constructed upon it an argument from design.[68] Their problem was thus only to account for the *preservation* of the primeval balance between the elements, to show how, erosion notwithstanding, dry land and sea both persevere 'in species.' By contrast, Ibn Sînâ does away with the idea of a particularizer altogether: take a world consisting of the four concentric spheres of the elements and let time pass; through the sole interplay of natural forces dry land and sea will ultimately emerge, as indeed will all other species, including man. The actually existing species and structure of the surface of the earth are the work of nature, as will be also those which will, in due course, emerge subsequent upon a flood completely immersing the dry land.[69] Ibn Sînâ thus in fact succeeded in rebuking the argument adduced against the Aristotelians by the author of *De elementis*.[70] Through his novel, chemically-founded, geological theory, Ibn Sînâ the natural philosopher buttressed an essential doctrine of Ibn Sînâ the metaphysician: it allowed him, somewhat like Laplace some eight centuries later, to dispense with the hypothesis postulating a mindful, well-*intentioned* Creator, from whom issued the order 'Let the water be gathered!'

Writing in the third decade of the fourteenth century, the Hebrew poet Immanuel of Rome was of the opinion that for this geology Ibn Sînâ deserved nothing short of eternal hell. Following in Dante's footsteps, Immanuel describes his voyage through Hades, and reports having sighted there, among others, Aristotle, Galen, al-Fârâbî, Plato and Hippocrates; Ibn Sînâ is the last on this list of illustrious free-thinkers:[71]

שָׁם אִבֶּן סִינָא, הָיָה לְלַעַג וּשְׂחוֹק

יַעַן אֲשֶׁר אָמַר כִּי הַוָּלֵד אָדָם לֹא מֵאָדָם אֶפְשָׁר לִזְמַן רָחוֹק

וְאָמַר כִּי לֵדַת הֶהָרִים הָיָה דֶּרֶךְ טִבְעִי,

מִי יִתֶּן וְנֶאֱלָם

נִמְשָׁךְ אַחַר אֱמוּנַת קַדְמוּת הָעוֹלָם.

NOTES

1. I. Kant, *Kritik der reinen Vernunft*, A 424—33; B 452—61.
2. For a fine collection of sources, *cf.* La Naissance du monde (= *Sources oriéntales*, 1) (Paris: Seuil, 1959).
3. Unfortunately (and surprisingly) there is no comprehensive and detailed study of this debate in its various contexts. For the earlier period, *cf.* J. Baudry, *Le Problème de l'origine et de l'éternité du monde dans la philosophie grecque de Platon à l'ère chrétienne* (Paris: Les Belles Lettres, 1931). A good study, albeit of limited scope, is W. Wieland, 'Die Ewigkeit der Welt (Der Streit zwischen

Johannes Philoponus und Simplicius),' in: *Die Gegenwart der Griechen im neueren Denken. Festschrift für H.-G. Gadamer zum 60. Geburtstag* (Tübingen: J. C. B. Mohr, 1960), 291—316. For the medieval period I know only of Ernst Behler, *Die Ewigkeit der Welt. Problemgeschichtliche Untersuchungen zu den Kontroversen um Weltanfang und Weltuntergang im Mittelalter. Erster Teil: Die Problemstellung in der arabischen und jüdischen Philosophie des Mittelalters* (Paderborn: F. Schöningh, 1965), which, however, is almost entirely derivative and, moreover, compiles its information mostly from outdated sources; Behler says some words on the context, and the stakes, of the debate in medieval philosophy on pp. 12—19. H. A. Davidson, *Proofs for Eternity, Creation and the Existence of God in Medieval Islamic and Jewish Philosophy* (New York/Oxford: Oxford University Press, 1987) provides an excellent overview of the arguments and of their history but does not seek to situate the debate in relation to other issues.

4. The subject of the following paragraph has been treated very extensively by P. Duhem in the chapters on 'L'équilibre de la terre et des mers' and on 'Les petits mouvements de la terre,' in *Le Système du monde*, Vol. IX (Paris: Hermann, 1958), pp. 79—323. In the few pages that follow my aim is not to add to Duhem's wealth of detail, but only to highlight how the geological problem is related to the metaphysical and physical contexts.

5. *Cf. De gen. et corr.* 2.10, 337a8; *Tim.* 58 A; and Friedrich Solmsen, *Aristotle's System of the Physical World* (Ithaca: Cornell University Press, 1960), p. 384 (n. 18), p. 394.

6. *Cf.* Solmsen, *Aristotle's System*, p. 407ff.

7. *Ibid.*, p. 420ff.

8. *Cf.*, notably, *Physics* 8.9 and Baudry, *Le Problème*, pp. 170—73.

9. *Cf. De gen. et corr.* 2.10—11; F. Solmsen, *Aristotle's System*, pp. 379—89; A. L. Peck, 'Appendix A' to his *Aristotle, Generation of Animals* (Cambridge, Mass.: Harvard University Press [The Loeb Classical Library], 1942), pp. 567—76.

10. *Cf.* Solmsen, *Aristotle's System*, p. 420ff.; and *idem*, 'Aristotle and Presocratic Cosmogony,' *Harvard Studies in Classical Philology*, *63* (1958): pp. 265—82 (reprinted in his *Kleine Schriften* [Hildesheim: Olms, 1968], *I*, pp. 356—73), at p. 273ff.; I. Düring, *Aristoteles* (Heidelberg: C. Winter, 1966), pp. 386—7.

11. *Meteor.* 1.14 and 2.3.

12. *Cf.* Duhem, *Système du monde*, IX, p. 91ff.

13. Solmsen, *Aristotle's System*, pp. 431ff.; *cf.* also W. K. C. Guthrie, *A History of Greek Philosophy*, Vol. 1: *The Earlier Presocratics and the Pythagoreans* (Cambridge: Cambridge University Press, 1962), pp. 387—90. On the eternity of the human species see the remarkable study by K. Oehler, 'Ein Mensch zeugt einen Menschen. Über den Missbrauch der Sprachanalyse in der Aristotelesforschung,' reprinted in his *Antike Philosophie und byzantinisches Mittelalter* (München: C. H. Beck, 1969), pp. 95—145.

14. *Meteor.* 1.14, 351a19ff.; *cf.* Solmsen, *Aristotle's System*, p. 436f.

15. *Cf.* Thomas S. Hall, 'Life, Death and the Radical Moisture,' *Clio Medica*, 6 (1971): pp. 3—23 at p. 6; Gad Freudenthal, 'The Theory of the Opposites and an Ordered Universe: Physics and Metaphysics in Anaximander,' *Phronesis 31* (1986): pp. 197—228, at p. 221.

16. *Cf.*, e.g., M. Pohlenz, *Die Stoa* (Göttingen: Vadenhoek & Ruprecht, 1948), *I*, pp. 77ff.; M. Lapidge, 'Stoic Cosmology,' in: J. M. Rist (ed.), *The Stoics* (Berkeley: University of California Press, 1978), pp. 161—85, at pp. 180ff.; D. E. Hahm, *The Origins of Stoic Cosmology* (Columbus, Ohio: Ohio University Press, 1977), pp. 185ff.

17. Philo, *De aeternitate mundi* 118—119, quoted after *Works*, vol. 9, trans. by F. H. Colson (Cambridge, Mass.: Harvard University Press [The Loeb Classical Library], 1941). The thesis that the arguments preserved by Philo are Theophrastus', who was already defending the eternity thesis against Zenon was the subject of some discussion. *Cf.* O. Regenbogen, 'Theophrastos,' in: Pauly-Wissowa, *Realencyclopädie*, Suppl. VII (Stuttgart, 1940), cols. 1539—40; Pohlenz, *Die Stoa, II*, 44; Baudry, *Le Problème*, pp. 219—20, pp. 236ff. The first to recognize the significance of these passages for the history of geology was Pierre Duhem; *cf.* his 'Léonard de Vinci et les origines de la géologie,' in his *Etudes sur Léonard de Vinci*, seconde série (Paris: Hermann, 1909), pp. 283—357, at pp. 286ff. The same passages are again discussed in *Système du monde, IX*, pp. 241ff.

18. The Stoic arguments are repeated by the Atomists, of whom, however, our evidence is scantier. *Cf.* Lucretius, *De rerum natura* 5.235ff. and Baudry, *Le Problème*, pp. 249ff.

19. Philo, *De aeternitate mundi*, 132—133. The passage seems to echo Aristotle, *Meteor.* 1.14, 351a27—29 referred to above. For the way this argument fits into Theophrastus' metaphysics *cf.* P. Steinmetz, *Die Physik des Theophrast* (Bad Homburg: Gehlen, 1964), p. 167. On the general context of the argument and on related arguments *cf.* H. A. Wolfson, 'Patristic Arguments Against the Eternity of the World,' in his *Studies in the History of Philosophy and Religion* (Cambridge, Mass.: Harvard University Press, 1973), *I*, pp. 182—206, on pp. 187ff.

20. Philo, *De aeternitate mundi*, 135—137.

21. Strabo, *The Geography*, 17.1.36, quoted after: *The Geography of Strabo*, with an English translation by H. L. Jones (Cambridge, Mass.: Harvard University Press [The Loeb Classical Library], 1932), vol. 8. On the attribution to Posidonius, *cf.* K. Reinhardt, *Poseidonius* (München: C. H. Beck, 1921), pp. 88ff. and *idem*, 'Poseidonius', in Pauly-Wissowa, *Realencyclopädie*, Vol. 43, pp. 665—6.

22. Strabo, *Geography*, 17.1.36.

23. For what follows, *cf.* Ikhwân al-Safâ *Rasâ'îl* (Beirut, 1957), **II**, pp. 91f.; F. Dieterici, *Die Naturanschauung und Philososphie der Araber im zehnten Jahrhundert. Aus den Schriften der Lauteren Brüder* (Berlin, 1861), pp. 99ff. For al-Bîrûnî, *cf.* his *The Determination of the Coordinates of Positions for the Correction of Distances Between Cities*, translated by Jamil Ali (Beirut: The American University of Beirut, 1967), pp. 23—5. For al-Ghazâlî, *cf.* Duhem, *Système du monde, IX*, p. 105. The argument was of course often rehearsed by Scholastic natural philosophers; *cf.*, e.g., Duhem, *Système du monde, IX*, pp. 126, 129, 131, 133.

24. Moreover, since the existence of land does not follow by natural *necessity*, its existence at one place rather than at another indicates that it has resulted from the voluntary action of a *particularizer* (*mukhaṣṣiṣ*); cf. H. A. Wolfson, *The Philosophy of the Kalam* (Cambridge, Mass.: Harvard University Press, 1976), p. 441. Quite evidently, the idea of God's providence, the doctrine of particularization, and the

argument from design are intimately related (*cf.* Maimonides, *The Guide of the Perplexed*, I, 74, 'The Fifth Method'); they are all discussed in Davidson, *Proofs for Eternity, Creation and the Existence of God* (n. 3, *supra*).

25. Quoted after Duhem, *Etudes sur Léonard de Vinci, II*, pp. 300f. *Cf.* also *Système du monde, II*, pp. 226—8 and *IX*, p. 256. The Arab original of *De elementis* (dating from the middle of the ninth century) seems to be lost and its author escapes identification to this day; cf. Charles B. Schmitt and Dilwyn Knox, *Pseudo-Aristotles Latinus. A Guide to the Works Falsely Attributed to Aristotle Before 1500* (Warburg Institute Surveys and Texts, Vol. XII) (London: The Warburg Institute, The University of London, 1985), p. 20.

26. *Ibid.*, p. 309; also in *Système du monde, IX*, pp. 256—7.

27. *Averroes on Aristotle's 'De generatione et corruptione'. Middle Commentary and Epitome*, translated by Samuel Kurland (Cambridge, Mass.: The Medieval Academy of America, 1958), p. 135.

28. Nor does Ibn Rushd describe here in any detail how the heavenly bodies are to 'preserve' the 'species' of dry places. He is a little more elaborate in his *Epitome of [Aristotle's] Meteorologica*. There he argues that the existence of dry land cannot be due to the sun's heat, for the southern hemisphere is warmer than the northern one and yet more submerged in water. Ibn Rushd believes that the drying effect is brought about by the rays of the sun immixed with those of the fixed stars, which are more numerous in the northern hemisphere. Levi ben Gerson (Gersonides, 1288—1344) severely criticizes Ibn Rushd's account, arguing that the existence of dry land is due to Providence and thus confirms the thesis of creation; *cf.* his *Milhamot ha-Shem* 6.1.13 (Riva di Trento, 1560, fol. 57b) and Ch. Touati, *La Pensée philosophique et théologique de Gersonide* (Paris: Minuit, 1973), pp. 185—7 (*cf.* n. 52 for references to some further authors who discussed the problem), as well as Davidson, *Proofs for Eternity, Creation and the Existence of God*, p. 231 (with note 108). (I have consulted Ibn Rushd's text as incorporated in Levi ben Gerson's Supercommentary on it in the manuscript of the Staatsbibliothek preussischer Kulturbesitz Berlin (West), Ms. Orient Fol. 1055, fol. 120b.) As has been noted by M. Joel [*Lewi ben Gerson (Gersonides) als Religionsphilosoph* (Breslau, 1862), p. 77], a position very similar to Gersonides' had been advocated by his contemporary, the astronomer Isaac Israeli, in his *Sefer Yesod ᶜOlam* 2.2 [(Berlin, 1846—1848), I, pp. 17bff.], written in 1310 (*ibid.*, pp. I, 1b, II, 31a). Levi's argument is rebuked by R. Ḥasdai Crescas (*Sefer Or ha-Shem* 3.1.3—4), but embraced by Crescas's student Joseph Albo (*Sefer ha-ᶜIqqarim*, 4.8), as well as by Shimeᶜon ben Zemah Duran (*Magen Avot* [Livorno, 1785], p. 9a) and by the fifteenth-century philosopher Abraham Shalom (*cf.* H. A. Davidson, *The Philosophy of Abraham Shalom* [Berkeley and Los Angeles: University of California Press, 1964), pp. 62, 74). To be sure, the entire issue is closely bound up with the controversial question whether or not the world, though created, is destructible; *cf.* on this Seymour Feldman 'The End of the Universe in Medieval Jewish Philosophy,' *Association for Jewish Studies Review, 11* (1986), pp. 53—77. Dante seems to have followed Ibn Rushd and to have further developed his account; *cf.* E. O. von Lippmann, *Beiträge zur Geschichte der Naturwissenschaften und der Technik*, Vol. 2 (Weinheim: Verlag Chemie, 1953), pp. 169—70.

70

29. For what follows, *cf.* Avicenne, *Le Livre de science*, translated by M. Achena and H. Massé (Paris: Les Belles Lettres, 1958), *II*, pp. 27, 31—3, 45f.

30. *Cf.* Aristotle, *Physics* 4.4, 212a5f., 20.

31. *Cf.* M. Steinschneider, *Die Hebraeischen Übersetzungen des Mittelalters* (Berlin, 1893), pp. 132—5.

32. Steinschneider (*ibid.*, p. 200, n. 676) concludes from internal evidence that it was written after 1221; Shmuel Ibn Tibbon was already dead in 1232 (*ibid.*, p. 132, n. 179).

33. Shmuel Ibn Tibbon, *Ma'amar Yiqqawu ha-Mayim*, ed. by M. L. Bisliches (Pressburg, 1837), p. 2.

34. Ibn Tibbon's *exposé* has been compared with Ibn Sînâ's own text in Georges Vajda, *Recherches sur la philosophie et la kabbale dans la pensée juive du Moyen Age* (Paris: Mouton, 1962), pp. 14—15. This chapter of Vajda's book has appeared in English as: 'An Analysis of the *Ma'amar Yiqqawu ha-Mayim* by Samuel b. Judah Ibn Tibbon,' *Journal of Jewish Studies, 10* (1959): pp. 137—49.

35. Hebrew: *kefi ha-mezi'ût ha-mehuyav le-seder ha-kôl*; in Arabic the last two words read: *liniẓâm al-kull. cf.* G. Vajda, *Recherches*, p. 14, where they are translated by 'l'ordre du Tout.'

36. Shmuel Ibn Tibbon, *Ma'amar Yiqqawu ha-Mayim*, pp. 7—8, I have somewhat corrected the printed text according to the manuscript Cod. Heb. 33 of the Bayerische Staatsbibliothek München (using the microfilm no. 1163 of the Institute of Microfilmed Hebrew Manuscripts in the Jewish National and University Library, Jerusalem), fol. 75ᵛf. This passage became known to a very wide public because it was quoted *in extenso*, without yet indicating its source, in the popular encyclopedia of R. Gershom b. Shlomo, *Sefer Shaʿar ha-Shamayim* 2.2 (Rödelheim, 1801), pp. 6ᵛ f.; (Warsaw, 1875), p. 13.

37. For their edition of the Arabic and Latin texts, accompanied by an English translation, *cf.* E. J. Holmyard and D. C. Mandeville (eds. and trans.), *Avicennae De congelatione et conglutinatione lapidum, Being Sections of the Kitâb al-Shifâ'* (Paris: Geuthner, 1927). That Ibn Sînâ is the author of the *De mineralibus* had been convincingly argued by Duhem, albeit without using the Arabic text, in 1909; *cf. Etudes sur Léonard de Vinci, II*, p. 302ff.

38. For what follows, *cf. ibid.*, pp. 71ff. (Arabic text), pp. 18ff. (English translation).

39. Aristotle, *De gen. et corr.* 2.2, 239b32f. All quotations from Aristotle are given after *The Complete Works of Aristotle. The Revised Oxford Translation*, edited by Jonathan Barnes (Princeton: Princeton University Press, 1984).

40. *De gen. et corr.* 2.8, 335a2f.,

41. Holmyard and Mandeville, *op. cit.*, p. 71 (Arabic), p. 18 (English).

42. *Ibid.*, p. 72 (Arabic), p. 19 (English). Holmyard and Mandeville render *lazij* by 'agglutinative'; I have preferred 'unctuous' so as to conform to medieval Latin usage.

43. The term and the notion go back at least to the ps.-Hippocratic treatise *On Fleshes* (fifth century B.C.), but tracing and dealing with this must remain outside the scope of the present overview.

44. *De long. et brev. vit.* 6, 467a6ff.

45. *Ibid.*, 5, 466b33ff.; *cf.* also 466b22f. and *De sensu* 2, 438a20f. for similar accounts.

46. *Generation of Animals* 5.3, 782b2ff.
47. *Meteor.* 4.7, 383b34.
48. *De long. et brev. vit.* 5, 466a23; *History of Animals* 3.19, 521a1. As has already become clear, the notion of unctuous moisture seeks to provide a physical or 'chemical' account of the cohesion of individual substances compounded of the four sublunary elements. But in Aristotelian philosophy the cohesion of substances is also subsumed under the notion of form, which, where living substances (plants and animals) are concerned, is their (vegetative) soul. It is therefore clear that the present *aperçu* artificially isolates one aspect of a general and fundamental problem in Aristotelian philosophy of nature, that of cohesion — or better: transtemporal stability [*cf.* M. Furth, 'Transtemporal Stability in Aristotelian Substances,' *Journal of Philosophy,* 75 (1978), pp. 624—46] — of material sublunary substances. Elsewhere I intend to deal with the entire problem in greater detail and try to show how the notion of unctuous moisture relates to the Aristotelian notion of soul, through that of vital heat.
49. *Problems* 21.12, 928a26ff.; *cf.* also 21.6.
50. The alchemists themselves accounted for their observations with their own distinctive sulfur-mercury theory, which the natural philosophers incorporated into their own system. Although the question of how this alchemical synthesis came about is highly germane to our topic and is crucial for an adequate understanding of the explanatory import of the notion of unctuous moisture, I will have to disregard it in the present paper. I have given a brief sketch of the historical developments involved in my paper 'Die elektrische Anziehung im 17. Jahrhundert zwischen korpuskularer und alchemischer Deutung,' in Chr. Meinel (ed.), *Die Alchemie in der europäischen Kultur- und Wissenschaftsgeschichte (Wolfenbütteler Forschungen Band 32)* (Wiesbaden: Otto Harrassowitz, 1986), pp. 315—326, and intend to come back to the topic in greater detail in the future.
51. R. J. Forbes, *Short History of the Art of Distillation* (Leiden: Brill, 1948).
52. *Cf.*, e.g., the 'Syriac and Arabic Alchemical Treatise,' in M. Berthelot, *La Chimie au Moyen Age* (Paris, 1893; reprinted Osnabrück/Amsterdam, 1967), *II*, pp. 184—5; J. Ruska, *Al-Râzî's Buch Geheimnis der Geheimnisse* (Berlin: J. Springer [*Quellen und Studien zur Geschichte der Naturwissenschaften und der Medizin, VII*], 1937), pp. 78ff.; 204ff.; 207ff.
53. Conrad Gesner, *The Treasure of Evonymus* (London, 1559), p. 2 (my italics). Similar descriptions can repeatedly be gleaned in al-Râzî's writings. *Cf.* Ruska, *Al-Râzî's Buch Geheimnis der Geheimnisse,* pp. 78ff., 204, 211, 216, 217, 218, 225.
54. J. Ruska, 'Der Salmiak in der Geschichte der Alchemie,' *Zeitschrift für angewandte Chemie, 41* (1928): pp. 1321—4.
55. For this and for what follows, *cf.* P. Kraus, *Jâbir ibn Ḥayyân: Contribution à l'histoire des idées scientifiques dans l'Islam.* Vol. II: *Jâbir et la science grecque* (= *Mémoires présentés à l'Institut d'Egypte,* Vol. 45) (Cairo, 1945 [reprinted: Paris: Les Belles Lettres, 1986]), pp. 5ff.; 41f.
56. *Cf.* also F. Rex, *Zur Theorie der Naturprozesse in der früharabischen Wissenschaft. Das 'Kitâb al-Ikhrâǧ', übersetzt und erklärt. Ein Beitrag zum alchemistischen Weltbild der Ǧâbir-Schriften* [= *Collection des travaux de l'Académie internationale d'histoire des sciences,* n° 22] (Wiesbaden: F. Steiner, 1975), p. 42.
57. Kraus, *op. cit.,* p. 10.

58. H. E. Stapleton, R. F. Azo and Hidâyat Husain, 'Chemistry in Iraq and Persia in the Tenth Century A. D.,' *Memoirs of the Asiatic Society of Bengal, 8* (1927): pp. 317—418, at p. 338f. [Added in proof: In antiquity, the natural philosophers most concerned with the problem of cohesion were the Stoics, who accounted for it in terms of their fundamental concept of *pneuma*. I have recently suggested that this concept may have gone into the Arab alchemists' notion of unctuous moisture: *cf.* my "The Problem of Cohesion Between Alchemy and Natural Philosophy: From Unctuous Moisture to Phlogiston," in Z.R.W.M. von Martels (ed.), *Alchemy Revisited. Proceedings of the International Conference on the History of Alchemy at the University of Groningen, 17—19 April 1989* (Collection de travaux de l'Académie internationale d'histoire des sciences, vol. 33) (Leiden: Brill, 1991), pp. 107—16.]

59. Ruska, *Al-Razi's Buch*, p. 126 (*cf.* n. 52 *supra*); *cf.* also p. 74.

60. Ibn Sînâ, *Kitâb al-Qânûn fi-l-Tibb*, Book VI, Treatise 18, Article 2, Chapter 6 (Rome, 1593), p. 434.

61. For more details *cf.* my paper referred to *supra*, note 50.

62. *Cf.* Aristotle, *Meteorologica* 2.8. In this chapter Aristotle uses the terms *pneuma* and *anemos* to designate the dry exhalation, the cause of earthquakes; *cf.* H. D. P. Lee, *Aristotle, Meteorologica* (Cambridge, Mass.: Harvard University Press [The Loeb Classical Library], 1952), p. 203. In the Arabic translation of the *Meteorologica* both terms are rendered as *rîh*; *cf.* C. Petraitis, *The Arabic Version of Aristotle's Meteorology* (= *Pensée arabe et musulmane*, tome 39) (Beirut: Dar el-Machreq, 1967), Arabic pagination, pp. 73ff., p. 138 *sub rîh.*

63. Holmyard and Mandeville, *op. cit.*, p. 77 (Arabic), p. 27 (English).

64. *Ibid.*, p. 78f. (Arabic), p. 29 (English).

65. For what follows, *cf.* Albertus Magnus, *Book of Minerals*, trans. by D. Wyckoff (Oxford: Clarendon, 1967), pp. 12—13. The first to emphasize the extent to which Albertus Magnus is dependent on Ibn Sînâ (whom he mentions by name) was P. Duhem; *cf.* his *Etudes sur Léonard de Vinci, II*, pp. 302ff., included almost *verbatim* in his *Système du monde, IX*, pp. 257ff. In these two works Duhem gives much information on the influence of Ibn Sînâ on the Latin Middle Ages.

66. As D. Wyckoff has noted, the account of 'viscous' in terms of parts hanging together like the links of a chain, goes back to Aristotle, *Meteorologica* 4.9, 387a11—19.

67. Shmuel Ibn Tibbon, *Ma'amar Yiqqawu ha-Mayim*, p. 8 (*cf.* n. 33 *supra*), corrected after the MS (*cf.* n. 36), fol. 76ʳf. G. Vajda, *Recherches* (*cf.* n. 34 *supra*), pp. 16—7, translates some passages and gives a summary of the rest. On Ibn Tibbon's views *cf.* also S. Rosenberg, "Remarks on the History of the Idea of a Restaurative Redemption in Medieval Jewish Philosophy" (in Hebrew), in *Ha-Raᶜyon ha-Meshihi be-Yisrael* (Jerusalem: The Israel Academy of Sciences, 1982), pp. 37—86, on pp. 50 ff. Ibn Sînâ's theory discussed here, in particular his 'outrageous' idea (*cf.* below in the text) that even man can come to be through spontaneous generation, so that creation is not a necessary condition for his existence, of course hinges on his own notion of the agent intellect, the giver of forms. The details of this theory are authoritatively exposed in H. A. Davidson, 'Alfarabi and Avicenna on the Active Intellect,' *Viator, 3* (1972): pp. 109—78, particularly pp. 156—9. Ibn

Rushd's view, or rather changing views, on this subject is studied in H. A. Davidson, 'Averroes on the Active Intellect as a Cause of Existence,' *Viator, 18* (1987): pp. 191—225.
68. *Cf.* note 28 *supra*.
69. In other words, Ibn Sînâ invokes efficient rather than final causes; *cf.* Sh. Pines, 'What was Original in Arabic Science?,' in A. C. Crombie (ed.), *Scientific Change* (London, 1963), pp. 181—205, on p. 189 (= *The Collected Works of Shlomo Pines*, vol. II: *Studies in Arabic Versions of Greek Texts and in Medieval Science* [Jerusalem: Magnes and Leiden: Brill, 1986], pp. 329—353, p. 337).
70. *Cf.* text at notes 25 and 26 *supra*.
71. In all likelihood Immanuel derived his knowledge of Ibn Sînâ's geology from Shmuel Ibn Tibbon's resumé. *Cf. Maḥbarot 'Immanuel ha-Romi* ed. by Dov Yarden (Jerusalem: Mosad Bialik, 1957), 28.90—98. Here is my translation, which has no pretensions whatsoever to do justice to the poetical qualities of the original:

Present there is Ibn Sînâ,
He has been put to ridicule and to laughter,
For he said that the generation of man not from man is possible at times,
And for he said that the generation of mountains is along the natural way.
Would that he had remained dumb!
For he followed the belief in the eternity of the world.

STOIC PHYSICS IN THE WRITINGS OF R. SAADIA GA'ON AL-FAYYUMI AND ITS AFTERMATH IN MEDIEVAL JEWISH MYSTICISM*

This paper is dedicated to my son Michael who, while I was writing it, one evening spontaneously explained to me that God is the air, omnipresent and therefore omniscient.

I. INTRODUCTION

Rav Saadia Ga'on (882-942), one of the very first medieval Jewish thinkers to advocate a synthesis between Greek-Arabic philosophy and traditional Jewish thought, was born and educated in Egypt but lived in Baghdad from ca. 922 onwards. In addition to being a philosopher, grammarian, and translator and commentator of the Bible, he was the outstanding spiritual leader of the Eastern Jewish communities of his day.[1] Saadia wrote two philosophic works: first, in 931, the *Tafsīr Kitāb al-Mabādi'* (*Commentary on "Sefer Yeẓira"*) and, two years later,

* *Acknowledgements*: A first version of this paper was presented at the colloquium on "Perspectives médiévales (arabes, latines, hébraïques) sur la tradition scientifique et philosophique grecque," organized conjointly by the Centre d'histoire des sciences et des philosophies arabes et médiévales (C.N.R.S., Paris) and the Société internationale d'histoire des sciences et de la philosophie arabes et islamiques, held in Paris on March 31 - April 3, 1993. I am very grateful to the colleagues and friends who kindly read a draft of this paper and made important helpful suggestions: Haggai Ben-Shammai (Jerusalem), Bernard R. Goldstein (Pittsburgh), Menachem Kellner (Haifa), Y. Tzvi Langermann (Jerusalem), Sarah Stroumsa (Jerusalem), Mauro Zonta (Pavia), and an anonymous referee for *Arabic Sciences and Philosophy*. I also express my gratitude to the Sidney M. Edelstein Center for the History and Philosophy of Science, Technology and Medicine at the Hebrew University, Jerusalem, for the material facilities put at my disposal during my regular stays in Jerusalem. The research underlying this paper was supported by an individual grant from the Memorial Foundation for Jewish Culture during the years 1991/92 and 1992/93, for which I am very grateful.
[1] The best general account is still Henry Malter, *Saadia Gaon. His Life and Works* (Philadelphia, 5702-1942).

114

in 933, his major work, the *Kitāb al-Amānāt wa-al-i'tiqādāt* (*The Book of Doctrines and Beliefs*[2]).

The main objective of this paper is to show that distinctly Stoic ideas can be identified in both of Saadia's works and that these ideas had important repercussions in medieval Jewish thought. Specifically, I will argue that the Stoic theory of *pneuma* looms behind Saadia's doctrine of God's created Glory, by which Saadia seeks to avoid anthropomorphism through a "realist," as contrasted with an "allegoric," interpretation of the accounts of God's revelations. I will also give some brief indications on the aftermath of Saadia's theory within Jewish mysticism and thus point to hitherto unnoticed repercussions of Stoic physics in medieval theology.

II. SOME CHARACTERISTICS OF STOIC PHYSICS

Before turning to Saadia, let me summarize some elements of the early Greek, notably Stoic, views of *pneuma*. My presentation will be exceedingly brief, limited to the few points that will prove to be most relevant.

Air played an important role in ancient Greek natural philosophy. Anaximenes, in the sixth century B.C.E., held air, or *pneuma*, to be the *archē* of the world, the unique substrate underlying the infinite variety of material substances. Anaximenes identified the universal element air with the godhead and held it to be the cause of cohesion of individual living beings and of the entire world: "As our soul, being air, holds us together and controls us, so does wind [or: breath; *pneuma*] and air enclose the whole world."[3] Thus the analogy between the role of breath or air in the living animal – the microcosm – and in the entire world – the macrocosm – which will be of great importance in our subsequent discussion, is invoked already by this early Presocratic philosopher.

[2] I follow the translation of the title proposed by the late Alexander Altmann. Cf. his *Saadya Gaon, Book of Doctrines and Beliefs* in: *Three Jewish Philosophers* (New York, 1969). The justification for this choice is given in the "Translator's Introduction," pp. 19-20.

[3] Text in H. Diels and W. Kranz, *Die Fragmente der Vorsokratiker*, 6th edn (Berlin, 1952), fr. 13B2. Translation quoted from: G.S. Kirk, J.E.R. Raven, and M. Schofield, *The Presocratic Philosophers*, 2nd edn (Cambridge, 1983), pp. 158-9. Cf. also W.K.C. Guthrie, *A History of Greek Philosophy*, vol. I: *The Earlier Presocratics and the Pythagoreans* (Cambridge, 1962), pp. 115-40.

A century later, Diogenes of Apollonia (fl. ca. 440-430 B.C.E.) carried Anaximenes' monistic theory further. Diogenes too holds air to be both the godhead and the only underlying *Urstoff* out of which all substances are generated. Diogenes further maintains that in its purest form, air is intellect or reason (*noesis*). This implies that nature consists of an intelligent substance, an assumption taken to explain nature's inherent teleology. The air endows animals and man at once with life and with reason: in the absence of air, life is extinguished, and reason ceases too. The differences in intelligence between the various living beings are held to be due to differences in the quantity and quality of the inhaled air.[4]

Following generations of natural philosophers did not ascribe any theological role to air, although, naturally, no one questioned the connection between air and life. Aristotle's mature physics posited a sharp distinction between the sublunar world – where air is simply one of the four elements – and the supralunar realm, whose matter is the eternal and unchanging "fifth body," which alone was ascribed theological dignity. Whereas in Diogenes' philosophy of nature, matter (viz. air) in the celestial and in the terrestrial realms differed only in purity, in Aristotle's scheme, the relationship between the supra- and the sublunar matter could at most be an analogy with respect to certain functions.[5] Aristotle's God was an Intellect and as such thoroughly transcendent.

Stoic philosophy, whose beginnings coincide with Aristotle's death, turned the tables again by creating a unified physical *cum* theological world-picture. By abandoning Aristotle's fifth element (and with it the separation of the universe into a supra- and a sublunar realm) and by positing a basic, divine, substance permeating the entire cosmos, the Stoics created a world-picture characterized by immanentism, monism, and materialism.[6] In

---

[4] Cf. Kirk, *Presocratic Philosophers*, pp. 434-42; W.K.C. Guthrie, *A History of Greek Philosophy*, vol. II: *The Presocratic Tradition from Parmenides to Democritus* (Cambridge, 1965), pp. 362-81.

[5] This is the relationship between the celestial matter and the vital heat, or the *pneuma*, contained in the semen. Cf. Aristotle, *De generatione animalium* 2.3, 736b37. This statement is unique in the Aristotelian *corpus*. I discuss it in my *Aristotle's Theory of Material Substance* (Oxford, 1995), pp. 114-19.

[6] Cf. Robert B. Todd, "Monism and immanence: The foundations of Stoic physics," in John M. Rist (ed.), *The Stoics* (Berkeley, 1978), pp. 137-60. Out of the rich literature devoted to early Stoicism I mention only the following: Max Pohlenz, *Die Stoa*, 6th edn, 2 vols. (Göttingen, 1984); G[érard] Verbeke, *L'évolution de la doctrine du*

fact, the fundamental and most distinctive postulate of the Stoic philosophy of nature is that the universe is thoroughly permeated by a tenuous substance called *pneuma*, a term whose initial meaning is wind or breath. It is through *pneuma* that matter, which in itself is amorphous and passive, is endowed with cohesion; indeed, that it coheres into definite substances (metal, plant, animal, etc.). The *pneuma* is also the cause of the cohesion of the entire universe.

The best insight into the *rationale* of the Stoic doctrine of *pneuma* can be gained by considering the model which presumably inspired it. The Stoic concept of *pneuma* apparently grew out of certain biological views which explained life functions in terms of a connate *pneuma*, i.e. a *pneuma* which does not enter the animal from outside (through respiration), but is the outcome of physiological processes. These biological theories held that the connate *pneuma* was the physiological agent fulfilling the multifarious functions ascribed to "soul," i.e. growth, motion, sense perception, reproduction, etc. In line with their distinctive monism, the Stoics radicalized these biological notions into the view that the *pneuma* was *identical* with the soul. The *pneuma*, they believed, has its center at the heart and, being continuous throughout the body, it governs all its operations. The Stoics made this biological theory into the cornerstone of their cosmology. They construed the entire cosmos as a living being (*zōon*), and argued that the universe is pervaded by a cosmic *pneuma*, fulfilling precisely the same roles in the universe as the connate *pneuma* does in the living body. And just as in the living being the *pneuma* is the soul controlling the animal, so the cosmic *pneuma*, which is endowed with reason (*nous*; *logos*), governs the cosmos wisely. Indeed, for the Stoics, the *pneuma is* the godhead (*theos*), which can direct every aspect of the world because it is immanent within it – the *pneuma* permeates all matter and every substance. (To support this

---

pneuma du stoïcisme à S. Augustin (Paris / Louvain, 1945); Shmuel Sambursky, *The Physics of the Stoics* (London, 1957; repr. Princeton, 1987); David E. Hahm, *The Origins of Stoic Cosmology* (Columbus, Ohio, 1977); James Longrigg, "Elementary physics in the Lyceum and Stoa," *Isis*, 66 (1975): 211-29; Michael Lapidge, "*Archai* and *stoicheia*: A problem in Stoic cosmology," *Phronesis*, 18 (1973): 240-78; *id.*, "Stoic cosmology," in Rist (ed.), *The Stoics*, pp. 161-85; Paul Hager, "Chrysippus' theory of pneuma," *Prudentia*, 14 (1982): 97-108; A.A. Long and D.N. Sedley, *The Hellenistic Philosophers*, vol. I (Cambridge, 1987), pp. 266-343; Richard Sorabji, *Matter, Space, and Motion* (New York, 1988), especially pp. 79-105.

claim, the Stoics elaborated a revolutionary theory according to which two substances can interpenetrate one another and thus co-exist in actuality at the same place.[7])

The *pneuma*, which permeates matter, confers upon it a certain tension (*tonos*), by virtue of which it becomes a specific substance, concomitantly endowing it with cohesion, and thus with continuity over time. On the lowest grade of tension, called *hexis*, the *pneuma* endows matter only with the cohesion found in inanimate substances: examples are stones, or bones and sinews in the living body. The next two degrees of *tonos*, which are superposed on the first, are *phusis* and *psuche*, in which the *pneuma* gives rise to plants and animals, respectively. The highest degree is that of the human and divine intellects – these are the highest concentrations of *pneuma*. The *pneuma* is at times construed as consisting of the two active elements: fire (not the ordinary fire, but rather a formative fire, *pur technikon*, construed as analogous to the vital heat in the living body), and air, which together act upon the two passive elements – water and earth.

It is the *pneuma*-godhead, then, that endows matter with forms (in analogy with the functions ascribed in medieval philosophy to the active intellect[8]). In other words: the *pneuma* generates, or creates, the world. The Stoics indeed held that every created world reaches its end in due course, namely, when the element fire overpowers the other elements, bringing about a conflagration (*ekpurōsis*); subsequently, the *pneuma*, being formative, creates the world anew. It seems that the *pneuma*'s formative power, particularly its capacity to confer cohesion upon matter, is due to the component air within it; in fact, the terms "*pneuma*" and "air" are often used interchangeably.[9]

Stoicism, then, produced a worldview in which physics and theology were a unified whole. The divine *pneuma* was immanent in all matter, and through it the world could persevere (just as the individual animal persists by virtue of its connate *pneuma*). Being Reason, the *pneuma* governs every aspect of the world with wisdom, whence Stoic determinism.

---

[7] For a detailed discussion cf. Sorabji, *Matter, Space, and Motion*, pp. 79-105.

[8] For an exhaustive overview cf. Herbert A. Davidson, *Alfarabi, Avicenna, and Averroes on Intellect: Their Cosmologies, Theories of the Active Intellect, and Theories of the Human Intellect* (New York, 1992).

[9] Cf. Sorabji, *Matter, Space and Motion*, pp. 85-9.

III. STOIC PHYSICS IN SAADIA GAON'S "COMMENTARY ON SEFER YEZIRA"

The *Sefer Yezira* (*SY*; *Book of Creation*), a short mystic text in Hebrew of unknown origin and date, ascribed by tradition to Abraham the Patriarch, was already surrounded by an aura of antiquity and sanctity in Saadia's time.[10] In addition to a translation (*tafsīr*) of *SY*, Saadia composed a commentary (*sharḥ*) on it. The question why a rationalist philosopher like Saadia should at all wish to write a commentary on the *SY* has recently been discussed and satisfactorily answered.[11] Here, we will be interested only in Saadia's doctrine of God's Glory as expounded in that work.[12]

To thwart anthropomorphism, Saadia elaborated a theory according to which those phrases of the Scripture which may seem to attribute material manifestations to God, in fact refer not to God Himself, but rather to a distinct, created, entity. Some hesitancy notwithstanding, Saadia's considered view is that the accounts of Divine appearances should not be interpreted allegorically.[13] Instead, they describe real physical mani-

---

[10] Before Saadia, *SY* was hardly mentioned by Jewish scholars. Joseph Dan has noted that the abrupt rise to prominence of *SY* in the work of Saadia Gaon in the tenth century involves a change of attitude and is an intriguing cultural phenomenon that calls for an explanation. Cf. J. Dan, "R. Yehuda ben Barzilai ha-Barzeloni's Commentary on *Sefer Yezira*: Its character and purposes" (Hebrew), in M. Oron and A. Goldreich (eds.), *Massu'ot. Studies in Kabbalistic Literature and Jewish Philosophy in Memory of Prof. Ephraim Gottlieb* (Jerusalem, 1994), pp. 99-119, and "The religious significance of *Sefer Yezira*" (Hebrew), *Jerusalem Studies in Jewish Thought*, 11 (1993): 7-35.

[11] Haggai Ben-Shammai, "Sa'adya's goal in his *Commentary on Sefer Yezira*," in Ruth Link-Salinger *et al.* (eds.), *A Straight Path. Studies in Medieval Philosophy and Culture. Essays in Honor of Arthur Hyman* (Washington, D.C., 1988), pp. 1-9; Raphael Jospe, "Early philosophical commentaries on the *Sefer Yezirah*: Some comments," *Revue des études juives*, 146 (1990): 369-415, esp. pp. 370-81.

[12] The work is available in two editions: *Commentaire sur le Séfer Yesira ou Livre de la création, par le Gaon Saadya de Fayyoum*, publié et traduit par Mayer Lambert (Paris, 1891); Yosef Qafih (ed.) *Sefer Yezira [Kitāb al-Mabādi']: im Perush ha-Gaon Rabbenu Sa'adiah b. R. Yosef Fayyumī z. l.* (Jerusalem, 1972). A French translation is included in the former, a translation into modern Hebrew in the latter. In what follows the references to the Arabic text of Saadia's *Commentary on SY* are given according to both Lambert's (L) and Qafih's (Q) editions, indicating page and line numbers; these are followed by the page number of Lambert's translation into French (T).

[13] Alexander Altmann, "Saadya's theory of revelation: Its origin and background," in his *Studies in Religious Philosophy and Mysticism* (London, 1969), pp. 140-60, on pp. 145-7.

festations, in which, however, not God Himself was perceived, but rather a distinct, created entity. Saadia maintains that this entity is the one which traditional Jewish sources alternately designate by the terms *Kavod* (Divine Glory), *Shekhinah* (Divine Presence), or *Ruaḥ ha-Qodesh* (Holy Spirit).[14] Saadia further identifies this entity with what he names "the second, subtle air" (*al-hawā' al-thānī al-laṭīf*),[15] to be distinguished, of course, from the "visible air," i.e., the atmospheric air, one of the four elements. The expression *Ruaḥ 'Elohim Ḥayyim* (the Spirit of the Living God) in *SY*, Saadia argues, refers precisely to this second, subtle air,[16] which is "doubtless something created."[17] According to Saadia, God's manifestations took place and were perceived by the prophets within and through this entity.[18] The sources and contents of the theological components of this theory have received thorough study,[19] and will not detain us here, where we concentrate on Saadia's view of the "second air" and its relationships to God and to the world.

Saadia suggests that an anthropomorphic construal of the Deity can be avoided through a thought-process proceeding by "analogy."[20] The first step in this cognitive process is to realize that the relationship of the Creator to the world is the same as

---

[14] L 72:6-7, 73:5; Q 108:13-15, 109:18; T 94-5. Cf. also *Doctrines and Beliefs* 2:10, in: S. Landauer, *Kitāb al-Amānāt wa'l-I'tiqādāt von Sa'adja b. Jūsuf al-Fajjūmī* (Leiden, 1880), pp. 99-100, also in *Sefer ha-Nivḥar be-Emunot u-ve-De'ot*, ed. and trans. (into modern Hebrew) Y. Qafiḥ (Jerusalem, 1970), pp. 103-4. On the three traditional notions cf. e.g. *Encyclopedia Judaica*, 16 vols. (Jerusalem, 1972), vol. XIV, coll. 364-8 and 1349-54.

[15] L 72:5 f.; Q 108:12 f.; T 94.

[16] L 72:8; Q 108:17; T 94.

[17] L 72:11; Q 108:22 f.; T 94.

[18] L 72:14 ff.; Q 108:26 ff.; T 94 f. Similarly *Doctrines and Beliefs* 2:5; ed. Landauer, 87 f.; ed. Qafiḥ, 92.

[19] Altmann, "Saadya's theory of revelation" (n. 13); Georges Vajda, "Sa'adyā commentateur du *Livre de la Création*," *Annuaire de l'École pratique des hautes études (Sciences religieuses)* 1959-1960, pp. 3-35, reprinted in G.E. Weil (ed.), *Mélanges Georges Vajda. Études de pensée, de philosophie et de littérature juives et arabes* (Hildesheim, 1982), pp. 37-69, on pp. 53-7. To dissipate possible misunderstandings, it may be useful to observe that Saadia's theory of divine visions as being brought about through the second air does not seem to have anything in common with the well-known Neoplatonist doctrine of *ochēma-pneuma* (the equivalence of the terms *pneuma*=air notwithstanding). On this latter theory cf. Robert Christian Kissling, "The *ochema-pneuma* of the Neo-Platonists and the *De insomniis* of Synesius of Cyrene," *American Journal of Philology*, 43 (1922): 318-30; E.R. Dodds, "The astral body in Neoplatonism," Appendix II to his *Proclus, The Elements of Theology*, 2nd edn (Oxford, 1963), pp. 313-21.

[20] "*bi-hādhihi al-amthāl*": L 71:12-13; Q 107:23-24; T 93.

the relationship of life (*ḥayāt*) to the living being.[21] "By analogy" (*'alā al-tamthīl*), therefore, God is the "life of the world."[22] This preliminary conclusion already implies the important corollary that God is immanent in the world: "Just as we see life present in every part [of an animal] as well as in the entire animal, so also He, be He exalted, is present in every part [of the world] and in the world as a whole."[23] But clearly God cannot be life *simpliciter* and, indeed, the above is only a first approximation, a step on the way to the precise formulation of Saadia's position.

In the second, and last, step, we pass from "life" to the level of intellect (*martabat al-'aql*).[24] We are now told that God is not the life of the world, but rather its *intellect* (*'aql al-'ālam*).[25] What, then, is God's relationship to the material world? Saadia elaborates his immanentist stance as follows. In all animals, the body is the abode (*maḥall*) of life; but in a rational animal, i.e. in man, in whose body the elements are blended in a perfect equilibrium, life in turn becomes the abode of intellect.[26] This position is also reflected in Saadia's *Tafsīr* of Genesis 2:7. Where the original Biblical text states that when God gave man the spirit of life, he became "a *living* soul" (*nefesh ḥayyah*), Saadia's translation states that the spirit of life *ipso facto* endowed man with "a *rational* soul" (*nafs nāṭiqa*).[27] In man,

[21] L 70:5-6; Q 106:4-7; T 91.
[22] L 70:7; Q 106:7; T 91.
[23] L 70:8-10; Q 106:9-11; T 92.
[24] L 70:10; Q 106:12; T 92.
[25] L 70:10-11, 18; Q 106:12-13, 24; T 92.
[26] L 70:12; Q 106:14-5; T 92.
[27] *Œuvres complètes de R. Saadia Josef al-Fayyoūmī*, publiées sous la direction de J. Derenbourg. Volume premier: Version arabe du Pentateuque (Paris, 1893), Arabic part, 7:9-10; French translation on p. 4. In his *Tafsīr* Saadia explicitly comments on this shift of meaning: translating "life" literally, he argues, would imply making Eve, of whom it is said that she was *em kol ḥay* (literally: "the mother of every living creature"; Gen. 3:20), into the mother of the lion, the ox, the ass, etc. Since experience disproves this, "we say that there is a hidden word in this phrase." Cf. *Saadya's Commentary on Genesis*, edited with Introduction, Translation and Notes by Moshe Zucker (New York, 1984), p. 18 (original), p. 191 (Hebrew translation); see also p. 296. This topic was taken up by later authors; cf. M. Perez, "A further fragment from the *Kitāb al-Tarjih* by R. Yehuda Ibn Bileam," *Proceedings of the American Academy of Jewish Research*, 57 (1990-91), Hebrew Section, p. 10, nn. 51-52. (I am grateful to Prof. H. Ben-Shammai for the last two references.) It should be noted that although Saadia here uses the distinctive philosophical term *nāṭiqa*, his translation doubtless echoes an ancient tradition going back to Onqelos' translation of the Bible, where our phrase is translated as *wa-hawat be-adam le-ruaḥ memalelā* (= speaking spirit). (The last observation was communicated to me independently by Dr. Y. Tzvi Langermann and by Prof. Moshe Hallamish, to both of whom I am very grateful. As I belatedly discovered, this observa-

therefore, the body is the abode of life, which in turn is the abode of intelligence, each one of these three being finer, or subtler, than the former. This idea provides the basis for Saadia's view of the "second air." According to Saadia, the three entities existing in the microcosm – body, life, intellect – have analogical counterparts in the macrocosm. These are: the world, the second air, and God. The second air, which is "simple, subtle" (*basiṭ, laṭīf*),[28] has the same relation to the world as life has to the body of the living being; and the relationship of the second air to God is the same as the relationship of life to man's intellect. The basic idea is thus expressed in the identity:

body : life : intellect = world : second air : God (= the world's Intellect)

The analogy posited by Saadia between the micro- and the macrocosm, between man and the world, obviously comes very near the one underlying Stoic thought: just like the Stoics, Saadia postulates the existence of a cosmic all-pervading air, to which he ascribes physical *cum* theological functions. There are two differences, however. First, for Saadia, the microcosmic counterpart of the cosmic air is "life," whereas the Stoics held that the macrocosmic *pneuma* had its microcosmic parallel in the inborn *pneuma*. Second and more important, for the Stoics, the godhead was forthrightly *identical* with the material, cosmic *pneuma*. As against this, Saadia, in continuity with a central (although not exclusive[29]) Jewish tradition, presumably also under the influence of Aristotelian or Neoplatonic philosophies,[30] construes God as

---

tion had already been made in A. Schmiedel, "Parallelen zwischen Onkelos und Saadja," *Monatsschrift für Geschichte und Wissenschaft des Judentums*, 46 (1902): 358-63, on p. 361; elsewhere Schmiedel makes the general observation that "Saadja im Grossen und Ganzen in Onkelos seinen kundigen Wegweiser erblickte"; cf. *id.*, "Saadja und Onkelos," *Monatsschrift für Geschichte und Wissenschaft des Judentums*, 46 (1902): 84-8, on p. 88.) Interestingly, Onqelos' rendering may itself reflect Stoic influence; cf. [Jacob Immanuel] Neubürger, "Onkelos und die Stoa," *Monatsschrift für Geschichte und Wissenschaft des Judentums*, 22 (1873), 566-8, on p. 567.

[28] L 70:15; Q 106:19; T 92.

[29] Cf. J. Abelson, *The Immanence of God in Rabbinical Literature* (London, 1912).

[30] Cf. on this the late Shlomo Pines' remarks in his "Quotations from Saadya's Commentary on *Sefer Yeẓira* in Maimonides' *Guide of the Perplexed*," Appendix III to his "Points of similarity between the exposition of the doctrine of the Sefirot in the *Sefer Yeẓira* and a text of the Pseudo-Clementine Homilies. The implications of this resemblance," *Proceedings of the Israel Academy of Sciences and Humanities*, vol. VII, no. 3 (Jerusalem, 1989), pp. 63-142, on pp. 127-32, esp. pp. 129-31. Pines does not consider the role of the air in the triad nor, consequently, the possibility that Saadia integrated Stoic elements in his scheme; cf. also n. 43 below.

transcendent: *qua* "Intellect of the world," the godhead is an entity totally separate from all matter, including the created "second air." Saadia clarifies this idea with the following simile: "The second, special [or: peculiar] subtle air is to God, be He praised and exalted, what the throne is to the king."[31] God's transcendence notwithstanding, however, the functions which Saadia ascribes to the second air come very close to those the Stoics had ascribed to their divine *pneuma*: Saadia holds that the Creator's will spreads throughout the second air and moves it ("as life moves the body"), and that God governs the world through the air as intermediary.[32] Thus, although the second air is not itself Reason, it is the abode of the "Intellect of the world," and it is the medium through which God exercises His wise governance of the world and through which He is immanent in it. This view clearly preserves central elements of the Stoic physics *cum* theology.

The Stoic connection is further corroborated by the fact that some of Saadia's physical arguments in favor of his theory of the subtle air are of distinctly Stoic origin. Thus, in order to prove that the subtle air is indeed all-pervading, Saadia writes: "Just as there is no bone, cartilage [or: sinews[33]], or any other hard part of the body, which is not permeated by the governance of the intellect – for life [i.e. the substrate of intellect] is present in any part of it –, so also in the world there is no mountain, sea, or any other gross thing, in which the Creator is not present, for the air is all-pervading."[34] The first part of this argument carries a clear echo of the central Stoic idea that the *pneuma* is the cause of the cohesion of all substances. It is especially noteworthy that bones and sinews are precisely the instances adduced by some Stoics to exemplify the *hexis* of the *pneuma* in the living body.[35]

---

[31] L 72:21 f.; Q 109:9-11; T 95.

[32] L 70:15-16; Q 106:18-21; T 92. Cf. also L 74:2; Q 110, *sharḥ* of the second *halakhah*; T 96.

[33] So reads a medieval Hebrew translation preserved in Judah b. Barzilai ha-Barzeloni's *Commentary on "Sefer Yeẓira"*; cf. *Commentar zum Sepher Jezira von R. Jehuda b. Barsilai aus Barcelona*, hrsg. von S.J. Halberstram nebst ergänzenden Noten von D. Kaufmann (Berlin, 1885), 177:37. Another medieval Hebrew translation confirms the rendering "cartilage": cf. *ibid.*, 340:33.

[34] L 71:1-4; Q 107:3-8; T 92. To be sure, the allusion to sea as an instance of something gross (or dense; *kathīf*) is somewhat surprising; this however is the reading carried by both printed editions (*baḥr*), and it is confirmed by the medieval Hebrew translations (cf. also n. 37).

[35] Cf. Lapidge, "Stoic cosmology," p. 171; Long and Sedley, *Hellenistic Philosophers*, p. 284.

In the sequel of the passage,[36] however, we encounter an idea which is the precise opposite of the Stoic one just considered. Saadia now reasons as follows: the fact that air is present in water is obvious, for water is thin and fluid[37]; but how is it possible that air be in stones and mountains, which are hard? Saadia's answer is that were it not for the air permeating these hard bodies, they would be absolutely solid and would never disintegrate. Underlying this argument is obviously the assumption that air confers upon substances softness and fluidity, rather than solidity and cohesion. Saadia thus seems to accept the Stoic idea that air is all-pervading, but to be unsure whether it is the cause of cohesion and solidity of substances, or, on the contrary, the cause of their thinness and softness.

How are we to account for this hesitancy and even self-contradiction? The answer, it seems to me, is that while Saadia has adopted the basic postulate of Stoic physics, some arguments adduced by its critics have crept into his reasoning. Indeed, Galen had already argued that it is unreasonable to think that a subtle body such as *pneuma* should be capable of bringing about the cohesion of solid bodies such as earth: how can a tenuous body endow matter with hardness, a property it does not itself possess? It would rather seem, Galen reasoned, that hard substances cohere by themselves, and that the presence of the *pneuma*, which is thin, endows matter with softness and fluidity.[38] Much the same arguments were put forward by Alexander of Aphrodisias.[39] These ideas, it seems, found their way into Saadia's argumentation. The claim that the hardness of bones and of cartilage (or sinews) is due to the presence within them of the subtle air and, as a result, of the intellect inhering in it is derived from the central Stoic postulate concerning the roles of the pervasive cosmic *pneuma*. In this context, Saadia explicitly formulates the principle on which the entire Stoic view of *pneuma* hinges, namely, that the "peculiar, subtle second air" is "subtler than everything subtle and yet

---

[36] L 71:5-7; Q 107:8-15; L 92 f.

[37] This idea may also underlie the allusion to the sea in the argument quoted in the preceding paragraph (cf. n. 34).

[38] Cf. Sambursky, *Physics of the Stoics*, pp. 35, 119-20.

[39] *Ibid.*, pp. 35, 120-21; and Robert B. Todd, *Alexander of Aphrodisias on Stoic Physics. A Study of the De mixtione with Preliminary Essays, Text, Translation and Commentary* (= *Philosophia antiqua*, vol. XXVIII) (Leiden, 1976), pp. 135, 217.

stronger than everything strong."[40] It would seem, incidentally, that the very fact that Saadia explicitly formulates this idea indicates that he was aware of the theoretical difficulty besetting the Stoic theory that had been pointed out by its critics. Yet when Saadia came to answer the rather cogent objection: "how can air be present within stones and mountains?",[41] confusion set in and he followed the critics of Stoicism, arguing that the presence of air within water is attested by its fluidity and – by extrapolation – that if there were no air in stones they would never disintegrate.

This confusion, it seems to me, need not cast doubt on the thesis that Saadia's notion of "second air" goes back to Stoicism. It has been demonstrated that Saadia's use of the sources at his disposal was not very careful.[42] Moreover, Stoic physics was ill understood during Antiquity; its ingenuity has begun to be appreciated only recently, largely thanks to the late Shmuel Sambursky. Saadia, therefore, was presumably unable to achieve a detailed and profound understanding of the Stoic theory of cohesion. It is also possible that the confusion was already in the source(s) on which Saadia depended.

We can thus conclude that underlying Saadia's theory of the "second, subtle air" is the Stoic theory of *pneuma*.[43] Saadia clearly affirms that the second air is the cause of cohesion of hard substances – although tenuous, it is "stronger than everything strong." Also in his Commentary on Genesis Saadia writes that through the air, "God's power becomes more mani-

---

[40] L 73:9-10; Q 109:25-6; T 95. As Qafih notes *ad loc.*, the same phrase – now applied to the Deity – occurs in *Doctrines and Beliefs*, Treatise 2, Introduction (ed. Landauer 77:16-17; ed. Qafih, 80:27-29); it is discussed at some length also at 2:8 (ed. Landauer 91:17-92:10; ed. Qafih, 95:32-96:21). Cf. below p. 125.

[41] L 71:6; Q 107:11; T 92.

[42] This has been brilliantly shown by H.A. Davidson apropos of Saadia's synopsis of theories of soul; cf. his "Saadia's list of theories of soul," in Alexander Altmann (ed.), *Jewish Medieval and Renaissance Studies* (Cambridge, Mass., 1967), pp. 75-94, especially pp. 88-94.

[43] That Saadia's views derive from Stoicism has been perceived and briefly noted by S. Horovitz in his "Ueber den Einfluss der griechischen Philosophie auf die Entwicklung des Kalam," *Jahres-Bericht des jüdisch-theologischen Seminars Fraenckel'scher Stiftung [1909]* (Breslau, 1909), pp. 1-91, on pp. 42-3. His opinion has been endorsed in Malter, *Saadia Gaon*, pp. 188-9. The late Professor Shlomo Pines pointed to distinctive Neoplatonic elements in Saadia's thought, but did not discuss the origin of the notion of "second air." Cf. Pines, "Points of similarity between the exposition of the doctrine of the Sefirot in the *Sefer Yeẓira* and a text of the Pseudo-Clementine Homilies," Appendices I and II on pp. 115-32. As far as I can see, there is no conflict between Prof. Pines' theses and those presented here (cf. n. 30).

fest than through anything else."[44] The second air relates to the
world as life relates to the living body, a stance that implies –
although Saadia does not formulate this corollary explicitly – that
the cosmic air is the cause of cohesion of the entire material
world.[45] Saadia's air, just like the Stoic *pneuma*, is all-pervading
and so God is immanent in the world, His transcendence
notwithstanding. Unlike the Stoic *pneuma*, Saadia's "second
air" is created and not endowed with reason, yet God's Wisdom
and governance are exercised through it.

IV. STOIC PHYSICS IN SAADIA'S "DOCTRINES AND BELIEFS"

In *Doctrines and Beliefs*, although presumably written a mere
two years after the *Commentary on SY*, the theory of the second
air does not explicitly appear. Nonetheless, I will now briefly
attempt to show that Stoic physics is indubitably present in
Saadia's major philosophical work as well. Henry Malter has in
fact observed that notwithstanding the absence of any explicit
reference to the theory of the "second air" in *Doctrines and
Beliefs*, that theory seems to underly the theory of divine immanence in that work too.[46] His contention, I will now try to show,
can be corroborated by further arguments.

To begin with, in *Doctrines and Beliefs*, Saadia explicitly
adduces and defends the postulate that "what is more subtle is
also more powerful," which underlies his theory of the "second
air" as developed in the *Commentary on SY*. As stated in
Doctrines and Beliefs, this principle holds, first, regarding the
elements: the more subtle element is more powerful – e.g. fire is
more powerful than air. But the applicability of the principle
goes further: "soul" is subtler than body, "wisdom" (*ḥikma*) is
subtler than soul, and yet, although being more tenuous, soul
directs the body, and wisdom the soul. Similarly, in the macrocosm, God, Who is "subtler than everything and more powerful

[44] M. Zucker (ed.), *Saadya's Commentary on Genesis*, p. 29 (original), p. 214
(Hebrew translation).
[45] In fact, Saadia maintains that it is the circular motion of the "great orb" of the
air (or fire) that holds together the world, or at least that keeps the Earth at its place
in the center; cf. *infra* at nn. 54, 56-57.
[46] Cf. Malter, *Saadia Gaon*, p. 188 (n. 440). In *Doctrines and Beliefs* 2:8, especially,
Saadia draws a comparison between the relationship of the soul and the body to
God's relationship to the world.

than everything," is (or: is associated with) Wisdom.[47] We have seen that this is precisely the idea that defines the role of the second air in the *Commentary on SY* (where, however, Saadia speaks of "intellect" and not of "wisdom").

Further, in *Doctrines and Beliefs* Saadia maintains, in complete conformity with his stance in the *Commentary on SY*, that the divine manifestations perceived by the prophets were real appearances. In the *Commentary on SY*, the medium producing these manifestations is referred to both by the expression "second air" and by the Hebrew term *ruah*, while in *Doctrines and Beliefs* only the equivalent Arabic term *rūh* is used. The theory itself, however, appears in both works, as does the affirmation that the *ruah/rūh* is a created entity.[48]

But the Stoic connection in *Doctrines and Beliefs* goes beyond the mute presence of the doctrine of the "second air," for Saadia there draws also on other Stoic physical teachings. Thus, in *Doctrines and Beliefs* Saadia rejects the Aristotelian celestial fifth element, subscribing to the view, propounded notably by the Stoics (and by John Philoponus), that the world is made up of four elements only. The Aristotelian argument to the effect that the heavens must consist of a fifth substance because the natural motion of the superior sublunar element (fire) is not circular is rejected by Saadia. This reasoning is erroneous, he maintains:

the cause of his [Aristotle's] mistake lies in his argument that, if the heavens had been [composed of the element of] fire, its motion would have been upward like that of fire. We declare, however, that the natural motion of fire itself is circular. The proof thereof is [provided by] the motion of the heaven, which is pure fire, as is clearly proved to us by the perceptible heat of the sun. ... You thus see how, on account of this slight mistake, this person [Aristotle] has compelled himself to affirm the existence of a fifth element which cannot be [rationally] apprehended and had to explain the perceptible heat of the sun by ascribing it not to its own substance, but to air.[49]

Both the rejection of the fifth element, one of the important innovations of Aristotle's physics, and the stance that the natural

---

[47] *Doctrines and Beliefs* 2:8, ed. Landauer 91:21-92:3; ed. Qafih, 96:5-9.

[48] *Doctrines and Beliefs* 2:5, ed. Landauer 88:3; ed. Qafih, 92:8.

[49] *Doctrines and Beliefs* 1:3, "eighth theory"; ed. Landauer 59:2-6, 14-17; ed. Qafih, 61:27-33, 62:11-15. Translation (slightly modified) quoted after Samuel Rosenblatt, *Saadia Gaon, The Book of Beliefs and Opinions* (New Haven / London, 1948), pp. 70-1. We are not here concerned with the fact that Saadia's argument involves a *petitio principii*.

motion of the celestial fire is circular instead of rectilinearly upward, are positions associated mainly (although not exclusively) with Stoicism.[50] Consistent with Saadia's adherence to the Stoic view of the celestial region as consisting of fire is his acceptance (in the *Commentary on SY*) of another Stoic

---

[50] An anonymous referee for this journal perspicaciously suggested the following possibility (for which I am very grateful): "For Plato too there are only four elements, and Philoponus agreed. Could *they* be the sources, rather than the Stoics?" Now although in different works Philoponus is inconsistent on the nature of the celestial matter and of its natural motion, in his lost *De aeternitate mundi contra Aristotelem*, notably, he indeed seems to have developed the idea that when fire is at its natural place it rotates naturally (cf. Christian Wildberg, *John Philoponus' Criticism of Aristotle's Theory of Aether* (*Peripatoi*, Bd. 16) [Berlin/New York, 1988], pp. 130-1). Philoponus was known to Arab philosophers, and some of his ideas found acceptance among them (for a recent statement cf. Ahmad Hasnawi, "Alexandre d'Aphrodise *vs* Jean Philopon: Notes sur quelques traités d'Alexandre 'perdus' en grec, conservés en arabe," *Arabic Sciences and Philosophy*, 4 (1994): 53-109). *Prima facie*, therefore, it seems possible that in rejecting Aristotle's fifth element, Saadia followed a tradition going back to Philoponus no less than to Stoicism. Although this possibility cannot be ruled out, it yet seems that the evidence does not warrant it. For one thing, in his *Contra Aristotelem* Philoponus apparently held the celestial region to consist of a mixture of fire and all the other elements. (This assumption was crucial in order to rebut Aristotle's argument that an overwhelming quantity of celestial fire would overpower all other elements, transforming the substance of the entire cosmos into itself. Cf. Wildberg, *ibid.*, p. 174.) Saadia, by contrast, in the passage just quoted, considers heaven to consist of "pure fire" only. Further, Philoponus consistently holds that like all other substances the heavens possess heaviness and lightness (Wildberg, *ibid.*, pp. 147 ff., 159; Paul Moraux, "Quinta essentia," in Pauly-Wissowa, *Realencyclopädie*, XXIV(1) [Stuttgart, 1963], coll. 1171-1263, on col. 1244), an assumption obviously not shared by Saadia. Moreover, to buttress his four-element cosmology, Philoponus argues that the circular motion of the heavenly bodies does not entail that their matter must be of a fifth kind, unlike the sublunar four elements, for air and fire, too, have a natural circular motion (cf. Wildberg, *John Philoponus' Criticism*, pp. 132-4, 161-3, 183; Moraux, "Quinta essentia," col. 1244). Now, although this argument indeed recalls the one we saw was adduced by Saadia, it should be noticed that, like the Stoics, Saadia ascribes a circular motion to fire alone, and not to the air as well. (Admittedly at some places Philoponus too ascribes a circular movement to fire only; cf. Wildberg, *John Philoponus' Criticism*, pp. 129-31. On the circular motion of fire in Stoic physics cf. Longrigg, "Elementary physics," p. 222; Michael Wolff, "Hipparchus and the Stoic theory of motion," in Jonathan Barnes and Mario Mignucci (eds.), *Matter and Metaphysics* [Napoli, 1988], pp. 473-545, esp. pp. 504-6, 540-2.) It may be added that Philoponus at times considered that the heavens' circular motion is both natural and caused by soul (Wildberg, *John Philoponus' Criticism*, pp. 161-3), a conception of which there seems to be no trace in Saadia. In view of these discrepancies between Philoponus' and Saadia's views, and considering that Saadia's argument of the "second air" which penetrates everything is undoubtedly of Stoic origin, it seems plausible to suppose that his view of the heavenly matter derives from the same source. But perhaps Saadia integrated a vaguely-defined anti-Aristotelian position on the issue of the heavenly matter, one that grew out of the views of both the Stoics and Philoponus (who anyway was indebted to the former); arguably Saadia's pronouncements on the subject are too brief to allow us to decide the matter. For a concise history of views on the fifth element cf. Paul Moraux, "Quinta essentia,"

principle, namely the distinction between two kinds of fire – the ordinary, sublunar fire, and the celestial fire.[51]

Another Stoic element in Saadia's thinking can be identified through the following consideration. In *Beliefs and Opinions*, Saadia maintains that the Earth is kept at rest in its place through the natural circular motion of the sphere of fire.[52] In

---

esp. coll. 1231 ff. and Longrigg, "Elementary physics"; the medieval debate is discussed in Georges Vajda, "La philosophie et la théologie de Joseph Ibn Caddiq," *Archives d'histoire doctrinale et littéraire du Moyen Âge*, 17 (1949): 93-181 (reprinted in G.E. Weil, *Mélanges Georges Vajda* [Hildesheim, 1982], pp. 423-511), on pp. 110-11 (= 440-1). The rejection of Aristotle's fifth element may go back to Theophrastus. For a recent assessment of his view cf. R.W. Sharples, "Theophrastus on the heavens," in J. Wiesner (ed.), *Aristoteles Werk und Wirkung Paul Moraux Gewidmet*, vol. I (Berlin/New York, 1985), pp. 577-93.

[51] *Commentary on SY*, Introduction, "Sixth method"; Q 29:3-4; L 9:15-16; T 24. Saadia here argues against those who posited fire as the *archē*, out of which the world was created. His argument is based on the *premise* (which he adopts) that "fire is twofold – celestial [lit. belonging to the orbs; *al-falakiyya*] and terrestrial [*al-arḍiyya*]." A similar distinction is admittedly adduced by Philoponus, but his view in turn "no doubt shows influence of the Stoic distinction between *pur phusikon* and *pur atechnon*" (Wildberg, *John Philoponus' Criticism*, p. 168; cf. also pp. 177, 179; Moraux, "Quinta essentia," col. 1243). Moreover, for Philoponus the two types of fire are at bottom "merely instances of one and the same element" (Wildberg, *John Philoponus' Criticism*, pp. 168-9; cf. also p. 134), a view that is consistent with his stance that the other three elements are also present in the celestial region (*ibid.*, p. 172). Saadia's view on the fiery nature of the supralunar realm is quoted with approval (albeit without ascribing it to Saadia) in R. Bahya Ibn Paqūda's *The Duties of the Hearts* (composed in Arabic between 1050 and 1150): "Some philosophers were of the opinion that the heavenly orbs and the upper beings [i.e. the heavenly bodies] are of the element fire. This is similar to what David said: "Who makest winds Thy messengers, the flaming fire Thy minister" (Ps. 104:4) – which is confirmatory evidence for that view; there is no fifth element as Aristotle held." Cf. *Ḥovot ha-Levavot* 1:6, in: Rabbenu Bahyay ben Yosef Ibn Paqūda, *Sefer Ḥovot ha-Levavot*, trans. by Juda ibn Tibbon, ed. A. Zifroni (Tel Aviv, 1954), p. 118; English translation (modified) quoted after Moses Hyamson (ed. and trans.), *Duties of the Heart* by R. Bachya ben Joseph ibn Paquda (1925-1947), 2 vols. (Jerusalem, 1978), vol. I, pp. 75-7. Dr. M. Zonta (University of Pavia) kindly informed me that in his *Kitāb al-Ḥadīqa*, building upon a summary of Aristotle's theory of elements as exposed in *De caelo*, Moshe ibn Ezra (ca. 1055 - after 1135) writes that "in his book *Beliefs and Opinions* R. Saadia Gaon, may his memory be blessed, rejected this [Aristotle's] view"; cf. Jerusalem, Jewish National and University Library, MS 8° 5701, fol. 50b.

[52] *Doctrines and Beliefs* 2:8, ed. Landauer 92:6-7; ed. Qafih, 96:14-15. Saadia also rejects the Aristotelian doctrine of natural motions. For him, the natural motion of an element, the one belonging to it essentially, is that which it has at its *origin* (the term here is *ma'din*): thus, rest is the natural state of a stone which has reached its origin, just as circular motion is the natural motion of fire at its own origin, i.e., in the heavens. In contrast, the rectilinear motion which the stone and fire acquire when removed from their respective origins is *accidental*. Cf. *ibid.* 1:3, "Eighth Theory"; ed. Landauer, 59:6-14; ed. Qafih, 61:33-62:10. This position is, of course, non-Aristotelian, but it is also not Stoic; cf. Longrigg, "Elementary physics," on pp. 223-7.

the *Commentary on SY*, Saadia makes the very same statement, albeit with respect to air, whose motion in this work is said to move the "great sphere."[53] Lastly, in his *Commentary on Genesis*, he ascribes the fact that the Earth is maintained "in the middle" to the action of wind (*rīḥ*).[54] It thus appears that Saadia followed the Stoics also in these two respects: First, like the Stoics, he blurred the distinction between the elements air (or wind) and fire (both of which were taken to be constituents of the *pneuma*).[55] Second, and more important, he adheres to a "vortex theory" as an explanation of the Earth's immobility at the center, a view that was explicitly rejected by Aristotle,[56] but which can plausibly be ascribed to the Stoics.[57] Saadia's physics, we can conclude, incorporates significant building blocks of characteristically Stoic origin.

Lastly, let me very tentatively suggest the possibility that Saadia's theory of the soul, too, draws on Stoic motifs. Saadia holds that the human soul is created and that its "quality ... is that of a pure substance similar to the substance of the celestial spheres. When the soul receives the light it becomes more brilliantly illuminated than the [celestial] sphere when it receives its light; its substance becomes more refined than that of the celestial spheres. It is because of this that it is endowed with intellect. ... Were the soul ... of the same nature as the celestial bodies it would, like them, be lacking in Reason. It thus follows that the soul is a substance even finer, clearer, purer and simpler than the celestial spheres."[58] Now the question concerning the source(s) of Saadia's view of soul has been discussed by a number of scholars. Recently, Professor H.A. Davidson has

[53] *Commentary on SY*, Introduction, Q 27:24-5; L 8:15-16; T. 23.

[54] Zucker (ed. and trans.), *Saadya's Commentary on Genesis*, p. 29 (Arabic), pp. 214-5 (Hebrew). A few lines further on Saadia says that it is "the motion of the [heavenly] sphere" that keeps the Earth at the center (*ibid.* pp. 30 and 216, respectively). This view was held also by the Karaite Yefet ben 'Elī (second half of the tenth century); cf. Haggai Ben-Shammai, *The Doctrines of Religious Thought of Abū Yūsuf Ya'qūb al-Qirqisānī and Yefet ben 'Elī*, Ph.D. thesis, The Hebrew University (Jerusalem, 1977), vol. I, pp. 158-61, 163.

[55] Cf. above p. 117 and Verbeke, *L'évolution de la doctrine du pneuma*, pp. 68-71, 82, 89-90 (Chrysippus) and 91, 93, 99 (Panaetius).

[56] *De caelo* 2.13, esp. 295a17 ff. It is reported by Averroes in his epitome of *De caelo*. I consulted the Hebrew translation in Berlin, Staatsbibliothek, MS Orient 1055, fol. 77^r.

[57] Cf. Wolff, "Hipparchus and the Stoic theory of motion," esp. pp. 523-33 and 541-2.

[58] *Doctrines and Beliefs* 6:3, ed. Landauer 193:17-19, 194:4-7; ed. Qafih, 199:9-13, 19-22; translation quoted after Altmann, *Saadya Gaon, Book of Doctrines and Beliefs*, pp. 145-6; partially quoted in Davidson, "Saadia's list," p. 85, n. 44.

drawn attention to the "most striking ... resemblance of Saadia's statement to that of the Ikhwān al-Ṣafā', who describe the soul as created and as of a 'celestial, spiritual' character."[59]

Without going into the matter in any depth, two brief remarks seem apposite. First, it should not be overlooked that, as we have seen, Saadia denied the existence of a fifth, celestial, element, which the Ikhwān al-Ṣafā' upheld. This means that for Saadia (but not for the Ikhwān), the matter of the human soul is the purest part of the most subtle of the four elements (presumably of celestial fire).[60] Since, as noticed above, the Stoics often failed to distinguish between *pneuma* and its two constituents (fire and air), Saadia's theory may well be a variant of the Stoic view of soul. Second, let me suggest the possibility that the relevant doctrine of the Ikhwān al-Ṣafā' itself may have integrated significant Stoic elements. The Ikhwān refer to a "spiritual force" (*quwwa ruḥāniyya*) emanating from the sun and permeating the entire world, including all material substances, to which the continued existence of the cosmos is due. This "spiritual force" they construe as being analogous to the vital heat permeating the body and maintaining it in life.[61] The late Shlomo Pines has convincingly argued that the term *ruḥāniyya* here is a rendering of the Greek *pneuma*.[62] Now, although in most contexts the term, usually in the plural, refers to *pneumata*, which are spiritual beings – angels or demons – as construed in Neoplatonic magic and theurgy, it seems possible that the most comprehensive "spiritual force" is, at least in part, a heir to the Stoic *pneuma*. The fact that this "spiritual force" is conceived after the model of the vital heat, a typical Stoic *topos*, would seem to be consonant with this interpretation. Also, the fact that this "force" is held to emanate from the sun seems to lend credence to this suggestion, seeing that some Stoics (e.g., Cleanthes) took the *hegemonikon* of the world to be

---

[59] Cf. Davidson, "Saadia's list," p. 85, n. 44, who also gives a succinct overview of previous scholarly opinions on the question.

[60] On this interpretation, Saadia's rejection of the theory that the soul is fire (*Doctrines and Beliefs* 6:1, "Third theory," ed. Landauer 190:4-6; ed. Qafiḥ 195:11-15) is directed against the view that the substance of the soul is *terrestrial* fire.

[61] Ikhwān al-Ṣafā', *Rasā'il* (Cairo, 1928), vol. II, pp. 124 ff.; translation in F. Dieterici, *Die Naturanschuung und Naturphilosophie der Araber im zehnten Jahrhundert. Aus den Schriften der lautern Brüder* (Berlin, 1861), p. 155; translated and discussed by Shlomo Pines in his "On the term *ruḥaniyyot* and its origin, and on Judah Halevi's doctrine" (Hebrew), *Tarbiẓ*, 57 (1988): 511-40, on p. 515.

[62] Pines, "On the term *ruḥaniyyot*," pp. 521 ff.

situated in the sun.[63] In sum, then, it would seem possible – although this remains hypothetical – that Saadia's theory of soul, whatever its immediate sources of inspiration, incorporates Stoic elements that are consistent with other points of doctrine he borrowed from Stoicism.[64]

A number of Stoic elements, then, both theological and physical, seem to be worked into Saadia's *Commentary on SY* and his *Doctrines and Beliefs*.[65] Saadia's proximity to the Stoic view can be gauged from the fact that an uncompromising critic of his views, R. Moshe ben Ḥasdai (Taku), writing in the first half of the thirteenth century, had no qualms in ascribing to Saadia the view that "the Air *is* the Godhead."[66] We may therefore conclude that Saadia occupies an important place in the largely still unchartered history of the reception of Stoicism generally, and of Stoic physics in particular, within the theology, philosophy, and science of medieval Arabic culture.[67]

---

[63] Cf. e.g. Long and Sedley, *The Hellenistic Philosophers*, p. 284.

[64] A somewhat related suggestion has been put forward in H. Ben-Shammai, "Al-Qirqisānī's theory of the generation of fire and related theories concerning the change of the elements into one another," in his "Studies in Karaite atomism," *Jerusalem Studies in Arabic and Islam*, 6 (1985): 243-93, on pp. 287-90: Ben-Shammai relates Saadia's views to some Presocratics, especially Anaximenes, whereas I suggest a Stoic ascendency.

[65] Jacob Guttmann [*Die Religionsphilosophie des Saadia* (Göttingen, 1882), pp. 26, 49, 75-6] believed that when Saadia wrote the *Commentary on "Sefer Yeẓira,"* two years before composing his *Doctrines and Beliefs*, he had not yet acquired the extensive philosophical knowledge displayed in the latter. Our conclusions concerning the consistent drawing on Stoic ideas in both works seem to disconfirm that contention.

[66] R. Moshe ben Ḥasdai (Taku), *Ketav Tamim*, ed. R. Kirchheim in *Oẓar Nechmad*, 3 (1860): 58-99, on p. 96. Similarly, he ascribes to Saadia the view that "the Creator ... is an Air which is subtler than anything subtle..." (*ibid.*, p. 64). This, of course, is not Saadia's true position, but it is significant that R. Moshe thought it was.

[67] S. Horovitz's pioneering study "Ueber den Einfluss des Stoicismus auf die Entwickelung der Philosophie bei den Arabern," *Zeitschrift der deutschen morgenländischen Gesellschaft*, 57 (1903): 177-98 has been subjected to severe criticism: cf. e.g., Shlomo Pines, "Études sur Awḥad al-Zamān Abu'l-Barakāt al-Baghdādī," reprinted in *The Collected Works of Shlomo Pines*, vol. I: *Studies on Abu'l-Barakāt al-Baghdādī. Physics and Metaphysics* (Jerusalem/Leiden, 1979), pp. 1-95, on p. 47, n. 188. The monographic study devoted to the subject is far from exhausting it: Fehmi Jadaane, *L'influence du stoïcisme sur la pensée musulmane* (Beirut, 1968). Cf. also Paul Kraus, *Jābir ibn Ḥayyān: Contribution à l'histoire des idées scientifiques dans l'Islam*, vol. II: *Jābir et la science grecque* (= *Mémoires présentés à l'Institut d'Égypte*, vol. 45) [Cairo, 1945; repr. Paris, 1986 and Hildesheim, 1989], pp. 168 ff.; S. van Riet, "Stoicorum veterum fragmenta arabica. À propos de Nemesius d'Émèse," in P. Salmon (ed.), *Mélanges d'Islamologie* (Leiden, 1974), pp. 254-63; Gad Freudenthal, "Clandestine Stoic concepts in mechanical philosophy: The problem of electrical attraction," in J.V. Field and Frank A.J.L. James (eds.), *Renaissance and Revolution:*

132

I conclude with two remarks. First, it has to be stressed that the question concerning the sources through which Saadia became acquainted with Stoic ideas remains open. Cursory attempts to identify works from which Saadia may have borrowed the theory of "second air" or have learnt of the Stoic view of the celestial realm have yielded no results.[68] Consequently, our above conclusions leave us with the problem concerning the identity of the source material that was available to Saadia.

Second, Saadia has repeatedly been characterized as a syncretistic thinker, whose philosophical system is constructed out of an almost haphazard collection of ideas.[69] In view of the above, it would seem that this description needs some qualification. We have seen that Saadia makes consistent use of the same theory of "second air" in both his philosophical works and that, more generally, he draws on Stoic ideas in several contexts. Assuming that Saadia has not simply taken over *en bloc* these ideas from his contemporaries, this suggests the possibility that, at least at some points, he may have selected the "building blocks" of his philosophical edifice with more forethought, discernment, and consistency than is apparent.[70]

Humanists, Scholars, Craftsmen and Natural Philosophers in Early Modern Europe (Cambridge, 1993), pp. 161-72. Some further bibliographical references are given in Josef van Ess, "The logical structure of Islamic theology," in G.E. von Grunebaum (ed.), *Logic in Classical Islamic Culture* (Wiesbaden, 1970), pp. 21-50, on p. 31 f. (n. 55; for calling this reference to my attention I am grateful to Dr. Sarah Stroumsa). In contrast, the reception given to Stoic thought within Latin medieval thought has been the subject of increasing attention. Cf. the masterly overview given in Michael Lapidge, "The Stoic inheritance," in P. Dronke (ed.), *A History of Twelfth-Century Western Philosophy* (Cambridge, 1988), pp. 81-112. Cf. also *id.*, "A Stoic metaphor in late Latin poetry: the binding of the cosmos," *Latomus*, 39 (1980): 817-37 and several of the papers included in Margaret J. Osler (ed.), *Atoms, Pneuma, and Tranquility* (Cambridge, 1991).

[68] Notably the examination of the Arabic translation of Pseudo-Plutarch's *Placita philosophorum*, ed. by A. Badawi in his *Arisṭūṭālīs fī l-nafs*, 2nd edn (Kuwait/Beirut, 1980) as well as by H. Daiber in his *Aetius Arabus. Die Vorsokratiker in arabischer Überlieferung* (Wiesbaden, 1980), which was the source for Saadia's exposition of the theories of soul (cf. above, n. 42). I thank Dr. Y. Tzvi Langermann and Dr. M. Zonta for their suggestions on this subject.

[69] E.g. Guttmann, *Die Religionsphilosophie des Saadia*, pp. 30-1.

[70] I am grateful to Prof. Menachem Kellner for having drawn my attention to this point.

V. REPERCUSSIONS OF SAADIA'S STOIC IDEAS
IN JEWISH MEDIEVAL THOUGHT

Saadia's *Doctrines and Beliefs* and his *Commentary on SY* were paraphrased in Hebrew as early as the eleventh century. The very poetical and flowery style of these anonymous paraphrases helped them to be accepted in circles which, as a rule, were quite hostile to philosophic thought. It thus happened that Saadia's "second air," and the associated doctrine of God's immanence, became constituents of the theology of some early, pre-kabbalistic Jewish mystic circles, particularly among those usually referred to by the collective designation *Ḥasidey Ashkenaz* (German Pietists).[71] In view of our foregoing discussion, it now becomes manifest that the mystics' view of divine immanence and their conception of *Kavod* as an intermediate between God and the world can ultimately be traced back to Stoicism.

Most immediately relevant to us here is the Hasidic doctrine of God's immanence.[72] An early and representative work of this circle, the "Song of Unity," eloquently expresses this immanentism: "Everything is in Thee, and Thou art in everything; Thou fillest every thing and dost encompass it; when everything was created, Thou wast in everything; before everything was created Thou wast everything."[73] Indeed, the idea of God's omnipresence, which has been shown to be borrowed directly from the old Hebrew paraphrase of Saadia's *Doctrines and Beliefs*, appears in all the writings of the German Hasidim. Moreover, some Hasidim associate God's immanence with Saadia's doctrine of the second air. Thus R. Eleazar of Worms (ca. 1165 - ca. 1230) quotes Saadia's arguments concerning the all-pervasiveness of the subtle air, and draws on them to establish God's omnipresence, which is mediated by that air.[74] Air also plays an essential theological role in the *Sefer ha-Ḥayyim* (*The Book of Life*), composed about the year 1200 by an anonymous author who

---

[71] The mystics' indebtedness to Saadia has already been pointed out by scholars in the nineteenth century and has been highlighted particularly by Gershom Scholem and Joseph Dan: cf. Gershom G. Scholem, *Major Trends in Jewish Mysticism* (New York, 1961); Joseph Dan, *The Esoteric Theology of Ashkenazi Ḥasidism* (Hebrew) (Jerusalem, 1968).

[72] Dan, *Esoteric Theology*, pp. 48, 85-7, 100-13, 140-2, 164-70, 171 ff.

[73] Quoted after Scholem, *Major Trends*, p. 108.

[74] Eleazar of Worms, *Perush 'al Sefer Yeẓirah* (Przemyal, 1883), 2ª, 3ª-ᵇ.

134

belonged to the circles of *Ḥasidey Ashkenaz*.[75] The Hasidim also adopted from Saadia the view that it is not God in person Who reveals Himself to man, but rather the created Glory.[76]

This superficial overview suffices to make clear that essential components of the theology of *Ḥasidey Ashkenaz* – most notably, the doctrine of God's (mediated) immanence in the world – have their ultimate source in Stoic philosophy. To be sure, unlike the Stoics, but like Saadia whom they follow, the Hasidim maintain a mediated and not a direct Divine presence in the material world. Yet, the doctrine of immanence itself, and its frequent association with an omnipresent "air" as the intermediary between God and the world, are *in fine* of Stoic origin.

Much the same holds of the doctrine of God's immanence upheld by the poet-philosopher Shlomo ibn Gabirol (ca. 1020-ca.1057). In fact, the late Professor Shlomo Pines has conclusively shown that several formulations in which Ibn Gabirol states the view that God and his Will, which is His Power, are immanent to the entire physical world are literally borrowed from Saadia's *Commentary on SY*.[77]

Lastly, let me mention here a book that has so far gone almost unnoticed by historians. I refer to *Sefer ha-Maskil (The Book of the Perspicacious Sage)*, written by R. Shlomo Simḥa ben Eliezer at Troyes in 1294, and extant in a unique manuscript.[78] In a short paper published in 1983, Prof. Israel Ta-Shma drew attention to this work, to which I have recently devoted a detailed study.[79]

Let it be said from the outset that the sources of *Sefer ha-*

[75] London, British Library, MS Heb. 1055. A transcription of this manuscript was published by the Department of Hebrew Literature, The Faculty of Humanities, The Hebrew University (Jerusalem, 1973).

[76] Scholem, *Major Trends*, pp. 111-13.

[77] Pines, "Quotations from Saadya's *Commentary on the Sefer Yeẓira* in a poem by Ibn Gabirol and in the *Fons vitae*," Appendix II to his "Points of similarity between the exposition of the doctrine of the Sefirot in the *Sefer Yeẓira* and a text of the Pseudo-Clementine Homilies," pp. 122-6.

[78] Moscow, State Library, MS Günzburg 508. I consulted the microfilm no. 16881 of the Institute for Microfilmed Hebrew Manuscripts, The Hebrew University, Jerusalem.

[79] I. Ta-Shma, "*Sefer ha-Maskil*: An unknown Hebrew book from the thirteenth century," (Hebrew), *Jerusalem Studies in Jewish Thought*, 3 (1982/3): 416-38 (English summary on pp. XIII-XIV); Gad Freudenthal, "'The air blessed be He and blessed be His name' in *Sefer ha-Maskil* by R. Shlomo Simḥa of Troyes: Some characteristics of a Stoically-inspired midrashic-scientific cosmology of the thirteenth century," Part One: *Da'at* (Bar-Ilan University, Israel) no. 32-33 (1994): 187-234 (in Hebrew; English abstract, pp. LXVII-LXVIII); Part Two: *ibid.*, no. 34 (1995): 87-129.

Maskil and its intellectual context are still largely an enigma. The work contains a highly idiosyncratic mixture of cosmology, mysticism, *midrash*, and *halakha*, which seem to place it in the vicinity of *Ḥasidey Ashkenaz*. It does not reveal even the faintest acquaintance with science or philosophy: the author presumably never even heard of the *Guide of the Perplexed* and has read none of the numerous scientific and philosophic texts that had been translated into Hebrew in the preceding century and a half. And yet the book expounds a theology that can be qualified as Neo-Stoic.

R. Shlomo holds,[80] that God's Spirit, *Ruaḥ Elohim*, which according to Genesis hovered over the waters before the creation, is "the wondrous and perfect Principle [*'iqqar*], be He blessed for ever and for ever. And this Spirit is necessarily the Air."[81] During creation, the upper and the lower worlds were formed out of this primeval Air. In the upper world, the Air is an infinite brightness; in fact, the firmament is a veil separating us from that heavenly Splendor, and the sun is simply a tiny hole in that veil, allowing an infinitesimal proportion of the celestial Light to reach us. In the terrestrial world, the primal Air gave rise to matter generally and to the atmospheric air in particular, of which R. Shlomo distinguishes a number of kinds. Notwithstanding their different qualities, the upper Air and the lower Air are one and the same, they are branches of the unique Principle. The godhead is *identical* with this cosmic Air: "the concealed and the perceptible and the hidden Air ... is the Creator, is the Maker; He is in everything and everything is in Him, He is the Foundation of everything and superior to everything."[82] The Godhead is thus immanent to His world – He is all-pervading and therefore omniscient, and governs everything: "The King of the Kings of the Kings, the Holy One Blessed be He, is omnipresent and everything is in Him; and His Spirit is everywhere – above and below, inside the earth and above the heaven."[83] Consequently, "there is no place empty of Him, for everywhere His *Kavod* and His *Shekhina* are fully present, and this equally at all times and in all seasons, at this place just as in all other places."[84] Moreover, the unique cosmic

---

[80] The following précis is given on the basis of my paper cited in the previous note.
[81] *Sefer ha-Maskil*, MS Günzburg 508, fol. 14a:40 f.
[82] *Ibid.*, fol. 13b:19-21.
[83] *Ibid.*, fol. 9b:36-7.

136

Air extending throughout the world in fact sustains it – it holds together the entire world and all that is in it. For R. Shlomo, the Air is, just as it had been for Diogenes of Apollonia and for the Stoics, a conscious Being endowed with Reason. Indeed, the Spirit (*ruaḥ*) which man harbors as long as he is alive, and which gives him consciousness and intelligence, is a ramification of the unique Principle.

These few indications are sufficient, I believe, to point out the remarkable affinity between Stoic philosophy and R. Shlomo's theology. To be sure, R. Shlomo has no inkling of the physical ideas of the Stoics, but his theology, although undoubtedly less sophisticated than its Stoic ancestor, is within a hairsbreadth of it. Now the identity of the sources which nourished R. Shlomo's thought still loom in the dark, but presumably his intellectual matrix owes much to *Ḥasidey Ashkenaz*. Therefore, although R. Shlomo does not seem to have been directly acquainted with any of Saadia's writings, presumably some of Saadia's ideas nonetheless reached him indirectly. *Sefer ha-Maskil* can therefore be considered as the most outstanding and impressive instance of Stoic influence which, in all likelihood, reached medieval Jewish thought through the writings of the tenth-century Ga'on of Baghdad.

Stoic physics was one of the great achievements of ancient natural philosophy. Its fructifying effects on early modern science have only recently begun to be appreciated.[85] We should now realize that Stoicism was also a source of inspiration to traditions far removed from scientific thinking, indeed from rational thought altogether. Stoicism provided some such traditions with models, images, and metaphors which proved to be psychologically gratifying and intellectually satisfying.

[84] *Ibid.*, fol. 10a:42-4.

[85] Peter Barker and Bernard R. Goldstein, "Is seventeenth-century physics indebted to the Stoics?", *Centaurus*, 27 (1984): 148-64; Osler, *Atoms*, Pneuma, *and Tranquility*.

L'héritage de la physique stoïcienne dans la pensée juive médiévale (Saadia Gaon, les dévots rhénans, *Sefer Ha-Maskil*)[1]

Notre propos dans ce bref article est de montrer que la pensée juive médiévale a intégré des éléments provenant de la philosophie stoïcienne et plus précisément de sa théorie du *pneuma*. Cette affirmation va à l'encontre d'une idée reçue, car on a l'habitude de penser que, s'agissant de la réception de la philosophie grecque, les penseurs juifs du Moyen Âge ont intégré surtout des idées aristotéliciennes et, à un degré moindre, néo-platoniciennes. Or cette vision des choses est fondée tacitement sur la prémisse selon laquelle la transmission des idées philosophiques s'est effectuée, au Moyen Âge, par le biais des traductions de textes canoniques[2]. Il s'avère cependant — et dans ce qui suit nous nous efforcerons de démontrer cette thèse — que les idées empruntent parfois des chemins « souterrains », invisibles. En l'occurrence, le Moyen Âge ne disposait d'aucun ouvrage stoïcien et n'avait du stoïcisme qu'une connaissance très fragmentaire, notamment par le biais de citations et de rapports doxographiques[3]; et pourtant, on peut discerner dans la pensée juive médiévale certaines idées, ayant trait notamment à l'existence d'un Air

---

1. Pour l'essentiel, l'article qui suit résume en français les résultats de nos recherches présentés auparavant en hébreu dans notre article « 'L'Air béni soit-Il et béni Son nom' dans le *Sefer ha-Maskil* de R. Shlomo Simḥa de Troyes : quelques caractéristiques d'une cosmologie midrashique-scientifique d'inspiration stoïcienne du XIIIᵉ siècle » (en hébreu), *Da'at*, Israël, Université de Bar-Ilan, n° 32-33, 1994, p. 187-234 (première partie) et n° 34, 1995, p. 87-129 (deuxième partie), ainsi qu'en anglais dans l'article « Stoic Physics in the Writings of R. Sa'adia Ga'on al-Fayyumi and Its Aftermath in Medieval Jewish Mysticism », *Arabic Sciences and Philosophy*, 6, 1996, p. 113-136.

2. Sur les traductions hébraïques médiévales, on consultera l'ouvrage monumental de Moritz STEINSCHNEIDER, *Die Hebraeischen Übersetzungen des Mittelalters und die Juden als Dolmetscher*, Berlin, 1893; réimprimé Graz, 1956. Cf. aussi nos articles « Les sciences dans les communautés juives médiévales de Provence : leur appropriation, leur rôle », *Revue des études juives*, 152, 1993, p. 29-136 et « Science in the Medieval Jewish Culture of Southern France », *History of Science,* 33, 1995, p. 23-58, et surtout Mauro ZONTA, *La filosofia antica nel Medioevo ebraico*, Brescia, Paideia, 1996.

3. Pour une analyse complète des connaissances actuelles sur la « voie diffuse » de la transmission des idées stoïciennes dans le monde musulman médiéval cf. Dimitri GUTAS,

(*pneuma*) subtil originel par le biais duquel s'effectue l'immanence divine dans le monde, dont nous suggérerons qu'elles sont l'aboutissement d'une tradition de pensée stoïcienne. Nous nous attarderons ici surtout sur le *Sefer ha-Maskil* (« le Livre du sage »), un ouvrage d'orientation mystique, écrit en 1294 à Troyes, et qui assimile Dieu à l'Air originel dont le monde tout entier est issu ; il présente ainsi des parentés frappantes avec la pensée stoïcienne.

I

Rappelons très sommairement quelques idées grecques relatives au *pneuma*. Commençons par Anaximène, un présocratique du VIe siècle[4]. À la question fondamentale des penseurs de Milet : quelle est la substance invariable sous-jacente à toutes choses, Anaximène répondit par sa doctrine selon laquelle l'élément universel (*arché*) est l'air (Anaximène utilise tantôt le terme de *pneuma* tantôt celui de *aer*). L'air est la matière dont les substances se forment par des processus de condensation et de raréfaction ; cet air est infini et Anaximène l'identifie avec Dieu. L'air est aussi la cause de la cohésion de tout ce qui existe — de chaque être vivant et du monde tout entier : « de même que notre âme, qui est air, nous soutient, de même le vent et l'air entourent le monde tout entier »[5]. Anaximène pose ainsi déjà l'analogie entre le rôle de l'air dans le vivant (microcosme) et son rôle dans le monde (macrocosme), qui nous occupera dans la suite.

La doctrine d'Anaximène a été développée par Diogène d'Apollonie (c. 440-430) un siècle plus tard[6]. Diogène soutient, lui aussi, que l'air, qui est Dieu, est l'élément unique du monde. Il considère que dans sa forme

« Pre-Plotinian Philosophy in Arabic (Other than Platonism and Aristotelism): A Review of the Sources », *in* Wolfgang HAASE (éd.), *Aufstieg und Niedergang der römischen Welt ANRW*). Teil II: *Principat*, Bd. 36: *Philosophie, Wissenschaften, Technik, 7*, teilband: *Philosophie*, Berlin et New York, Walter de Gruyter, 1994, p. 4939-4973, notamment p. 4959-4962. (Je remercie M. A. Elamrani d'avoir attiré mon attention sur ce travail.)

4. Cf. G. S. KIRK, J. E. R. RAVEN et M. SCHOFIELD, *The Presocratic Philosophers*, Cambridge, Cambridge University Press, 2e éd., 1983, p. 143-162 ; W. K. C. GUTHRIE, *A History of Greek Philosophy*, t. 1: *The Earlier Presocratics and the Pythagoreans*, Cambridge, Cambridge University Press, 1962, p. 115-140.

5. KIRK, RAVEN et SCHOFIELD, *The Presocratic Philosophers, op. cit.*, p. 158.

6. Cf. KIRK, RAVEN et SCHOFIELD, *The Presocratic Philosophers, op. cit.*, p. 434-452 ; W. K. C. GUTHRIE, *A History of Greek Philosophy*, t. 2: *The Presocratic Tradition from Parmenides to Democritus*, Cambridge, Cambridge University Press, 1965, p. 362-381.

la plus pure, l'air est raison ou intellect (*noesis*). La nature consiste donc en un élément doué de raison, ce qui explique la téléologie dont elle fait preuve. L'air donne aux animaux et aux hommes et la vie et l'intelligence, de sorte qu'en son absence la vie et l'intelligence cessent simultanément. Ce sont les différences de qualité et de quantité de l'air dont dispose tout être vivant (y compris l'homme) qui déterminent les différences de leur intelligence.

Les générations suivantes des penseurs grecs n'ont accordé à l'air aucun rôle théologique, bien que, logiquement, la relation entre l'air et la vie n'a jamais été mise en doute. En effet, la physique aristotélicienne a nettement séparé le monde sublunaire (dans lequel l'air n'est qu'un des quatre éléments) du monde supralunaire, dont la matière est le « cinquième élément », qui est tout à fait étranger aux quatre éléments sublunaires[7]. Dans ce cadre de pensée, l'élément céleste, qui seul est éternel et non sujet à des changements, est aussi le seul à posséder une signification théologique[8]. Tandis que chez Diogène la différence entre l'air des cieux et l'air des substances terrestres n'était qu'une différence de degré de pureté, dans le système de pensée aristotélicien la relation entre le cinquième élément et une des matières sublunaires ne saurait être une identité, mais tout au plus une relation d'analogie[9]. Chez Aristote, Dieu était un intellect et, partant, était tout à fait transcendant.

La philosophie stoïcienne, dont les débuts coïncident avec la mort d'Aristote, a produit un retournement de la situation en créant une philosophie physico-théologique dans laquelle Dieu est immanent au monde. Nous nous intéresserons ici surtout à la première période de la philosophie stoïcienne, notamment au système de Chrysippe (c. 281 à 208); nous ne mentionnerons que quelques points qui sont directement pertinents pour notre discussion.

La philosophie de la nature stoïcienne se caractérise par son immanentisme, son monisme, et son matérialisme[10]. L'idée maîtresse de la philo-

---

7. Paul MORAUX, « Quinta essentia », *in* PAULY-WISSOWA, *Realencyclopädie*, XXIV(1), Stuttgart, 1963, col. 1171-1263.

8. Cf. par exemple ARISTOTE, *De caelo* I, 9, 270a30 *sq.*, II, 1, 284a12 *sq.* et cf. W. K. C. GUTHRIE, *The Greeks and Their Gods*, Londres, Methuen, 1950.

9. Une telle analogie est postulée par Aristote dans un seul passage, dans lequel il soutient qu'il existe une analogie entre la matière céleste et le *pneuma* ou la chaleur innée se trouvant dans le sperme. Cf. ARISTOTE, *De gen. an.* II, 3, 736b37. Nous avons interprété la nature de cette analogie dans Gad FREUDENTHAL, *Aristotle's Theory of Material Substance. Form and Soul, Heat and Pneuma*, Oxford, Clarendon Press, 1995, p. 114-119.

10. Robert B. TODD, « Monism and Immanence: The Foundations of Stoic Physics », *in* John M. RIST (éd.), *The Stoics*, Berkeley, University of California Press, 1978, p. 137-160.

sophie de la nature stoïcienne est que le cosmos est traversé par une substance ténue appelée *pneuma*, un terme dont la première signification est « vent » ou « respiration ». Ce *pneuma* subtil donne sa cohésion à la matière passive et amorphe et il est aussi ce qui en fait des substances spécifiques (un métal, une plante, un animal, etc.). Le *pneuma* est également la cause de la cohésion et de l'unité du cosmos tout entier.

Afin de saisir les idées que les Stoïciens ont associées avec le concept de *pneuma*, il convient de les considérer à la lumière des conceptions biologiques admises alors, qui expliquaient les fonctions de la vie par référence au *pneuma* inné[11]. Selon ces conceptions, le *pneuma* est l'agent physiologique qui accomplit les fonctions de l'âme, à savoir la croissance, le mouvement, la perception, la reproduction, etc[12]. Dans l'esprit de leur monisme caractéristique, les Stoïciens ont franchi un pas supplémentaire, soutenant que le *pneuma* est non l'instrument de l'âme, mais *identique* à celle-ci : il a son centre dans le cœur et, étant continu à travers le corps animal tout entier, il domine ses opérations. Les Stoïciens ont fait de cette doctrine biologique le fondement de leur cosmologie. Ils percevaient le cosmos comme un animal (*zoon*) gouverné par un *pneuma* cosmique, qui y remplit précisément les mêmes fonctions que remplit le *pneuma* inné dans le corps animal. Par conséquent, de même que, dans l'animal, le *pneuma* est l'âme qui le dirige, de même le *pneuma* cosmique, qui est doué de raison, dirige le cosmos tout entier. En effet, pour les Stoïciens, le *pneuma* est Dieu (*theos*) qui dirige totalement le monde, une conception que nous avons déjà rencontrée chez Anaximène et chez Diogène. Dieu peut diriger le monde car il est totalement immanent à celui-ci — le *pneuma* se trouve partout et il traverse toutes les substances. (Afin de soutenir cette idée, les Stoïciens ont développé leur doctrine révolutionnaire qui admet un mélange total, dans lequel deux substances peuvent coexister simultanément dans le même lieu[13]).

Mentionnons encore la littérature suivante sur la physique stoïcienne : G[érard] Verbeke, *L'Évolution de la doctrine du pneuma du Stoïcisme à S. Augustin. Étude philosophique,* Paris, Desclée de Brouwer et Louvain, Éditions de l'Institut supérieur de philosophie, 1945; Max Pohlenz, *Die Stoa,* 2 vols., Göttingen, Vandenhoeck et Ruprecht, 6e éd., 1984; S. Samburský, *The Physics of the Stoics,* Londres, Routledge and Kegan Paul, 1957; réimpression: Princeton, Princeton University Press, 1987; Michael Lapidge, « Archai and Stoicheia: A Problem in Stoic Cosmology », *Phronesis,* 18, 1973, p. 240-278; *idem,* « Stoic Cosmology », in John M. Rist (éd.), *The Stoics, op. cit.,* 161-185.

11. Cf. Émile Brehier, *Chrysippe et l'ancien Stoïcisme,* Paris, PUF, 2e éd., 1951, p. 121.

12. Cf. pour un tableau général; Franz Rüsche, *Blut Leben und Seele. Ihr Verhältnis nach Auffassung der griechischen und hellenistischen Antike, der Bibel und der alten Alexandrinischen Theologen (= Studien zur Geschichte und Kultur des Altertums,* Ergänzungsband 5), Paderborn, F. Schöningh, 1930.

13. Une discussion détaillée dans Richard Sorabji, *Matter, Space, and Motion,* New York, Cornell University Press, 1988, p. 79-105.

Le *pneuma* qui pénètre le corps lui donne sa cohésion ainsi que son *tonos*, en vertu duquel la matière devient telle ou telle substance. Au niveau le plus bas du *tonos* (appelé *hexis*), le *pneuma* donne à la matière une cohésion dont résultent des substances dures, telles que des pierres dans la nature inanimée et des os ou des tendons dans le corps animal. Les degrés suivants du *tonos* du *pneuma* (*phusis, psuche*), correspondent aux végétaux et aux animaux. Le degré le plus élevé, où le *pneuma* est le plus concentré, est celui de l'intellect humain et divin. Pour certains Stoïciens, le *pneuma* est composé des deux éléments actifs — l'air et le feu (non le feu ordinaire, mais le feu créatif [*pur technikon*], qui est analogue à la chaleur vitale des êtres vivants) — qui agissent sur les éléments passifs — la terre et l'eau. Le *pneuma* divin est donc ce qui donne des formes à la matière (en analogie avec l'intellect agent de la philosophie médiévale), il génère le monde. Il semble que le pouvoir créatif du *pneuma* et en particulier sa capacité de donner de la cohésion (à des substances particulières et au monde en sa totalité) lui proviennent surtout de l'air qu'il contient; en fait, les termes de *pneuma* et d'« air » sont souvent employés comme synonymes.

Dans le cadre de la pensée stoïcienne, théologie et philosophie de la nature forment un tout. Le Dieu-*pneuma* est omniprésent et immanent à la nature — c'est ainsi qu'il dirige le monde. Se pose ainsi la question, déjà considérée par les premiers Stoïciens et qui revient chez certains penseurs juifs médiévaux : comment se fait-il que Dieu puisse « résider » en des lieux souillés ?[14]

Ce résumé, bref et sélectif, nous permettra d'identifier des éléments de la tradition stoïcienne dans des sources hébraïques et notamment dans le *Sefer ha-Maskil*.

II

Certains *midrashim*, composés vraisemblablement dans les premiers siècles de notre ère en Palestine, trahissent des influences stoïciennes. Un *midrash* met en scène un rabbin qui s'efforce de démontrer à son interlocuteur païen que le monde ne subsiste qu'en vertu d'un air ou un souffle (*ruah*) particulier qui le « supporte » (*sovel*). Le rabbin fait venir des chameaux chargés, puis les fait étrangler. Il invite ensuite son interlocuteur

---

14. Cf. POHLENZ, *op. cit.*, t. 1, p. 95, t. 2, p. 54; LAPIDGE, « Stoic Cosmology », art. cit., p. 170.

à les faire à nouveau tenir debout. Pourquoi cela ne serait-il pas possible, s'écrie-t-il devant la perplexité de son interlocuteur, puisque je ne leur ai enlevé aucune partie? « D'où tu apprends que le monde ne subsiste ['*omed*] qu'en vertu de l'air, qui est suprême et qui préexistait (à la création), car il est dit « le Souffle [*ruaḥ*] de Dieu flotta » (Gen. 1:2) »[15]. Un passage important du Talmud de Babylone fait état de plusieurs points d'analogie entre Dieu et l'âme (*neshamah* qui remplace ici l' « air » du texte *midrashique,* mais que d'autres passages identifient à ce dernier) : Dieu et l'âme (ou l'air) se ressemblent par le fait de remplir l'un le monde, l'autre le corps; par le fait de pouvoir voir, tout en étant invisibles; par le fait de nourrir le monde ou le corps; et enfin par le fait d'être purs[16]. Une autre version de ce passage ajoute l'idée typiquement stoïcienne selon laquelle de même que l'âme fait subsister le corps, Dieu fait subsister le monde[17].

Le passage talmudique, de même que les *midrashim* qui véhiculent la même idée, exprime des conceptions typiquement stoïciennes. La principale d'entre elles est l'analogie entre le rôle de l'âme dans le corps et celui de Dieu dans le monde, d'où la notion de l'immanence divine dans le monde. Il est vrai que la notion stoïcienne, selon laquelle le monde ne subsiste qu'en vertu de l'air, n'est pas exprimée explicitement dans le passage talmudique, mais elle se trouve dans des versions parallèles du même passage, selon lesquelles Dieu — ou plutôt Son Souffle— « soutient [*sovel*] le monde ». En effet, le caractère stoïcien du passage est obscurci par le fait que le terme d'âme remplace celui de souffle (*ruaḥ*), que l'on trouve dans les *midrashim*[18]. Reste que le passage talmudique exprime clairement l'idée de l'immanence divine; encore s'agit-il d'une immanence

---

15. *Midrash Temura ha-Shalem,* chap. 5, imprimé *in* S. A. WERTHEIMER, *Batey Midrashot,* nouvelle édition par A. WERTHEIMER, t. 2, Jérusalem, 1953, p. 193-194. (Une version légèrement différente a été imprimée *in* A. YELLINEK (éd.), *Beit ha-Midrash,* Jérusalem, 1938, t. 1, p. 109.) Une autre version de ce *midrash* se trouve in *Midrash Tanhuma, Parashat Bereshit,* § 5, Jérusalem, 1969, p. 7b.

16. *Berakhot 10b;* cf. *Devarim Rabba 2/37* sur l'équivalence d' « âme » et d' « air ».

17. *Midrash Tehillim ha-mekhune Shoḥer tov 103:4,* S. BUBER (éd.), réimpression New York, 1948, p. 217a. Dans un *midrash* publié par YELLINEK, on trouve cette formule : « de même que l'air [*ruaḥ*] supporte [*sovel*] le monde tout entier, de même l'âme de l'homme supporte [*sovel*] son corps », *Beit ha-Midrash, op. cit.,* ch. 5, p. 57. Dans la poésie hébraïque, l'image selon laquelle Dieu « supporte » le monde revient fréquemment à toutes les époques. Cf. L. ZUNZ, *Die synagogale Poesie des Mittelalters,* Francfort-sur-le-Main, 2ᵉ éd., 1920, p. 509-510 (« Beilage 26 »).

18. Juda Bergmann avait déjà émis l'idée selon laquelle le passage talmudique reflète des idées stoïciennes. Cf. [Juda] BERGMANN, « Die stoische Philosophie und die jüdische Frömmigkeit », in *Judaica. Festschrift zu Hermann Cohens siebzigstem Geburtstage,* Berlin, Bruno Cassirer, 1912, p. 145-166. En revanche le caractère stoïcien de ce passage n'a pas été

non matérialiste, contrairement à l'immanentisme matérialiste des Stoïciens qui est à son origine. En fait, ce passage talmudique fut le point de départ de la théologie immanentiste, tout à fait non matérialiste, que l'on trouve chez les Dévôts rhénans. Nous y reviendrons dans ce qui suit.

Les idées stoïciennes apparaissent très clairement chez Rav Saadia Gaon, particulièrement dans son *Commentaire du « Livre de la Création »*, composé en arabe à Baghdad en 931[19]. On sait que Saadia développe une conception selon laquelle les différentes manifestations divines furent des apparitions réelles et qu'il rejette l'idée qu'il faille les interpréter allégoriquement[20]. Saadia avance l'idée selon laquelle ce ne fut pas Dieu lui-même qui apparut aux prophètes, mais Sa Gloire (*Kavod*), qu'il identifie d'une part avec la Présence divine (*Shekhinah*) et le Trône divin et, d'autre part, avec le Souffle divin (*ru'aḥ Elohim*), qui est une chose créée[21]. Ce souffle (ou air), que Saadia nomme « le second air subtil », est bien entendu distinct de [l'« air visible » (atmosphérique). Ici nous nous intéressons surtout à la manière dont Saadia conçoit la relation entre Dieu, le second air et le monde. Il ouvre la discussion en avançant qu'afin d'éviter l'anthropomorphisme, il faut procéder par un processus cognitif qui progresse par « analogies ». Il convient d'abord de réaliser que la relation du Créateur à son monde est la même que la relation entre la vie (*ḥayât*) et l'être vivant; « par analogie » on peut donc dire que Dieu est « la vie du monde »[22]. De cette étape du raisonnement se dégage déjà la conséquence que Dieu est immanent au monde : « comme nous voyons que la vie existe dans chaque partie et dans chaque tout de l'animal, de même Lui — qu'il soit exalté — se trouve dans chaque partie et dans chaque tout du monde »[23].

---

remarqué par le grand savant Saul LIEBERMAN; cf. son article « How Much Greek in Jewish Palestine ? », *in* Alexander ALTMANN (éd.), *Biblical and other Studies*, Cambridge, Mass., Harvard University Press, 1963, p. 123-141.

19. Pour ce qui suit, nous nous appuyons notamment sur notre article « Stoic Physics in the Writings of R. Sa'adia Ga'on al-Fayyumi and its Aftermath in Medieval Jewish Mysticism », art. cit. (n. 1). La meilleure introduction générale à la pensée de Saadia demeure Henry MALTER, *Saadia Gaon. His Life and Works*, Philadelphie, The Jewish Publication Society of America, 5702-1942.

20. Alexander ALTMANN, « Saadya's Theory of Revelation: Its Origin and Background », dans ses *Studies in Religious Philosophy and Mysticism*, Londres, Routledge and Kegan Paul, 1969, p. 140-160 aux p. 145-147.

21. Nous utilisons l'édition du texte arabe et sa traduction en hébreu moderne : Yosef QAFIH (éd.), *Sefer Yeṣirah [Kitâb al-Mabâdi] ʿim Perush ha-Gaon Rabbenu Saʿadiah b. R. Yosef Fayyumî z.l.*, Jérusalem, 1972. Nous avons aussi consulté *Commentaire sur le Séfer Yeṣirah ou Livre de la création, par le Gaon Saadya de Fayyoum*, publié et traduit par Mayer LAMBERT, Paris, Émile Bouillon, 1891.

22. Y. QAFIH, p. 106, l. 7; trad. Lambert, p. 91.

23. Y. QAFIH, p. 106, l. 9-11; cité d'après la traduction de Lambert, p. 92.

Mais, manifestement, Dieu ne saurait être identifié à la vie et, effectivement, ce qui vient d'être dit n'est qu'une « première approximation », une première étape sur le chemin cognitif proposé par Saadia. Dans la deuxième et dernière étape, notre pensée se hisse au degré suprême qui est celui de la connaissance par l'intellect. Nous apprenons maintenant que Dieu n'est pas la vie du monde, mais son intellect (ʿaql al-ʿâlam)[24]. Quelle est alors la relation entre Dieu et le monde ? Saadia développe sa conception immanentiste de la façon suivante : s'agissant des animaux, le corps est la demeure (maḥall) de la vie ; mais dans l'animal rationnel, c'est-à-dire l'homme, au sein duquel les éléments sont mélangés dans un équilibre parfait, la vie devient à son tour la demeure de l'intellect[25]. Saadia exprime d'ailleurs la même idée dans son Tafsîr (paraphrase en arabe) de Genèse 2:7 : là où le texte hébraïque affirme que lorsque Dieu donna le souffle à l'homme celui-ci devint « une âme vivante » (nefesh ḥayyah), la traduction de Saadia affirme que le souffle de vie donna à l'homme ipso facto « une âme rationnelle » (nafs nâtiqa)[26]. Dans l'homme, donc, le corps est la demeure de la vie, qui, à son tour, est la demeure de l'intellect, chacune de ces trois entités étant plus fine, ou plus subtile, que la précédente. Cette idée fournit la base pour la théorie de l'immanence divine de Saadia et clarifie sa conception du rôle de l'air. Selon Saadia, les trois entités qui existent dans le microcosme — le corps, la vie, l'intellect — ont des entités qui leur sont analogues dans le macrocosme : le monde, le second air, et Dieu. La relation au monde du second air, qui est « simple et subtil » (basîṭ, laṭîf), est la même que la relation de la vie au corps ; et la relation du second air à Dieu est la même que la relation de la vie de l'homme à son intellect. Cette idée s'exprime dans la relation suivante :

corps : vie : intellect = monde : second air : Dieu (= l'intellect du monde)

24. Y. Qafih, p. 106, l. 11-12 ; 24, trad. Lambert, p. 92.
25. Y. Qafih, p. 106, l. 14-15, trad. Lambert, p. 92.
26. Œuvres complètes de R. Saadia Josef al-Fayyoûmî, publiées sous la direction de J. Derenbourg, vol. 1, version arabe du Pentateuque, Paris, Ernest Leroux, 1893, partie arabe, p. 7, l. 9-10 ; trad. fr. p. 4. Cf. à ce propos également Saadya's Commentary on Genesis, edited with Introduction, Translation and Notes by Moshe Zucker, New York, The Jewish Theological Seminary of America, 1984, p. 18 (texte), p. 191 (traduction en hébreu moderne) ; cf. aussi p. 296. Il y a lieu de supposer que Saadia, bien qu'il utilise le terme distinctement philosophique de nâtiqa, prolonge une ancienne tradition qui remonte à la traduction de la Bible par Onkelos, qui traduit notre phrase par « wa-hawat be-adam le-ruaḥ memalelâ » (= esprit parlant) ; cf. A. Schmiedel, « Parallelen zwischen Onkelos und Saadja », Monatsschrift für Geschichte und Wissenschaft des Judentums, 46, 1902, p. 358-363, à la p. 361 ; ailleurs, Schmiedel observe que, de manière générale, « Saadja im Grossen und Ganzen in Onkelos seinen kundigen Wegweiser erblickte » ; cf. idem, « Saadja und Onkelos », Monatsschrift für Geschichte und Wissenschaft des Judentums, 46, 1902, p. 84-88. Pour notre propos, il est particulièrement intéressant de noter que la traduction d'Onkelos peut elle-même refléter des influences stoïciennes ; cf. [Jacob Immanuel] Neubürger, « Onkelos und die Stoa », Monatsschrift für Geschichte und Wissenschaft des Judentums, 22, 1873, p. 566-568.

L'analogie postulée par Saadia entre l'homme et le monde, entre le micro-cosme et le macrocosme, s'apparente manifestement à la pensée stoïcienne. A l'instar des Stoïciens, Saadia postule l'existence d'un air cosmique omni-présent, auquel il attribue des fonctions physiques et théologiques. Cepen-dant, tandis que les Stoïciens postulaient une véritable *identité* entre Dieu et le *pneuma* cosmique, Saadia, lui, accorde au second air un rôle plus humble. En effet, en continuité avec une tradition juive établie[27] et probablement aussi sous l'influence des traditions philosophiques, Saadia postule une con-ception transcendante de Dieu : en tant que « intellect du monde », Dieu est séparé d'avec la matière, y compris d'avec le second air. Malgré cette dif-férence, cependant, les rôles que Saadia attribue au second air s'apparentent à ceux que les Stoïciens avaient attribués au *pneuma* divin : Saadia affirme que la volonté divine se répand dans le monde à travers le second air qu'elle met en mouvement (« comme la vie meut le corps »), de sorte que c'est par le biais du second air que Dieu gouverne le monde[28]. Ainsi, bien que le second air ne soit pas Raison, il est la demeure de « l'Intellect du monde » et il est la substance matérielle à travers laquelle Dieu exerce son gouver-nement sur le monde. Cette conception garde manifestement des éléments centraux de la physique et de la théologie stoïciennes. Certains passages dans lesquels Saadia avance que des substances telles que des pierres, des os, et des tendons sont dures à cause de l'air qui les traverse s'inspirent visible-ment de la physique stoïcienne. Il en va de même pour d'autres explications de phénomènes naturels, avancés par Saadia dans son grand ouvrage philo-sophique *Doctrines et Croyances*[29]. Tout cela tend à confirmer la thèse de l'inspiration stoïcienne de la doctrine saadienne du second air[30]. Notons,

27. Etablie mais non exclusive : sur des courants de pensée qui niaient la transcendance divine cf. J. ABELSON, *The Immanence of God in Rabbinical Literature,* Londres, Macmillan, 1912.
28. Y. QAFIH, p. 106, l. 18-21, trad. p. 92. Cf. aussi Y. QAFIH, p. 110, *sharḥ* de la deuxième *halakhah,* trad. p. 96.
29. Pour davantage de détails cf. notre « Stoic Physics in the Writings of R. Saʿadia Gaʾon al-Fayyumi », art. cit. (n. 1), p. 122-132.
30. Le premier à noter l'origine stoïcienne de la doctrine du second air fut S. HOROVITZ dans son article « Ueber den Einfluss der griechischen Philosophie auf die Entwicklung des Kalam », *Jahres-Bericht des jüdisch-theologischen Seminars Fraenckel'scher Stiftung [1909],* Breslau, 1909, p. 1-91 et p. 42-43. MALTER, *Saadia Gaon,* p. 188-189, prend à son compte cette thèse. Le regretté professeur Shlomo PINÈS a récemment attiré l'attention sur des com-posantes néo-platoniciennes dans la pensée de Saadia. Cf. son article « Points of Similarity between the Exposition of the Doctrine of the Sefirot in the *Sefer Yezira* and a Text of the Pseudo-Clementine Homilies. The Implications of this Resemblance », *Proceedings of the Israel Academy of Sciences and Humanities,* vol. VII, n° 3, Jérusalem, 1989, p. 63-142, notamment p. 122-132. Il me semble qu'il n'y a pas de contradiction entre ma thèse et celle du professeur Pinès (qui n'aborde pas la question de l'origine de la notion du second air).

pour terminer, que nous ignorons presque tout de la manière dont les idées stoïciennes ont pu être connues de Saadia[31].

III

Le *Commentaire du « Livre de la Création »* de Saadia a été paraphrasé en hébreu très tôt, dès le XIe siècle[32]. Ces paraphrases anonymes ont véhiculé les idées de Saadia sur le second air dans un langage certes poétique, mais qui ne laisse pas d'être précis. Il se trouve ainsi que les idées de Saadia sur le second air et la doctrine connexe sur l'immanence divine ont été reprises dans certains cercles mystiques pré-kabbalistiques, en particulier par ceux que l'on désigne par le nom collectif de *Ḥasidey Ashkenaz* (les Dévôts rhénans)[33]. En effet, Gershom Scholem et Joseph Dan ont montré, à la suite de chercheurs du siècle dernier, que la distinction, admise par tous les courants de pensée des Dévôts rhénans, entre la divi-

31. L'étude pionnière de S. Horovitz « Ueber den Einfluss des Stoicismus auf die Entwickelung der Philosophie bei den Arabern », *Zeitschrift der deutschen morgenländischen Gesellschaft*, 57, 1903, p. 177-198 a été vivement critiquée; cf. p. ex. Shlomo Pines, « Etudes sur Awhad al-Zamân Abu'l-Barakât al-Baghdâdi », réimprimé in *The Collected Works of Shlomo Pines*, vol. I, *Studies on Abu'l-Barakât al-Baghdâdî. Physics and Metaphysics*, Jérusalem, Magnes Press et Leyde, Brill, 1979, p. 1-95 et p. 47, n. 188. La monographie consacrée à ce sujet est loin de l'épuiser : Fehmi Jadaane, *l'Influence du Stoïcisme sur la pensée musulmane*, Beirut, 1968. Cf. également Paul Kraus, *Jâbir ibn Hayyân: Contribution à l'histoire des idées scientifiques dans l'Islam;* vol. II, *Jâbir et la science grecque* (= *Mémoires présentés à l'Institut d'Égypte*, vol. 45, Le Caire, 1945 [réimpressions : Paris, Les Belles Lettres, 1986 et Hildesheim, Olms, 1989], p. 168 *sq.;* S. van Riet, « Stoicorum veterum fragmenta arabica. A propos de Nemesius d'Emese », *in* P. Salmon (éd.), *Mélanges d'Islamologie*, Leyde, 1974, p. 254-263; Gad Freudenthal, « Clandestine Stoic Concepts in Mechanical Philosophy: The Problem of Electrical Attraction », *in* J. V. Field et Frank A. J. L. James (eds.), *Renaissance and Revolution: Humanists, Scholars, Craftsmen and Natural Philosophers in Early Modern Europe*, Cambridge, Cambridge University Press, 1993, p. 161-172. D'autres références bibliographiques sont données *in* Josef van Ess, « The Logical Structure of Islamic Theology », *in* G. E. von Grunebaum (éd.), *Logic in Classical Islamic Culture*, Wiesbaden, Otto Harrassowitz, 1970, p. 21-50 et p. 31 *sq.* (n. 55; je remercie Mme le professeur Sarah Stroumsa d'avoir attiré mon attention sur cette référence.) Pour une synthèse récente des recherches sur la transmission du stoïcisme au Moyen Âge arabe, cf. D. Gutas, art. cit. (n. 3). En revanche, la réception des idées stoïciennes en Occident est devenue récemment le sujet d'une attention grandissante. Cf. l'excellente étude de Michael Lapidge, « The Stoic Inheritance », *in* P. Dronke (éd.), *A History of Twelfth-Century Western Philosophy*, Cambridge, Cambridge University Press, 1988, p. 81-112. Cf. également *idem*, « A Stoic Metaphor in Late Latin Poetry: the Binding of the Cosmos », *Latomus*, 39, 1980, p. 817-837, ainsi que plusieurs études dans Margaret J. Osler (éd.), *Atoms, Pneuma and Tranquility*, Cambridge, Cambridge University Press, 1991.

32. Cf. Malter, *Saadia Gaon*, p. 355-359.

33. Pour une première approche, cf. Gershom G. Scholem, *Major Trends in Jewish Mysticism*, New York, Schocken, 3e éd., 1961, p. 80-118; Joseph Dan, *Hasidism in Medieval Germany* (en hébreu), Tel Aviv, 1992.

nité cachée (le Créateur), qui ne se révèle pas, et la divinité qui se révèle (la Gloire divine, la Présence divine), remonte à Saadia. Il en va de même pour la notion de l'immanence divine, qui est une des pierres angulaires de leur pensée[34]. Il est vrai que chez les Dévôts rhénans l'immanence divine est directe, tandis que chez Saadia le second air (la Gloire divine) lui sert d'intermédiaire, mais l'idée même de l'immanence divine a été suggérée aux Dévôts rhénans par Saadia. A la lumière de la discussion précédente nous pouvons maintenant constater que cette conception immanentiste est, en dernier lieu, d'origine stoïcienne. Certes, l'immanentisme des Dévôts rhénans est immatérialiste, contrairement au matérialisme de leurs aînés stoïciens, mais l'immanentisme lui-même est d'origine stoïcienne. La notion saadienne de l'air en tant qu'entité physique ayant une signification théologique a, elle aussi, été reprise par certains mystiques[35].

Les idées des Dévôts rhénans fournissent probablement le contexte dans lequel a vu le jour le *Sefer ha-Maskil*, qui sera au centre de notre intérêt dans ce qui suit.

Écrit à Troyes en 1294 par R. Shlomo Simḥa ben Eliezer, un descendant de Rachi dont on ignore presque tout, le *Sefer ha-Maskil* n'a été conservé que dans un seul manuscrit[36]. Dans un bref article publié en 1983, le professeur Israël Ta-Shma de l'Université hébraïque de Jérusalem a attiré l'attention sur ce texte[37], auquel nous avons, pour notre part, consacré une étude détaillée[38]. Le *Sefer ha-Maskil* présente un mélange assez singulier de cosmologie, mysticisme, *midrash* et *halakhah*. Ses sources restent presque entièrement inconnues ; l'ouvrage ne trahit pas la moindre connaissance de la science et de la philosophie juive médiévales — probablement R. Shlomo n'a-t-il même pas entendu le nom du *Guide des égarés* et n'a-

---

34. Cf. Gershom SCHOLEM, *op. cit.*, p. 108-116 (p. 374, n. 91 pour des références aux travaux du siècle dernier) et Joseph DAN, *The Esoteric Theology of Ashkenazi Hasidism* (en hébreu), Jérusalem, Mosad Bialik, 1968.

35. Cf. p. ex. R. Eleazar de WORMS, *Persush al Sefer Yetsira*, Przemyal, 1883, p. 2ª, 3ª⁻ᵇ ; Anonyme, *Sefer ha-Hayyim*, Londres, British Library, MS Heb. 1055, *passim*. Une transcription de ce manuscrit a été publiée par le Département de littérature hébraïque, Faculté des Lettres, Université hébraïque de Jérusalem, 1973.

36. Moscou, Bibliothèque de l'État, MS Günzburg 508. Nous avons utilisé le microfilm n° 16881 de l'Institut de microfilms de manuscrits hébraïques à la Bibliothèque nationale et universitaire de Jérusalem, que nous remercions pour l'avoir mis à notre disposition. Tous les renvois au *Sefer ha-Maskil* dans ce qui suit se réfèrent aux folios de ce manuscrit.

37. I. Ta-SHMA, « *Sefer ha-Maskil:* An Unknown Hebrew Book from the Thirteenth Century » (en hébreu), *Jerusalem Studies in Jewish Thought*, 3, 1982/1983, p. 416-438.

38. Voir *supra*, n. 1. La deuxième partie de cette étude consiste en une édition scientifique annotée d'extraits de *Sefer ha-Maskil*, concernant surtout la cosmologie ; les folios et numéros de lignes du manuscrit de Moscou y sont indiqués.

t-il pas eu connaissance des nombreux textes philosophiques et scientifiques traduits en hébreu dans le siècle et demi précédant la rédaction de son livre[39]. Le grand intérêt du livre pour notre propos réside dans sa théologie particulière, qui peut être qualifiée de néo-stoïcienne.

L'idée maîtresse au cœur du *Sefer ha-Maskil* est que Dieu, le Principe (*Iqqar*) comme l'appelle R. Shlomo, est identique avec l'Air cosmique qui remplit le monde entier, supérieur et inférieur. En fait, soutient R. Shlomo, bien que l'air se révèle à nous sous une multitude de formes, il est, au fond, unique. Cette idée se clarifie en considérant la cosmogonie que développe R. Shlomo. A l'origine, avant la création, le Souffle (*ruah*) de Dieu flottait au-dessus des eaux : ce Souffle, dit R. Shlomo, « est le Principe merveilleux et excellent, béni soit-il, pour toute éternité. Ce Souffle est nécessairement l'Air. [...] Ainsi, l'Air merveilleux et excellent flottait au-dessus des eaux »[40]. Au deuxième jour de la création, l'Air originel a été séparé en deux parties — supérieure et inférieure. De la partie supérieure ont été faits les cieux ; de la partie inférieure ont été créés la terre, les mers et les fleuves. Dorénavant une partie de l'Air originel se trouvait dans les cieux, l'autre ici-bas. Ainsi, tout ce qui existe a été créé à partir de l'Air originel.

L'Air supérieur, céleste, est une Splendeur éternelle. R. Shlomo avance l'idée, assez étrange, selon laquelle nous voyons une infime partie de cette Splendeur par le petit trou qui se déplace dans le ciel et que l'on appelle communément « soleil » : par cette « fenêtre » passe la lumière qui illumine et qui chauffe notre monde ici-bas. L'Air supérieur n'est donc rien d'autre que la lumière que nous voyons « à travers » le soleil, seulement en plus fort et en plus magnifique, à la mesure du fait que la surface du ciel est plus grande que celle de la « fenêtre » qu'est le soleil. Il s'ensuit que « chaque jour nous voyons [un segment] de la Gloire du Principe, béni soit-il »[41]. D'ailleurs, à la fin des temps Dieu enlèvera le ciel qui nous sépare de l'Air d'en haut et l'Air-lumière occupera de nouveau sa place originelle et remplira de sa Splendeur le monde tout entier.

A côté de l'Air supérieur, il y a aussi « l'air au sein duquel nous vivons », l'air « du monde habité ». R. Shlomo l'appelle aussi l' « air d'obscurité », car il n'a ni lumière ni chaleur propres. Cependant, et c'est là un point capital, malgré les différences importantes entre l'Air supérieur et l'air d'ici-bas, ces deux espèces d'air sont, au fond, une seule et même

---

39. Pour une vue d'ensemble cf. la littérature citée *supra*, n. 2.
40. *Sefer ha-Maskil,* f° 14a, l. 40 *sq.*
41. F° 15b, l. 15.

chose; l'un et l'autre sont des ramifications du Principe suprême : « l'air parmi nous, l'Air au-dessus du ciel, l'air au sein du ciel et l'air des profondeurs — tout cela n'est qu'un seul Air »[42]. Cet unique air est Dieu Lui-même : « l'Air caché et l'Air dévoilé [...] c'est le Créateur, c'est le générateur; il est partout et tout est en Lui; il est le fondement de tout et supérieur à tout »[43].

Cette notion de l'Air cosmique et son identification avec Dieu sont la base d'une idée qui nous intéresse tout particulièrement. Selon R. Shlomo, « le monde tout entier est au sein de l'Air », ou, plus précisément, le monde « est suspendu » au sein de l'Air[44]. R. Shlomo utilise l'image suivante pour illustrer son idée : « le monde tout entier est au sein [de l'Air], comme les entrailles sont au sein de l'homme »[45]. Ailleurs, R. Shlomo écrit que le monde est entouré d'un « abîme » comme l'œuf l'est d'une coquille, précisant que le monde avec l'abîme se trouvent au sein de l'Air merveilleux[46].

L'Air, le Principe, avec ses diverses ramifications est donc omniprésent tout en restant un. C'est le fondement de la philosophie immanentiste de R. Shlomo : « Le Roi des rois, le Saint béni soit-il, est partout et tout est en Lui; son Souffle est partout — en haut et en bas, à l'intérieur de la terre et au-dessus des cieux »; « il n'y a aucun lieu qui soit vide de Lui, car Sa Gloire et Sa Présence sont partout et restent identiques à elles-mêmes toujours et dans toutes les saisons »[47].

Sur la base de sa théorie immanentiste, R. Shlomo développe des solutions novatrices à plusieurs problèmes d'ordre théologique et plus précisément herméneutique. Le plus important parmi ces problèmes est certainement celui de l'anthropomorphisme : comment interpréter les nombreux passages bibliques qui semblent attribuer à Dieu des traits humains ou un comportement humain? Le problème, on le sait, a grandement occupé les penseurs juifs médiévaux et il a sans doute été un des vecteurs qui ont ouvert dans la pensée juive traditionnelle la brèche ayant permis à la philosophie « étrangère » de s'introduire en son sein. L'identification de Dieu, ou du Principe, avec l'Air cosmique omniprésent permet à R. Shlomo d'avancer une solution inédite à ce problème.

---

42. F° 13b, 1. 26-27.
43. F° 13b, 1. 19-21.
44. F° 14a, 1. 20; p. 43a, 1. 43-44.
45. F° 14a, 1. 23.
46. F° 14b, 1. 21 *sq*; f° 15b, 1. 3 *sq*.
47. F° 9b, 1. 36-37; f° 10a, 1. 42-44.

R. Shlomo est tout à fait au courant des tentatives, associées notamment au nom de Maïmonide, d'éviter l'anthropomorphisme au moyen d'interprétations allégoriques de certains versets des Écritures[48]. Il critique en effet sévèrement « les hommes qualifiés de grands sages et de philosophes, dont est issue l'erreur, qui risque de rendre mécréants et de faire punir les fils d'Israël, » consistant à « affirmer que 'Dieu dit' et 'Dieu parla' ne sont que des allégories [mashal] ». Les tenants de cette opinion fausse, dit R. Shlomo, avancent l'argument selon lequel « une parole ne saurait être émise que par quelqu'un qui a une bouche, tandis qu'il ne convient pas d'attribuer au Béni-soit-Il aucune corporéité, ni aucune forme de membres » et ils concluent que ces versets « problématiques » sont à interpréter « allégoriquement » (we-dimmu hadavar le-mashal)[49].

Qui sont les tenants de l'interprétation allégorique que connaît R. Shlomo? Il n'est pas aisé de répondre à cette question, car notre auteur ne mentionne aucun nom et ne donne apparemment pas de citations directes. Il nous semble cependant possible d'affirmer que sa critique vise avant tout le Livre de la connaissance de Maïmonide. En effet, lorsque R. Shlomo formule la position qu'il rejette, il semble faire écho à des formulations maïmonidiennes. Ainsi, il se réfère à l'opinion selon laquelle les expressions bibliques qui en apparence sont anthropomorphiques ne sont « qu'une allégorie [mashal] » et ont été employées « seulement eu égard à l'entendement de Moïse et eu égard à l'entendement des fils d'Israël »[50]. Ces formulations rappellent celles de Maïmonide, qui affirme que les expressions en apparence anthropomorphiques « ne sont employées qu'eu égard à l'entendement des hommes », de sorte qu' « il ne s'agit que d'une allégorie et toutes les expressions que nous venons de citer ont également valeur d'allégorie »[51]. Le Livre de la connaissance est parvenu au nord de la France déjà dans le premier tiers du XIIIe siècle, presque un siècle après qu'y aient commencé, sous l'influence des écrits de R. Saadia Gaon, les discussions sur le problème de l'anthropomorphisme. Rien d'étonnant donc à ce que R. Shlomo, qui écrit en 1294, traite ce problème et, plus spécifiquement, à ce qu'il combatte l'opinion du Livre de la

48. Cf. Isaak HEINEMANN, « Die wissenschaftliche Allegoristik des jüdischen Mittelaters », Hebrew Union College Annual, 23(1), 1950-1951, p. 611-643.

49. F° 1b, l. 16-19.

50. F° 1b, l. 17, 20.

51. MAÏMONIDE, Sefer ha-Mada, Hilkhot Yesodey Tora 1:9; trad. fr. (modifiée) in MAÏMONIDE, Le livre de la connaissance, traduit de l'hébreu et annoté par Valentin Nikiprowetzky et André Zaoui, étude préliminaire de Salomon Pinès, Paris, PUF, 1961, p. 31-32.

connaissance[52]. Il semblerait pourtant que R. Shlomo n'avait pas du *Livre de la connaissance* une connaissance directe : non seulement il ne mentionne jamais le nom de Maïmonide, mais chaque page de son livre témoigne clairement du fait qu'il n'avait pas la moindre notion de la cosmologie aristotélicienne, qui est exposée dans les chapitres 2 à 4 du *Livre de la connaissance*, suite, précisément, à l'exposé de la doctrine maïmonidienne sur le problème de l'anthropomorphisme. *A fortiori*, on peut affirmer avec certitude que R. Shlomo ne connaissait pas le *Guide des égarés*. Il semblerait ainsi que R. Shlomo connaissait les idées de Maïmonide sur l'anthropomorphisme indirectement, par exemple par l'intermédiaire du livre *Arugat ha-bosem* de R. Abraham b. Yehiel (composé autour de 1234), qui rapporte textuellement les propos de Maïmonide sur ce sujet[53].

R. Shlomo rejette donc la solution maïmonidienne au problème de l'anthropomorphisme apparent des Écritures et s'insurge contre toute tentative d'interpréter les Écritures autrement que littéralement. Il ne se rallie cependant pas pour autant à la solution « réactionnaire », défendue notamment par R. Moshe Taqu[54] (vers 1225), qui consiste à défendre l'anthropomorphisme, c'est-à-dire à attribuer à Dieu des traits corporels. L'originalité de R. Shlomo réside précisément dans sa tentative d'éviter à la fois l'allégorisation et la corporification de Dieu.

R. Shlomo affirme clairement que « nous savons que Dieu, béni soit-il, n'a ni figure ni forme »[55]. A l'instar d'autres auteurs qui se sont penchés sur la question de l'anthropomorphisme apparent des Écritures, R. Shlomo fonde cette thèse en faisant observer que Dieu apparaît sous une multitude

---

52. R. Shlomo décrit le Maharam (R. Meir ben Baruch) de Rothenburg (1215-1293) comme un de ses maîtres (Ta-Shma, art. cit., p. 419) et nous savons que le Maharam, de même que d'autres personnalités de son entourage, connaissait le *Livre de la connaissance*, pour lequel il avait la plus grande estime. Cf. Isadore TWERSKI, « The Beginning of Mishneh Torah Criticism » in A. ALTMANN (éd.), *Biblical and Other Studies*, Cambridge, Mass., 1963, p. 161-182. (Réimprimé in Isadore TWERSKI, *Studies in Jewish Law and Philosophy*, New York, Ktav, 1982, p. 30-51.)

53. E. A. URBACH (éd.), *Sefer Arugat ha-bosem de R. Abraham b. Yehiel*, t. 4, en hébreu, Jérusalem, 1963, p. 198.

54. R. Moshe TAQU, *Ketav tamim*, reproduction du ms Paris, Bibliothèque nationale de France, héb. 711 avec une introduction de Joseph Dan, Jérusalem, Université hébraïque, 1984.

55. F. 9a, l. 35. Cette formule apparaît certes chez Maïmonide, mais elle figure également dans la paraphrase du *Livre des croyances et des opinions* de Saadia et elle avait été répandue chez les Devôts rhénans avant même qu'ils ne se familiarisent avec le *Livre de la connaissance*. Cf. MAÏMONIDE, *Livre de la connaissance* 1:9, trad. *op. cit.*, p. 32. Cette phrase est citée p. ex. dans URBACH (éd.), *Sefer arugat habosem*, *op. cit.*, t. 1, p. 199. Que cette phrase soit tirée de la paraphrase de l'écrit de Saadia a été établi par URBACH in *ibid.*, p. 136 *sq.* (n. 17) et p. 199 (n. 4).

de formes différentes : « tantôt le Béni-soit-Il apparaît dans le feu, tantôt il apparaît dans un nuage; une fois il apparaît comme un jeune guerrier, une autre comme un vieillard miséricorde [...] »[56]. Cette multitude d'apparitions de Dieu a créé la confusion chez les « pauvres d'esprit », qui se demandent comment les Écritures peuvent se contredire. R. Shlomo se propose d' « enrichir » l'esprit de ces pauvres et de leur permettre de se dégager de leur perplexité. Il rejette cependant de façon explicite et décidée la solution consistant à interpréter les Écritures allégoriquement : « il faut que nous adhérions aux paroles de notre Dieu, qu'Il soit béni et exalté, et que nous établissions Ses propos avant tout littéralement (*bimelitsa qayemet*) »[57]. S'appuyant sur une idée qui n'est pas sans rappeler Descartes, R. Shlomo soutient qu'il est exclu que Dieu veuille induire les hommes en erreur en employant le langage autrement que de façon littérale. Cette prémisse fonde le principe herméneutique selon lequel Dieu « a ordonné l'ensemble de Ses lois littéralement et Il ne s'est pas exprimé en utilisant seulement des énigmes et des paraboles »[58]. Selon R. Shlomo, toute parole divine doit avant tout être interprétée littéralement; c'est uniquement après cela qu'il est permis d'adjoindre à l'interprétation littérale aussi une interprétation allégorique, à condition toutefois que cette dernière ne contredise pas le sens premier du texte[59]. Comme d'autres adversaires de la méthode allégorique, R. Shlomo craint que l'allégorisation n'aboutisse à une invalidation des commandements divins : « si la Tora n'était qu'une interprétation [lit. opinion] et non une parole [à un sens déterminé], qui pourrait prouver que l'idolâtrie et d'autres transgressions sont interdites? En effet, le mécréant pourrait toujours affirmer : « c'est de cette façon que j'avais compris telle parole ou tel commandement »[60].

Pour R. Shlomo, la méthode allégorique renferme donc le danger de saper les fondements de la Tora. Comment l'écarter tout en évitant la corporification de Dieu? A cette question cruciale R. Shlomo offre une solution originale à deux volets :

1) R. Shlomo explique que les expressions bibliques qui semblent attribuer des traits humains ou matériels à Dieu sont employés eu égard à ce que l'homme peut entendre. De même que Maïmonide et d'autres philosophes, il cite le verset d'Isaïe « A qui vous comparerez Dieu et quelle

56. F° 9a, l. 30. Cf. notre édition de ce passage (art. cit., n. 1), p. 89, pour l'identification des sources de ces images.
57. F° 1b, l. 33-34.
58. F° 4b, l. 16.
59. F° 1b, l. 34-35.
60. F° 1b, l. 20-22; cf. également f° 1a, l. 40-45.

forme Lui attribuez-vous ? »[61], en développant toutefois cette idée dans une direction opposée à celle de Maïmonide. Pour réfuter l'anthropomorphisme, les philosophes avaient soutenu que les mêmes termes ont un sens différent selon qu'ils sont appliqués à Dieu ou aux hommes. R. Shlomo, lui, avance l'idée contraire, selon laquelle les expressions bibliques ont toujours un seul et même sens, qu'elles soient appliquées à Dieu ou à l'homme. Comment peut-il aller ainsi, si l'on veut éviter la corporification de Dieu ? La réponse, surprenante, de R. Shlomo consiste à affirmer que le risque de l'anthropomorphisme provient non du fait que l'on ne comprend pas correctement ces expressions là où elles sont appliquées à Dieu, mais, à l'inverse, du fait que l'on ne les comprend pas correctement là où elles sont appliquées à l'*homme*. R. Shlomo soutient en effet que des verbes désignant une action de l'âme, tels que « parla », « vit », etc., s'appliquent toujours, que leur sujet grammatical soit Dieu ou un homme, à l'*air*. L'homme, explique-t-il, ne voit et ne parle que grâce à son souffle (*ruaḥ*). En effet, l'homme mort possède tous ses membres, mais ne parle ni ne voit plus : c'est seulement « tant que le souffle est dans le cœur que l'homme voit par ses yeux et parle par sa bouche »; d'où cette conclusion que « c'est le souffle qui parle »[62]. Là où, grammaticalement, des verbes tels que « voir », « parler », etc. ont pour sujet l'homme, ils portent en fait sur son souffle, qui, puisqu'il est air, est une ramification du Principe. Il s'ensuit que ces verbes peuvent être appliqués à Dieu avec *la même signification* qu'ils revêtent lorsqu'ils sont appliqués à l'homme, et ceci sans que l'on encoure le danger de l'anthropomorphisme. Cette idée, conclue R. Shlomo, réfute l'argument des philosophes qui affirment qu' « on ne saurait attribuer au Béni-soit-Il aucune parole, car celui qui émet une parole a une bouche, alors que chez le Béni-soit-Il il n'y a aucun aspect corporel »[63]. De même, sont ainsi établies d'une manière solide, les interdictions contenues dans les dix commandements, qui ont été prononcées d'une voix réelle qui a été entendue d'une extrémité de la terre à l'autre et que tous les hommes ont réellement entendue[64].

2) Il reste le problème de la diversité des apparitions divines, que les tenants de l'interprétation allégorique ont invoquée pour prouver qu'aucune des descriptions de Dieu n'est à entendre littéralement. R. Shlomo, quant à

---

61. ISAÏE 40:18, cité f° 9a, l. 3 ; MAÏMONIDE cite un verset semblable (40:25) in *Livre de la connaissance 1:8*, trad. fr., p. 31.
62. F° 1b, l. 22-25.
63. F° 48a, l. 46-48.
64. F° 48b, l. 5-7.

lui, propose une solution tout autre à ce problème. Là où les philosophes avaient affirmé que Dieu « n'a ni figure ni forme », R. Shlomo ajoute le mot significatif « permanentes »[65]. Selon lui, l'essence divine, l'Air, peut revêtir des formes différentes, temporaires, sans que cela porte atteinte à son unité. Ainsi, comme nous le verrons maintenant, les apparitions qu'ont vues les prophètes étaient non pas « des figures réelles, permanentes », mais des figures « temporaires » ou « passagères »[66], qui malgré leur fugacité furent pourtant une partie intégrante de Dieu.

Les apparitions divines se font, selon R. Shlomo, par des anges, dont notre auteur distingue, selon leur degré de perdurabilité, trois catégories. Il y a d'abord les anges « permanents », comme Michaël et Gabriel, qui sont constitués par l'eau et le feu[67] et qui sont préposés à des fonctions spécifiques invariables. Les anges de cette classe ne nous concernent pas directement, précisément parce qu'ils n'apparaissent pas aux hommes sous des formes variées. La deuxième catégorie d'anges est composée d' « anges temporaires ». Un ange de ce groupe est « un souffle qui s'est séparé d'avec le Principe, le Souffle merveilleux, béni soit-Il »[68]. Ces anges-là sont créés en vue d'une mission spécifique et leur existence ne perdure qu'aussi longtemps que dure cette mission; une fois leur mission terminée, ils sont consumés par le feu[69]. Une dernière catégorie d'émissaires divins est constituée par les « souffles séparés » : à l'instar des anges de la deuxième catégorie, ils sont des « ramifications » du Principe, mais contrairement à ceux-là, ils reviennent à leur origine une fois leur mission accomplie. C'est au moyen d'anges des deux dernières catégories, qui sont des souffles dérivés du Principe premier, que Dieu accomplit Sa volonté. Aussi, « Son Souffle est partout — en haut et en bas, à l'intérieur de la terre et au-dessus des cieux. Où que ce soit — le Souffle qui s'y trouve ira accomplir la volonté divine, puis reviendra à sa place »[70]. Ainsi, la solution que suggère R. Shlomo aux problèmes de la diversité des apparitions divines consiste à dire que le Principe, étant un Souffle, a des ramifications diverses qui revêtent des apparitions différentes, sans que cela porte atteinte à l'unité fondamentale de Dieu.

---

65. F° 9a, l. 41.
66. F° 9a, l. 37.
67. L'association de l'ange Michaël avec l'eau et de Gabriel avec le feu renvoie à une conception ancienne que R. Shlomo prend à son compte. Cf. Reuven MARGOLIOUTH, *Mal'akhey Elyon*, Jérusalem, Mosad ha-Rav Kook, 1945, p. 24, 118.
68. F° 23a, l. 16-17; de même, f° 9b, l. 18.
69. F° 9b, l. 15-16.
70. F° 9b, l. 36-37.

IV

La proximité entre la doctrine du *Sefer ha-Maskil* et la philosophie stoïcienne est manifeste. L'idée centrale de la philosophie stoïcienne, que l'on trouve déjà chez Anaximène et chez Diogène d'Apollonie — à savoir que Dieu est identique à un Air cosmique, qui, malgré la diversité de Ses manifestations, reste Un — est au cœur même du *Sefer ha-Maskil*. Il est vrai que R. Shlomo porte son attention avant tout sur le côté théologique de la doctrine et se désintéresse des aspects physiques. Mais il énonce néanmoins clairement le principe typiquement stoïcien selon lequel l'air qui entoure le monde ici-bas « soutient le monde et tout ce qui est en lui »[71]. C'est d'ailleurs précisément la raison pour laquelle R. Shlomo appelle l'air « fondement » (*yesod*) : « c'est seulement ce qui soutient qui est appelé 'fondement' »[72]. Selon R. Shlomo, il y a neuf « fondements » (huit cieux et la terre) « qui sont plantés dans le [dixième] fondement [c'est-à-dire l'Air], qui n'a ni terme ni limite » et dont dépend tout le reste de l'existant[73].

La parenté entre la doctrine de R. Shlomo et la pensée stoïcienne se manifeste aussi dans le fait que, comme Diogène d'Apollonie et les Stoïciens, mais contrairement aux Aristotéliciens, R. Shlomo, loin de tenir l'air pour une simple matière, lui attribue une conscience et de la connaissance. Il maintient que c'est grâce au fait que l'air est omniprésent que s'exerce parfaitement la providence divine. La notion d'un air omniprésent fonde l'idée que cet air « voit tout et sans lui le cœur [de l'homme] ne saurait rien »; davantage, l'air dans les cœurs des hommes « voit leurs pensées »[74], une idée que l'on trouve d'ailleurs déjà chez Saadia Gaon[75]. En fait, comme nous l'avons vu, toute perception et toute connaissance sont dues à l'air : « nous savons que les yeux ne voient pas, les oreilles n'entendent pas, la bouche ne parle pas, mais c'est l'air qui remplit le cœur de l'homme » qui fait tout cela[76]. La même idée se trouve déjà chez Diogène d'Apollonie et chez les Stoïciens[77].

---

71. F° 14b, l. 23.
72. F° 43a, l. 34.
73. F° 43a, l. 39-45.
74. F° 13b, l. 21 ; f° 15b, l. 38-39.
75. Saadia GAON, *Commentaire du « Livre de la création »*, éd. Qafih, p. 107, l. 23-35.
76. F° 49, l. 11-14 ; f° 25, l. 2-12. Cette idée a des origines dans le Talmud de Babylone, *Nida* 31a et cf. RASHI, *ad loc.*
77. Cf. KIRK, RAVEN, SCHOFIELD, *op. cit.*, p. 448 ; LAPIDGE, « Archai and Stoicheia », art. cit., p. 253 ; POHLENZ, *Die Stoa, op. cit.*, 1:88, 2:52.

Comment expliquer la proximité des idées du *Sefer ha-Maskil* et du Stoïcisme? Nous ne sommes pas en mesure de répondre de manière définitive à cette question. Il semblerait pourtant que la « théologie de l'Air » de *Sefer ha-Maskil* remonte finalement aux écrits de R. Saadia Gaon, particulièrement à son *Commentaire du « Livre de la création »*. R. Shlomo et Rav Saadia attribuent tous les deux à l'air des rôles primordiaux, quoique différents, et c'est en se fondant sur la notion de l'air qu'ils établissent, chacun à sa manière, l'immanence divine dans le monde. Considérons deux problèmes qu'ils évoquent et résolvent dans ce contexte :

1) Saadia est conscient du fait que l'immanence divine dans le monde pose le problème du contact de la divinité avec la souillure de la matière. Il résout ce problème en avançant que de même que la lumière du soleil n'est pas salie par la saleté du monde, de même la Gloire de Dieu, quoique immanente au monde, n'est pas affectée par sa saleté. A la suite de Saadia, les Dévôts rhénans se sont, eux aussi, penchés sur le problème : certains, comme par exemple R. Eléazar de Worms, ont suivi Saadia en pensant que « le Créateur est dans le tout mais sans y toucher »; d'autres, comme R. Elḥanan b. Yaqar, ont, à l'inverse, soutenu que l'immanence divine dans le monde est sélective[78]. R. Shlomo, lui aussi, est conscient du problème que pose l'idée de l'immanence divine et lui propose une solution. D'une part, il soutient que l'air d'ici-bas est appelé (entre autres appellations) « l'air du pardon » (*awir ha-meḥilah*) et ceci pour signifier que « tous les besoins de l'homme y sont faits, mais [les hommes] sont pardonnés ». D'autre part, il insiste qu'il faut traiter avec respect même la petite portion de l'air qui nous parvient par la fenêtre qu'est le soleil[79]. A l'instar des Dévôts rhénans, R. Shlomo tire des conséquences pratiques de sa doctrine de l'immanence divine[80].

2) Saadia Gaon s'est fondé sur sa notion du « second air » pour soutenir l'omniscience divine. Selon lui, cet air omniprésent permet à Dieu de connaître les pensées humaines : « il n'y a rien qui échappe à Sa science ». « Même si l'homme creusait dans la terre à une profondeur de milliers et de milliers de coudées — l'air y sera. [...] Tout le monde admet que Dieu est partout par l'intermédiaire de l'air visible »[81]. Nous avons déjà constaté que R. Shlomo énonce précisément la même idée : « ayant compris que le

78. J. DAN, *The Esoteric Theology of Ashkenazi Hasidism, op. cit.*, p. 178-183.
79. F° 14b, l. 43-47.
80. Cf. J. DAN, *The Esoteric Theology of Ashkenazi Hasidism, op. cit.*, p. 177.
81. Nous citons d'après une des paraphrases du *Commentaire sur le « Livre de la création »*, citée par R. Moshe TAQU in *Ketav tamim, op. cit.*, f° 11a.

secret du Principe est l'Air, tu pourras comprendre qu'il est partout et tout est dans l'Air; même au fond du cœur des hommes — Il y est et voit leurs pensées »[82].

Il semble donc vraisemblable qu'il existe un lien étroit entre le Commentaire du « Livre de la création » de Saadia et le *Sefer ha-Maskil* : l'idée que l'air revêt une signification théologique et cosmologique à la fois, qui est au cœur même de la pensée de R. Shlomo, lui a été suggérée, directement ou indirectement, par Saadia. Cette proximité ne doit cependant pas oblitérer la différence principielle entre les deux systèmes de pensée. Pour Saadia, Dieu est absolument transcendant et ce n'est que le second air, qui est créé, qui est immanent au monde; pour R. Shlomo, par contre, Dieu est *identique* à l'Air, et c'est Lui-même qui est immanent au monde.

R. Shlomo connaissait-il directement le *Commentaire sur le « Livre de la création »* de Saadia? Dans l'état actuel de nos connaissances sur la transmission des paraphrases et des traductions de ce *Commentaire* il est impossible de répondre à cette question. Contentons-nous, ici, de l'observation suivante. Certains auteurs attribuent à Saadia des idées qui ne se trouvent pas dans son texte original arabe, ni dans les traductions hébraïques connues, mais qui sont très proches des idées du *Sefer ha-Maskil*. Ainsi, R. Moshe Taqu attribue à Saadia l'idée que le Créateur *est* « un Air extrêmement subtil, et le monde tout entier s'y trouve, si bien qu'il [l'Air] se trouve dans les collines, dans les pierres, dans les arbres et dans les animaux »[83]. Le même auteur rapporte aussi que Saadia « a écrit dans le *Livre des croyances* que l'air est Dieu »[84]. Ce ne sont certes pas les véritables opinions de Saadia, mais manifestement R. Moshe Taqu croyait qu'elles le sont. Vu la parenté entre les idées attribuées ici à Saadia et celles du *Sefer ha-Maskil*, on peut avancer la possibilité que R. Shlomo ait développé sa doctrine à partir de certains textes qui prétendaient exposer la position de Saadia.

S'agissant du contexte intellectuel dont est issu *Sefer ha-Maskil*, il y a aussi lieu de mentionner de nouveau la théologie des Dévôts rhénans. Parmi les idées de Saadia que ces derniers ont adoptées, figure l'idée que la présence divine dans le monde se fait par l'intermédiaire de l'air. Quelques chercheurs ont déjà évoqué brièvement cette « doctrine des airs » des Dévôts rhénans, mais nous ne disposons pas encore d'une étude d'ensemble de ce sujet[85]. Contentons-nous donc des brèves observations suivantes.

---

82. F° 15b, l. 38-40.
83. R. Moshe Taqu, *Ketav tamim, op. cit.*, f° 8b, l. 3-5.
84. *Ibid.*, p. 52a, l. 21-22.
85. Cf. particulièrement J. Dan, *The Esoteric Theology of Ashkenazi Hasidism, op. cit.*, p. 95, 99-100.

Le *Sefer ha-Hayyim* (« le Livre de la Vie »), écrit vers 1200, renferme une doctrine élaborée de l'air, mais elle est très différente de celle de *Sefer ha-Maskil*[86]. R. Eléazar de Worms s'appuie sur la doctrine de l'air, développée par Saadia, et maintient que l'immanence divine dans le monde s'opère au moyen de l'air, qu'il identifie, comme le fait R. Shlomo, avec la lumière[87]. Cependant, puisque R. Eléazar tient l'air pour une chose créée, il ne semble pas être une source d'inspiration de *Sefer ha-Maskil*.

Mentionnons enfin qu'il y avait dans la Kabbale une tradition qui postulait l'existence d'un « Air originel », ou d'un « air subtil imperceptible », et sans doute existe-t-il une certaine relation entre celle-ci et la doctrine du *Sefer ha-Maskil*[88]. Cependant, cette doctrine kabbalistique fait corps avec la doctrine des sefiroth qui semble tout à fait inconnue du *Sefer ha-Maskil*. De plus, on ne trouve pas, dans la Kabbale, cette identification de Dieu avec l'Air, qui est au cœur même du *Sefer ha-Maskil*. Le *Sefer ha-Maskil* ne saurait donc être considéré comme dépendant de cette tradition kabbalistique.

Il convient encore de mentionner ici le *Livre de l'âme* du ps.-Galien, un ouvrage d'origine inconnue qui a été traduit de l'arabe en hébreu par Yehuda al-Ḥarizî au début du XIIIe siècle, une traduction qui a d'ailleurs servi de base pour une version latine[89]. Ce petit ouvrage met en scène un dialogue entre « Moria » et « Galien », au cours duquel il est démontré que l'air est indispensable à l'existence de tous les êtres animés, de même que pour celle du feu, ce qui permet aux mineurs de vérifier, au moyen d'une bougie allumée, la présence d'air dans un puits[90]. Après avoir abordé les sujets liés à la physique, la discussion passe à des questions de théologie :

Moria : J'ai donc appris que l'existence des êtres créés dépend de l'air.
Galien : Et l'existence de l'air dépend [à son tour] du Créateur, béni soit-Il.

---

86. *Sefer ha-Hayyim*, *op. cit.* (n. 34). (Une transcription en a été publiée par le Département de littérature hébraïque de l'Université hébraïque de Jérusalem [Jérusalem, 1973].) Cf. également, J. DAN, *The Esoteric Theology of Ashkenazi Hasidism*, *op. cit.*, p. 52, 143 *sq.*

87. R. Eléazar de WORMS, *Commentaire sur le « Livre de la création »*, *op. cit.* (n. 34). p. 2a, 3a-3b; *idem*, *Sefer Sodot ha-razia*, Jérusalem, 1991, p. 12, 42. Et cf. J. DAN, *The Esoteric Theology of Ashkenazi Hasidism*, *op. cit.*, p. 95, 97, 99-100.

88. Cf. notamment, George MARGOLIOUTH, « The Doctrine of the Ether in the Kabbalah », *Jewish Quarterly Review*, 20, 1908, p. 825-861; M. IDEL, *Kabbalah: New Perspectives*, New Haven et Londres, 1988, p. 147, 348 (n. 310, 311).

89. Texte hébraïque : A. YELLINEK (éd.), *Sefer ha-Nefesh*, Leipzig, 1852; corrections de Y. B. Schorr in *He-Ḥalutz 11*, 1840, p. 109-110. Texte latin : E. BERTOLA, « Un dialogo pseudo-galenico sui problemi dell'anima », *Rivista di filosofia neo-scolastica*, 60, 1968, p. 191-210 (réimprimé in E. BERTOLA, *Il pensiero ebraico*, Padoue, 1972). (Je remercie M. Mauro Zonta de m'avoir signalé la dernière référence.)

90. Cf. THEOPHRASTE, *De igne*, § 23-24.

Moria : Quel est la preuve indiquant que l'existence de l'air dépend du Créateur ? Galien : De même qu'il t'est devenu manifeste que les êtres créés dépendent, pour leur existence, de l'air, de même l'existence de l'air dépend du Créateur qu'Il soit béni. En effet, ton intellect appréhendra qu'il n'y a rien derrière l'air, sauf le Créateur, béni soit-Il et qu'il n'y a rien avant l'air sauf le Créateur. [...] Ainsi l'existence de l'air dépend du Créateur, béni soit-Il[91].

Dans la suite, le *Livre de l'âme* s'efforce de montrer que l'air est une chose créée : la preuve en est que l'on peut enfermer de l'air dans des récipients et des outres et ainsi le diviser et le détruire. « Par conséquent, poursuit 'Galien', tout ce qui est divisible est corruptible, tandis que tout ce qui est illimité et indivisible perdure pour toute éternité »[92]. Manifestement, ce sont ceux qui identifiaient l'air avec Dieu Lui-même que l'auteur du *Livre de l'âme* met en cause : il semblerait donc que ce petit ouvrage a son origine dans les discussions entre les stoïciens et leurs adversaires. En effet, l'idée stoïcienne, que nous avons retrouvée dans différents *midrashim*, selon laquelle l'air « soutient » le monde, est totalement absente du *Livre de l'âme*. Il convient d'ajouter que certains des propos de l'ouvrage ps-galénique ont été repris dans un autre ouvrage pseudo-épigraphique, à savoir le *Sefer ha-Nimṣa* (Le Livre de ce qui est facilement trouvable), attribué à Maïmonide[93].

Le *Sefer ha-Maskil* et le *Livre de l'âme* partagent l'accent mis sur le fait que l'air est indispensable à la vie, ainsi que l'exploitation de ce fait à des fins théologiques. Cependant, les doctrines des deux ouvrages sont diamétralement opposées : là où *Sefer ha-Maskil* postule que Dieu est identique avec l'Air primordial, le *Livre de l'âme* souligne, au contraire, que l'air est une chose créée et corruptible, qui dépend pour son existence de Dieu qui, Lui seul, est éternel. Il ne semble donc pas que l'ouvrage hébraïque de la fin du XIIIe siècle ait été influencé par l'ouvrage pseudo-galénique.

L'origine de la doctrine qui est au cœur de *Sefer ha-Maskil* demeure ainsi incertaine. D'une part, il y avait toute une tradition de pensée, remontant à certains *midrashim*, à Saadia et aux Dévôts rhénans et s'appuyant sur le *Livre de l'âme* du ps.-Galien, qui attribuait à l'air un rôle théologique. D'autre part, cependant, aucun de ces ouvrages ne présente le radi-

91. *Sefer ha-Nefesh, op. cit.*, p. 17-19.
92. *Ibid.*
93. Pour le texte hébraïque cf. l'édition de I. L. MAIMON in *Sinai,* 36, 1955, p. 201-211. Une traduction anglaise, pas tout à fait fiable, ainsi qu'une bibliographie, se trouvent *in* Fred ROSNER, *The Existence and Unity of God. Three Treatises Attributed to Maimonides,* Northvale, N. J. et Londres, Jasson Aronson, 1990.

calisme de *Sefer ha-Maskil*, qui consiste à établir une stricte identité entre Dieu et l'Air cosmique et à affirmer que la lumière du soleil n'est qu'une infime partie visible de Dieu. L'auteur de *Sefer ha-Maskil* a-t-il puisé sa doctrine dans des ouvrages qui restent à identifier? Ou est-il à l'origine de cette conception si proche de la théorie stoïcienne? Dans l'état actuel de la recherche, il n'est pas possible de trancher cette question. Observons en tout cas que l'auteur de *Sefer ha-Maskil* écrit avec l'enthousiasme de quelqu'un qui a fait une grande découverte, dont il est fier et qu'il désire communiquer à autrui. Ceci nous incite à penser que R. Shlomo est l'auteur de la doctrine qu'il expose, qu'il l'avait développée comme une continuation et une radicalisation des théories de l'air qui circulaient dans son milieu. Dans la mesure où toutes ces théories de l'air remontent en dernier lieu au stoïcisme, nous pouvons voir en R. Shlomo un héritier de cette tradition, dont le radicalisme s'apparente à celui de ses aïeuls du Portique.

V

La physique stoïcienne constitue une des réalisations les plus remarquables de la pensée antique[94]. Son influence sur le cours de l'évolution de la pensée ultérieure, certes moins grande que celle de la philosophie platonicienne ou péripatéticienne, est encore trop souvent méconnue par les historiens. Ce n'est en effet que tout récemment que l'on a commencé à étudier la survivance de la philosophie stoïcienne dans la pensée latine médiévale[95] et à mesurer son influence sur la science moderne[96]. Dans cet article nous avons voulu suggérer que la physique stoïcienne était une source d'inspiration pour des traditions de pensée bien éloignées de la tradition scientifique, en fait éloignées de la pensée rationnelle tout court. Le stoïcisme a en effet mis à la disposition de ces traditions des modèles, des images et des métaphores qui se sont montrés gratifiants au niveau psychologique et satisfaisants au niveau intellectuel. Dans la tradition

---

94. C'est le mérite de feu le professeur Shmuel SAMBURSKY d'avoir attiré l'attention sur la signification de la physique stoïcienne pour l'histoire des sciences. Cf. notamment son ouvrage *The Physics of the Stoics, op. cit.,* n. 9.

95. L'influence de la philosophie stoïcienne sur le Moyen Âge latin est admirablement bien étudiée *in* Michael LAPIDGE, « The Stoic Inheritance », art. cit., n. 30.

96. Peters BARKER et Bernard R. GOLDSTEIN, « Is Seventeenth-Century Physics Indebted to the Stoics? », *Centaurus, 27,* 1984, p. 148-164; Margaret J. OSLER (éd.), *Atoms, Pneuma and Tranquility, op. cit.,* n. 30; Gad FREUDENTHAL, « Clandestine Stoic Concepts in Mechanical Philosophy: The Problem of Electrical Attraction », art. cit.

hébraïque, il y avait une longue tradition d'inspiration stoïcienne, remontant à certains *midrashim* et explicitée par Saadia, selon laquelle le monde subsiste grâce à un Souffle divin cosmique, immanent au monde matériel. Cette tradition atteint son apogée en 1294 avec le *Sefer ha-Maskil* de R. Shlomo Simḥa de Troyes : l'identification de Dieu avec l'Air originel et l'immanentisme matérialiste qui en résulte font du *Sefer ha-Maskil* un exemple frappant d'une pensée inspirée du stoïcisme.

XV

THE MEDIEVAL ASTROLOGIZATION OF ARISTOTLE'S BIOLOGY: AVERROES ON THE ROLE OF THE CELESTIAL BODIES IN THE GENERATION OF ANIMATE BEINGS

I. INTRODUCTION

The coming-to-be of plants and animals is among the most wondrous and impressive phenomena Nature offers us. From the seemingly amorphous and unorganized matter of the world, incredibly complex living beings come to be, which are astonishingly well fitted to live in their respective environments. The ancient mythologies of the Middle East have accounted for this marvel with stories of creation by a supernal conscious Being. The Presocratic philosophers, too, offered accounts for the fact that matter is organized and not haphazard. They did so notably by positing a universal underlying matter, which they construed as divine and, therefore, as endowed with a teleological capacity of intelligent self-organization. Plato accounted for the existence of informed living beings through his theory of Ideas.

Aristotle, as is his wont, broke with these traditions. In his philosophy of nature, the sublunar matter was devoid of teleological capacities; nor could it be informed by the supralunar realm, consisting of the so-called "fifth element", which was construed as a "stranger" to the sublunar generation and corruption.[1] Concomitantly he dismissed Plato's Ideas. Both of these innovations are encapsulated in Aristotle's oft-repeated catch-phrase: "man generates man, and the sun as well". Klaus Oehler has persuasively argued that this phrase essentially signals that each individual man is generated from another individual man, without the intervention of any transcendent, Platonic, Ideas.[2] Indeed, according to Aristotle's biology, the off-

---

[1] Friedrich Solmsen, *Aristotle's System of the Physical World* (Ithaca, 1960), p. 289.
[2] Klaus Oehler, "Ein Mensch zeugt einen Menschen. Über den Missbrauch der

112

spring acquires its form, which is the explanandum of any embryology, from the male parent's semen, the female contributing the matter only. The contributing role of the sun has nothing to do with the old ideas on the vivifying effect of the celestial bodies or of their heat. Rather, the very modest role of the sun in Aristotle's scheme is that of continually mixing up the four sublunar elements: were it not for the sun's motion along the ecliptic and the resultant four seasons, the world would have reached a stable state, in which the four elements would have been separated and statically arranged in four concentric spherical layers; there would be absolutely no generation and corruption.[3]

In this paper I will be concerned with the reception of these Aristotelian ideas by Averroes.[4] Averroes considered himself as

---

Sprachanalyse in der Aristotelesforschung", in his *Antike Philosophie und byzantinisches Mittelalter* (München, 1969), pp. 95-145. See also Gad Freudenthal, *Aristotle's Theory of Material Substance. Form and Soul, Heat and Pneuma* (Oxford, 1995), pp. 38-9.

[3] Aristotle, *De gen. et corr.* 2.10.

[4] Works by Averroes are referred to in the following editions and translations, and with the following abbreviations: (1) *De gen. et corr.*: Epitome: Arabic: *Epitome del libro sobre la generación y la corrupción*, edición, traducción y commentario Josep Puig Montada (Madrid, 1992); Epitome and Middle Commentary (= MC): Hebrew translations (Moses ibn Tibbon and Qalonimos ben Qalonimos, respectively): *Averroes on Aristotle's "De generatione et corruptione"*. *Middle Commentary and Epitome*, Hebrew text edited by Samuel Kurland (Cambridge, Mass., 1958); English translations: *Averroes on Aristotle's "De generatione et corruptione"*. *Middle Commentary and Epitome*, translated by Samuel Kurland (Cambridge, Mass., 1958). (2) *Parva naturalia*: Epitome: Arabic: *Averrois Cordubensis compedia liborum aristotelis qui Parva naturalia vocantur*. Arabic text edited by Henry Blumberg (Cambridge, Mass., 1972); Hebrew translation (Moses ibn Tibbon): *Averrois Cordubensis compedia liborum aristotelis qui Parva naturalia vocantur*. Hebrew text edited by Harry Blumberg (Cambridge, Mass., 1954); English: Averroes, *Epitome of the "Parva naturalia"*, translated from the original Arabic and the Hebrew and Latin versions with notes and introduction by Harry Blumberg (Cambridge, Mass., 1961). (3) *Book of Animals*, Treatise 16 (= *De generatione animalibus*): MC: Hebrew translation (Jacob b. Makhir): Oxford, Bodleian Library Opp. 1641 Qu (= Neubauer 1381; = Institute for Microfilmed Hebrew Manuscripts, Hebrew University, Jerusalem [= IMHM] # 22405) (This text is not preserved in Arabic.) (4) *Metaphysics*: Epitome: Arabic: *Compendio de la metafisica*, texto arabe con traducción y notas de Carlos Quirós Rodríguez (Madrid, 1919; reprinted with an introduction by Josep Puig Montada, Cordoba etc. 1998); Hebrew translation (Moses Ibn Tibbon): MS Paris, Bibliothèque nationale de France (= BNF), héb. 918 (= IMHM # 31960); German translations: Ger. (B): *Die Epitome der Metaphysik des Averroes*. Übersetzt und mit einer Einleitung und Erläuterungen versehen von S. Van den Bergh (Leiden, 1924); Ger. (H): *Die Metaphysik des Averroes*, nach dem Arabischen übersetzt und erläutert von Max Horten [Halle/S., 1912] [*Abhandlungen zur Philosophie und ihrer Geschichte*, 36]. MC: Hebrew (Qalonimos ben Qalonimos): MS Paris, BNF, héb. 954

Aristotle's faithful Commentator, but in point of fact his metaphysical universe differed considerably from that of his master, who lived some fifteen centuries earlier. Arabic-language Peripatetism had integrated within its framework certain Neoplatonic ideas, notably that of emanation, completely absent from the Stagirite's original conceptual scheme. The Avicennian idea of the active intellect as a Giver of Forms, in particular, which straddles metaphysics, physics and biology, largely determined the terms in which the young Averroes construed the generation of forms in matter, specifically those of plants and animals. But Averroes' commitment to the idea of the active intellect as the Giver of Forms did not last.

In his authoritative book *Alfarabi, Avicenna, & Averroes, on Intellect*, Professor Herbert A. Davidson has shown that Averroes' thinking on the role of the active intellect in nature underwent an evolution, and that Averroes in different periods of his life construed the origin of the formal cause of inanimate and animate substances differently. Averroes' first position was that forms generally, and the forms of living beings in particular, are infused into matter by an "incorporeal mover", identified as the active intellect. But in later writings, Averroes recanted his initial position, now qualifying the earlier view as that taken by "many of the later philosophers". (We learn from Averroes himself that he thereby meant essentially Ibn Bājja.)

(= IMHM # 32605). Long Commentary (= LC): Arabic: *Tafsīr mā ba'd al-ṭabī'a*, ed. Maurice Bouyges, 2nd edition (Beirut, 1973); English translation of Book Λ, with pagination of Arabic indicated: Charles Genequand, *Ibn Rushd's Metaphysics. A Translation with Introduction of Ibn Rushd's Commentary on Aristotle's "Metaphysics", Book Lâm* (Leiden, 1984). French translation of the same, with Arabic pagination also indicated: Aubert Martin, *Averroès, Grand commentaire de la "Métaphysique" d'Aristote ... Livre lam-lambda* (Paris, 1984). (5) *Meteorologica: MC*: Hebrew: *The Middle Commentary of Averroes on Aristotle's Meteorologica. Hebrew Translation of Kalonymos ben Kalonymos*, edited with Introduction, Critical Apparatus, and Hebrew-Arabic Vocabulary by Irving Maurice Levey, Dissertation (Harvard University, 1947). (6) *De caelo*: Epitome: Arabic: *[Jawāmi' Kitāb] al-Samā' wa-al-'Ālam* in *Rasā'il Ibn Rushd* (Hyderabad, 1947); Hebrew (Moses Ibn Tibbon): MS Paris, BNF, héb. 918 (= IMHM # 31960). MC: Arabic: *Talkhīṣ al-Samā' wa-al-'Ālam*, ed. Jamāl al-Dīn 'Alawī (Fez, 1984); Hebrew (Shlomo ben Ayyub): MS Berlin 212 (Or. Qu. 811) (= IMHM # 1769). (7) *Tahāfut al-Tahāfut*, ed. Maurice Bouyges (Beirut, 1930). English translation with pagination of Arabic indicated: *Averroes' "Tahafut al-Tahafut" (The Incoherence of the Incoherence)*, translated from the Arabic with Introduction and Notes by Simon Van den Bergh (London, 1954). Some passages of works by Averroes are translated in R. Shem-Tov Ibn Falaqera's encyclopedic work *De'ot ha-Filosofim* (= "The Opinions of the Philosophers"; henceforth *DF*) and were occasionally used: MS Parma, De Rossi # 16420 (= IMHM # 13897).

In these works Averroes takes the view that no separate, incorporeal mover is involved in the generation of living beings.[5]

Davidson has shown that certain manuscripts of Averroes' Epitome of the *Metaphysics* bear witness to the Commentator's change of mind: in what originally were marginal annotations, Averroes replaced passages reflecting his older views by passages presenting the new ones. In a recent paper I have shown that the same holds of Averroes' Commentary on the so-called *Book of Animals*: the first "edition" of this work claimed that the forms of animate beings derive from the active intellect, while the latter version denied it.[6]

In this paper I will ask how Averroes accounted for the generation of animate beings once he rejected the theory of the active intellect. That theory in fact supplied a straightforward explanation for the emergence of order within amorphous matter and it will be our goal to discover what was Averroes' alternative explanatory device for the same facts.

II. THE CONSTANT ELEMENT IN AVERROES' VIEW OF THE GENERATION OF ANIMATE SUBSTANCES: THE NEED FOR A "MOVER FROM WITHOUT"

The question how a wholly unitary Entity can give rise to a plurality exercised Averroes more than once, but I will not go into this subject, which has been extensively studied by Davidson.[7] Suffice it to say that Averroes takes the First Cause to give rise to the forms of the spheres, i.e. their souls, and to the corresponding intelligences. He changed his mind on the causal dependence of these entities, but not on the fact that the First Cause is the cause of the existence of the incorporeal substances associated with the celestial bodies. (Averroes apparently remains mute on the origin of the celestial *bodies*, i.e. of the

---

[5] Herbert A. Davidson, *Alfarabi, Avicenna, & Averroes, on Intellect* (New York and Oxford, 1992), pp. 232-42. In 1859 the great scholar Salomon Munk had already signaled in general terms that Averroes changed his mind on several philosophical issues and that he at times added glosses to this effect in some of his writings. See Munk, *Mélanges de philosophie juive et arabe* (Paris, 1859), pp. 442-3.

[6] Gad Freudenthal, "Averroes' changing mind on the role of the active intellect in the generation of animate being", in Proceedings of the SIHSPAI Colloquium (Cordova, 9-11 December 1998), forthcoming.

[7] Davidson, *Intellect*, pp. 223-31.

fifth, celestial, element.[8]) The question that is of interest to us here is: assuming that the celestial bodies, their spheres, souls, and respective intelligences exist, how, according to Averroes, do the sublunar beings come to be? In other words: what is the causal chain linking the First Cause to the sublunar beings, specifically the animate ones?

To begin with, the existence of the elements has to be explained. In his Epitome of the *Metaphysics* Averroes maintains that the problem receives different answers according to the science in which it is posed and he in fact offers there three different accounts. (i) In physical science, he says in the Epitome of the *Metaphysics* (referring to the *De caelo*), the generation of the four elements is explained "mechanically", with reference to the "great motion" of the sphere of the fixed stars,[9] but metaphysics calls for a different explanation. On the contents of that "metaphysical" explanation, as described in the Epitome of the *Metaphysics*, Averroes changed his mind: (ii) In the early version of the text, he maintained that the four forms of the elements, like all other sublunar forms, go back to the active intellect as the Giver of Forms.[10] (iii) In the later version of the Epitome of the *Metaphysics*, Averroes did not replace the above passage with one reflecting his new view and thus did not explicitly specify his "metaphysical" position on this question after the demise of the theory of the active intellect. It stands to reason, however, that his new "metaphysical" position now

[8] Davidson, *Intellect*, pp. 224, 235. In the Epitome of the *Metaphysics* Averroes writes however that "the cause of the existence of the 'matters' of the heavenly bodies is their forms only"; Ar., p. 163 (§ 69); Heb., fol. 146ab:9-18; Ger. (B), p. 139.

[9] Davidson, *Intellect*, p. 236; Averroes, Epitome of the *Metaphysics*, Ar., p. 160 (§ 63); Heb., fol. 145ba:38-bb26; Ger. (B), pp. 136-7. Averroes there says that in "natural science" it has been shown that the motion of the great sphere produces heat, which in turn generates lightness, which is the form of fire. By the same token, the absence of motion at the center creates heaviness, which is the form of the element earth. The intermediary elements, water and air, come to be inasmuch they are heavy or light with respect to the two extreme elements. Another physical argument to the same effect is that the great sphere needs an immobile center on which to revolve and this is the earth at the center; the existence of earth entails that of its contrary, i.e. of fire, and, in a further move, that of the intermediate elements. Thus, the "spherical body is necessarily what produces the elements and preserves them". However, on that subject, too, Averroes seems to have changed his mind, for in the Middle Commentary on the *Meteorologica* he writes (p. 17) that "the heavenly body preserves the elements, but does not generate them, as people may imagine, especially with respect to fire". See also Ruth Glasner, "Gersonides's theory of natural motion", *Early Science and Medicine*, 1 (1996): 151-203.

[10] Davidson, *Intellect*, pp. 236-8.

ascribed the generation of the elements to the action (presumably the "mechanical" action) of the heavenly bodies, a solution that is in continuity with his late view of the generation of animate beings, as we shall see below.

On the subject of the homoeomerous substances, Averroes again has three positions: (i) According to physical science, he says in the Epitome of the *Metaphysics*, the proximate causes of the existence of homoeomerous substances are only the elements and the motions of the planets. In other words: the forms of the homoeomerous substances are "reduced" to the ratio of the mixture of the elements in them – they have no form beyond the "blend-form", and hence their existence, too, like that of the elements, is accounted for "mechanically".[11] From a "metaphysical" point of view, things again are different. In the early redaction of the Epitome of the *Metaphysics*, Averroes holds (ii) that the homoeomerous bodies, just as the elements and the animate beings, receive their forms from the active intellect. In the later redaction, after he had discarded the notion of the active intellect, Averroes contends (iii) that the homoeomerous bodies receive their forms from the "celestial bodies", in a manner analogous to that of the animate substances.[12]

We thus come to the last stage in the formation of sublunar substances – that of the formation of plants and animals. Averroes' early view, as expounded in the original redaction of the Epitome of the *Metaphysics*, ascribes the forms of plants and animals to a separate intellect. Averroes writes:

[T]he doctrine in that science [viz. natural science[13]] compelled us to introduce *a principle from without* [in order to account for] the existence of the plants and the animals only. For it is manifest that they possess faculties (such as the nutritive soul) which accomplish actions for the sake of some end. For this reason, [plants and animals] cannot be ascribed to the elements. Nor can they be ascribed to the generating individual. For the individual supplies in such circumstances [merely] the matter receiving [the form] or the instrument, as is the case with semen and menses. All this has already been made manifest in natural science.[14]

---

[11] Davidson, *Intellect*, pp. 236-8.

[12] Davidson, *Intellect*, p. 241.

[13] The text refers back to "natural science" mentioned p. 161, § 65, ll. 1-2 (of the paragraph). In the original redaction, "that science" came just 2 or 3 lines later; in the later redaction, the interpolation of some 13 lines introduced a greater distance between them. See also Ger. (B), p. 137 and the apparatus, p. 319, note 80(13).

[14] This passage is summarized in Davidson, *Intellect*, p. 237.

[Indeed,] when we consider the issue in the context of this science [meta-physics[15]] it becomes evident that the notion [ma'nā] through which these things [plants and animals] become intelligible, cannot be generated from the individual material form, insofar as it is individual. For if it were the case that the material forms, insofar as they are material, as a rule generate the forms in the material substances, then separate forms would not do this. But it has already been made manifest that the separate forms do generate forms in material substances. It therefore follows with necessity that these forms are not generated by the material forms.

This follows with necessity also from another consideration: the material individual thing generates an individual thing [ma'nā] which is like itself. Yet the intelligible form[16] generated [in matter] manifestly is not an individual thing. It therefore follows with necessity that it is the active intellect that gives the forms of the simple bodies [i.e. elements] and the other [i.e. composite] bodies. That which is generated by an individual essentially is another individual like him – or itself. This is why Aristotle says that a man is generated by man and by the sun. It is the individual that is generated essentially. The form [of the species], however, is generated accidentally. It is manifest, therefore, that what generates it [i.e. the intelligible form] is something other than the individual. Hence, this concrete individual, which comes to be essentially, is generated by this concrete [individual] sun and a concrete [individual] man. [By contrast,] the thing that comes to be acciden-tally – viz. the form of humanity – is generated by humanity abstracted from matter. This makes up the difference between the route taken by Aristotle and the route taken by Plato concerning the activity of forms. Keep this in mind, because this is how the other errors can be removed.[17]

In the present context, the salient point in this passage is that in Averroes' early view, substantiated by considerations deriv-ing both from natural science and from metaphysics, the gener-ation of animate beings cannot be ascribed to the elements alone. Rather, a "principle from without" must be introduced to account for the appearance of forms in matter. Without that assumption, the generation in matter of telos-oriented faculties cannot be accounted for, nor can the emergence of intelligible forms. We will now see that although Averroes modified his view on the identity of this external mover – the active intellect

---

[15] "This science" refers to the subject of the present treatise and stands in opposi-tion to "that science" (natural science) mentioned a couple of lines before (see note 13).

[16] I.e. "the form of the physical object, which can be abstracted from the object and grasped by the mind as an intelligible thought"; Davidson, *Intellect*, p. 237.

[17] Averroes, Epitome of the *Metaphysics*, Ar., pp. 162-3 (§§ 66-67); Heb., fol. 146aa:4-40; Ger. (B), pp. 137-8. The passage is summarized in Davidson, *Intellect*, pp. 237-8 (with n. 77).

or not – he remained steady in his belief that such a principle must be introduced.

Thus, elsewhere in the Epitome of the *Metaphysics* (in a passage going back to the early version but maintained also in the later one) Averroes similarly writes that the biological factors (i.e. the soul-heat deriving from the parent plant or animal) "have been proved to be insufficient *without a principle from without*".[18] That this indeed is a premise common to the early and the later position – so that it is not the need for such a "principle from without" that is under discussion, but solely its precise identity – comes to the fore in a sentence Averroes added in the later redaction of his text, in immediate sequel to the above statement: Averroes there identifies the "principle from without" as the "celestial bodies in Aristotle's view, which is the correct one, or the active intellect in the view of many of the later philosophers"[19] – including, we may add, the young Averroes himself. Clearly, the later Averroes perceives the role of the active intellect in the earlier view as equivalent to that of the celestial bodies in the later one. Again, a few lines later, in another late addition, Averroes writes, apropos of "spontaneous generation", that "the ultimate mover is, in Aristotle's system, the celestial bodies through the mediacy of soul-powers emanating from them, or else the active intellect as the later philosophers interpret him [Aristotle]".[20] Lastly, in the Commentary on the so-called *Book of Animals*, in a brief statement manifestly also from the later period, Averroes offers a synthetic view of the question. He declares that "what Galen calls 'formative power' [is] what Aristotle calls 'the soul' (which according to him is donated by the sun and the rest of the planets) and the others call 'the separate form.' And it is identical to what Plato calls 'the [universal] soul' and many of the Peripatetics call it 'the Giver of Forms'".[21]

All these pronouncements unambiguously show, I submit, that in Averroes' later opinion, the celestial bodies are the functional equivalent of the active intellect in the early account. In other

---

[18] Averroes, Epitome of the *Metaphysics*, Ar. p. 50 (§27), l. 12; missing in Ger. (B), p. 40; translation quoted from Davidson, p. 239.

[19] Averroes, Epitome of the *Metaphysics*, Ar., p. 50 (§ 27); Heb., fol. 126ba:13-14; Ger. (B), 40; quoted from Davidson, *Intellect*, p. 239.

[20] Averroes, Epitome of the *Metaphysics*, Ar., p. 50 (§ 28); Heb., fol. 126ba:14-17; missing in Ger. p. 40; translation quoted from Davidson, *Intellect*, p. 239.

[21] Averroes, Commentary on the *Book of Animals*, fol. 241b:11-13.

words: according to Averroes' later philosophy, the heavenly bodies provide the formal cause of the generated living beings.

The meaning of this claim is clarified by an important sentence in the Commentary on the *Book of Animals*, which I take to belong to the part of the text shared by the early and the late redaction. Averroes writes that the power to initiate life in seed or semen "is not sufficient to endow [a plant or an embryo] with life, without the concourse of the sun and the sphere.[22] *For an infinite [series of] causes must perforce be dependent on an eternal Cause*".[23] The argument is general (it remains valid whether one identifies the "mover from without" as the active intellect or differently) and establishes that the form of the offspring cannot derive solely from the soul-heat carried by the seed or semen. For if the living form of Zayd derived only from his progenitor, Amr, and that of Amr only from that of his own progenitor, then we would obtain an infinite series of progenitors. Averroes regards this as an impossible consequence. Thus in his Epitome of *On Generation and Corruption* he writes that the infinite series of generation and corruption can only be circular and not linear (*'alā al-istiqām*, lit. "in a straight line"): "Since the existence of the later individual implies that of its predecessor [i.e. progenitor]", Averroes writes, "we cannot postulate it to go on linearly indefinitely, for then the existence of the later individual would presuppose an infinity of causes that had preceded it. This is in and by itself essentially false".[24]

Averroes summarizes this stance in the *Tahāfut al-Tahāfut*:

[S]ince, according to them [the philosophers], the causes do not form an infinite series, they introduced a primary, permanent efficient cause. Some of them believed that the heavenly bodies are this efficient cause, some that it is an abstract principle, connected with the heavenly bodies, some that it is the First principle, some again that it is a principle inferior to it [...].[25]

---

[22] The latter phrase ("without the concourse of the sun and the sphere") may seem to indicate that the two sentences belong to Averroes's later period, when he identified the "eternal Cause" as the heavenly bodies. I think, however, that it reflects also the view of the earlier period, and that the reference to the sun and the sphere at that time simply expresses in general terms the Aristotelian belief that "the sun and the sphere" are the ultimate cause of generation and corruption in the sublunar world (Aristotle, *De gen. et corr.* 2.10). This is also the case in the long passage just quoted, which is patently from the earlier period.

[23] Freudenthal, "Averroes' changing mind", sentences [1]-[2] (= *DF*, fol. 246aa:31-32). Similarly Commentary on the *Book of Animals*, fol. 229a.

[24] Averroes, Epitome of *De generatione et corruptione*, Ar., p. 57; Heb., 125; English, p. 136. Cf. Maimonides, *Guide for the Perplexed* 1:73 (11th premise).

[25] Averroes, *Tahāfut al-Tahāfut*, p. 211. Cf. Davidson, *Intellect*, p. 251.

This passage again highlights that in his later philosophy, Averroes construes the role of the heavenly bodies in the economy of the natural world as precisely equivalent to that of the active intellect in the earlier view.

Let us now try to determine with precision how the heavenly bodies fulfill the role Averroes ascribes to them.

III. THE ROLE OF THE HEAVENLY BODIES IN THE GENERATION OF ANIMATE BEINGS

In the very beginning of his Commentary on Book 16 of the so-called *Book of Animals* (corresponding to Book 2 of Aristotle's *Generation of Animals*) Averroes writes:

The existents in motion are of two kinds: (i) those whose motion is eternal and whose individuals are neither generated nor corrupted; these are the most noble existents among the things in motion and they exist by necessity. These are the heavenly bodies. (ii) The second kind of things in motion is the one [whose individuals] now exist now are non-existent. These are those whose natures are susceptible to be more or less perfect. By contrast, it is in the nature of the first kind that [its individuals] be of the utmost perfection. For this reason, this [first] kind of existents is *the cause* of the existence of the second kind. [...] The existents in motion of the first kind extend help to those of the other kind, exert providence over them, and educe them from non-existence to existence, and bring them forth from a lifeless existence to a living existence and from a living existence to an ensouled existence, until they acquire the most perfect existence their nature is capable of receiving. And just as [the heavenly bodies] exert providence in that they move a deficient existence into a perfect existence, so also they exert providence by endowing [the sublunar existents] with eternal and permanent persistence, as far as their nature allows.[26]

In a late addition to the Epitome of the *Metaphysics*, Averroes is somewhat more explicit on the capacity of the celestial bodies to move sublunar substances "from a lifeless existence to a living existence and from a living existence to an ensouled existence". Averroes there writes that "according to Aristotle" – i.e. according to Averroes' own later view –

some of the blended [i.e. homoeomerous] substances are ensouled owing to the heavenly bodies. This is why Aristotle says that man is generated by man and the sun. The reason for his holding this view is that a man is generated

---

[26] Averroes, Commentary on the *Book of Animals*, fol. 225b:1-11, roughly corresponding to Aristotle, *Generation of Animals* 2.1, 731b24-33.

by a man like himself and that *because those [celestial] bodies are alive they can endow with life what is [down] here [in the sublunar world]. For only a body whose nature is to be ensouled can move matter to the animate [i.e. soul-] perfection*".[27]

In the Long Commentary on the *Metaphysics*, Averroes similarly writes that inasmuch as celestial bodies "are the principle [*mabda'*] of animated and non animated things" they themselves necessarily are "animated" [*mutanāfisa*] too, and their principles "must be body and soul".[28] In the same work he similarly states that "it is the sun and the other stars which are [the] principle of life for every being in Nature",[29] a stance which, as we saw, Averroes upholds also in late additions to the Epitome of the *Metaphysics*.[30] Averroes explains that if Aristotle invoked the sun alone in this context, this is simply owing to the fact that "it is the star whose action is most manifest in that respect" (viz. that of endowing life).[31]

A complementary idea is expressed in the Epitome of the *Metaphysics*, where, following Alexander of Aphrodisias, Averroes construes the amounts of heating and cooling by the sun and the moon as a display of divine providence. The heavenly bodies are not presented here as the formal cause of the sublunary existents, but they are still described as regulating the entire economy of the natural world.[32] Indeed, were the

---

[27] Averroes, Epitome of the *Metaphysics*, Ar., p. 161 (§ 65); Heb., fol. 145bb30-146aa2; Ger. (B), p. 137:30. The passage is summarized in Davidson, *Intellect*, p. 241. See also the two redactions of another passage in the Epitome of the *Metaphysics*, translated in Davidson, *ibid.*, p. 239.

[28] Averroes, LC on the *Metaphysics*, 1534:8-10.

[29] Averroes, LC on the *Metaphysics*, 1502:1-3.

[30] Cf. above, p. 118.

[31] Averroes, LC on the *Metaphysics*, 1502:7-8.

[32] "The existence of things down here, i.e. the fact that their species persist", Averroes writes, "cannot be due to chance, as the Ancients had claimed. This can be seen if one considers how the motions of the spherical bodies are in agreement with the existence of each and every thing coming to be in this [lower] world, and how the former preserve the latter. This can be observed first and foremost with respect to the sun and next also with respect to the moon. Indeed, concerning the sun it has been established that if its body had been greater than it is, or if it had been closer [to us], then it would have destroyed the plants and the animals through an excess of heat. Again, had it been farther away or smaller than it is, then [the plants and the animals] would have perished through cold. This is attested by the places that are uninhabited because of excessive heat or cold. Providence can also be clearly observed in the [existence of] the sun's inclined sphere [i.e. the ecliptic]. For if the sun did not have an inclined sphere, there would be no summer, nor winter, nor [variations in] heat and cold. Yet it is evident that the four seasons are necessary for the existence of the species of plants and animals. Providence is also very clearly observable in the

122

motions of the heavenly bodies to change even slightly, the entire natural order would be disrupted.[33] In the Commentary on the *Book of Animals* Averroes goes one step further when, following Aristotle, he states that the life periods of animals (periods of gestation, lifespan, etc.) are controlled by the sun and the moon.[34]

In sum it can be said that in his late philosophy, Averroes ascribes to the heavenly bodies precisely the capital role that in his early view he had ascribed to the active intellect: it is they which endow the sublunar existents with their forms, including in particular souls: "because those [celestial] bodies are alive they can endow with life what is [down] here [in the sublunar world]. For only a body whose nature is to be ensouled can move matter to the animate [i.e. soul-] perfection".[35]

But how, precisely, does Averroes construe the mode of operation of the ensouled heavenly beings in the generation of animals? How do they act – in parallel to the heat in the seed or the semen – to generate an animated living being? And how do their "souls", by virtue of which, as we just saw, the heavenly bodies

---

[sun's] daily motion, for without it there would be no day and night, but rather there would be half a year night and half a year day. If this were the case, the existents would have perished by day due to an excess of heat, and at night for an excess of cold. Similarly, the operation of the moon can be observed in the existence of rains and the ripening of fruits. And it is clear that were the moon bigger than it is or smaller or farther away, or if its light were not borrowed from the sun, it would not have that effect. Again, if it [did not have] an inclined sphere it would not accomplish different operations at different moments in time. ... And what we have said with respect to the sun and the moon must be believed also with respect to the other stars. Indeed, Aristotle said that the motion of the stars takes after the sun, and he said so because it can be seen that the stars seek to resemble the sun. Therefore a truthful general statement must be adopted, according to which all [heavenly] motions [were so designed as to] exercise providence over what is below them in this world". Averroes, Epitome of the *Metaphysics*, Ar., pp. 166-8 (§§ 74-77); Heb., fol. 146bb:7-147aa:21; Ger. (B), pp. 141-3; Ger. (H), pp. 201-3; Falaqera, *DF*, 288b ff. Pierre Thillet has pointed out that Averroes here paraphrases (the Arabic version of) Alexander of Aphrodisias' *On Providence*. See Pierre Thillet, *Alexandre d'Aphrodise, Traité de la providence*, Doctorat d'État, Sorbonne (Paris, 1979), vol. 4, Introduction, p. 65, 75-7; text, pp. 18-24. See now the printed edition of Alexander's treatise in: Silvia Fazzo and Mauro Zonta (ed. and trans.), *Alessandro di Aphrodisia, La Providenza* (Milano 1999), pp. 124-34.

[33] Averroes, Epitome of the *Metaphysics*, Ar., p. 144:20-24; Heb., 142bb:35-40; Ger. (B), p. 122.

[34] Averroes, Commentary on the *Book of Animals*, fol. 248b, corresponding to Aristotle, *Generation of Animals* 4.10.

[35] *Supra*, p. 121.

at all participate in the generation of living beings, relate to the souls of the generated beings?

To get as close as possible to Averroes' intention, let us first note that the heavenly bodies fulfill two distinct roles in nature: the first, relatively humble one, is to continually mix up the elements in the sublunar world (without this the elements would all have been separated and at rest at their respective natural places); the other, nobler role is to endow adequately mixed portions of matter with forms. Thus Averroes writes:

> The matter of [individuals of] these species [viz. those that reproduce through sexual generation] is generated through mixing up the elements by these celestial bodies, [an action] through which the elements become a mixture suitable for receiving the form [of the species]. But the power whose rank is that of an instrument carrying the formative power is the heat generated by the sun and the stars in matter suitable for this action, namely the aeriform body [pneuma] which is akin to the body of the semen which is generated in the animals which reproduce sexually.[36]

Let us now look into the second of these roles, i.e. the formal cause. Averroes' most explicit statement on the subject seems to be a passage of the Long Commentary on the *Metaphysics*, a work belonging to the period in which Averroes already had abandoned his belief in the active intellect. In this passage Averroes says that the heat in semen has a formative power (like the heat used in the arts) and that it is for this reason that it is called "soul-heat".[37] This heat, Averroes goes on to write, "which is endowed with form, *is in the seeds, generated by the possessor of the seeds and the sun*": the soul-heat in seeds or semen derives both from the progenitor and from the sun (and the other heavenly bodies). The role of the celestial bodies in sexual generation having come up, Averroes inserts a short digression to prove his point through the consideration of "spontaneous generation": in this phenomenon,[38] the absence of

---

[36] Averroes, Commentary on the *Book of Animals*, fol. 241b:3-6. Apropos of the last affirmation, note that Averroes also writes: "semen, in its aeriform part, is an instrument of the formative power" (Averroes, *ibid.*, fol. 230a:16-17).

[37] *Ḥarāra nafsiyya*; Averroes, LC on the *Metaphysics* 1501:15. Similarly, in the Commentary on the *Book of Animals*, Averroes writes that the formative soul-power resembles "art" and "is of the celestial nature" which, as the *De anima* makes clear, is called "intellect" (Commentary on the *Book of Animals*, fol. 229b:20-21; Falaqera, *DF*, fol. 246ba:22-23). Falaqera makes clear that this passage is a late addition; see Freudenthal, "Averroes' changing mind".

[38] Is it necessary to add that the reality of "spontaneous generation" was not questioned by either Aristotle or by Averroes and their contemporaries?

124

heat deriving from the progenitor makes the contribution to generation of the sun, the most conspicuous celestial body, particularly manifest. The contribution of the celestial bodies to sexual generation, Averroes writes,

is the reason why Aristotle says that man begets a man like himself, with the help of the sun. [Indeed, in generation taking place without any seed or semen] it [the heat effectuating generation] is generated in earth and water through solar heat mixed with the heat of the other stars.[39] Therefore it is the sun and the other stars that are the principle of life of every living being in nature. And the heat of the sun and the other stars, which is generated in earth and water, is that which generates the animals which are generated in putrefied matter and, in general, everything that is generated without seed. There is no soul in actuality generated from the ecliptic and the sun, as Themistius claims.[40] All this has been explained in the *Book of Animals*. He [Aristotle] ascribed this action to the sun because it is the star whose action is most manifest in that respect.

As for the heats generated by the heats of the stars, which produce each distinct species of animals – [each one of these "heats"] *is potentially this or that species of animal*. The power [*taqdīr*] present in each of the "heats" depends on the amount of the motions of the stars [i.e. their velocities] and their relative positions with respect to their closeness or distance. This power originates from the divine and intelligent work [*mihna*], which is analogous to the single form of the single primary art, to which other arts are subordinated. Clearly, therefore, one must understand that when nature produces something which is highly structured without being itself intelligent, it is inspired [*mulhama*] by active powers which are nobler than it and which are called "intellect".[41]

The passage is not as clear as one might wish, but the following can still be concluded. Averroes chooses to discuss the subject of the so-called "spontaneous generation" – this indeed is where the vivifying powers of the celestial bodies come to the fore with particular clarity. Considering "spontaneous generation" allows one to draw general inferences and in a parenthetical sentence Averroes even goes as far as to contend that "the sun and the other stars [...] are the principle of life of *every* living being in nature". This statement is in keeping with those considered

[39] See Aristotle, *De generatione animalium* 3.11, 762a19ff. and Freudenthal, *Aristotle's Theory*, p. 123.

[40] The Arabic text of the passage from Themistius is quoted *in extenso* by Averroes in the Long Commentary on the *Metaphysics*, 1492-4; see also the Hebrew version of Themistius' text in *Themistii in Aristotelis Metaphysicorum librum Λ paraphrasis*, ed. Samuel Landauer (Berlin, 1903), Hebrew pagination, pp. 7:28-8:26. Annotated French translation in: Rémi Brague, *Thémistius, Paraphrase de la Métaphysique d'Aristote (Livre Lambda)* (Paris, 1999), pp. 63-4.

[41] Averroes, LC on the *Metaphysics*, 1501:16-1503:1.

earlier to the effect that the heavenly bodies fulfill the role of the "principle from without" which is indispensable for generation. Note that the conspicuous vivifying effects of the sun are only the visible part of the iceberg – the other stars (i.e. planets) have similar, although less perceptible effects. The passage also signals that, at least in "spontaneous generation", the stars act as formal causes: Averroes suggests that the "heats" of the various planets are different and their effects vary as a function of their velocities and relative positions. The result is that the identity of the species produced at any given moment by the "heats" of the planets depends on the latter's velocities and positions. We have here a clear instance of what I have elsewhere called *the "astrologization" of the Aristotelian philosophy of nature* in the Middle Ages.[42]

The planets, then, not only are indispensable for the emergence of souls in matter, but through their positions and velocities they determine the identity of the "intelligible forms" emerging in sublunar matter. The heavenly bodies accomplish this through their "heats". This claim, however, raises new questions. Specifically, we were not told by virtue of what causality the "heats" of the celestial bodies contribute to the generation of animals. Moreover, we have seen above that the celestial bodies were said to contribute to generation (both sexual and "spontaneous") by virtue of their being ensouled ("for only a body whose nature is to be ensouled can move matter to the animate [i.e. soul-] perfection"[43]): but how do the statements concerning the operations of the "heats" of the heavenly bodies square with this postulate? How, in other words, are the "heats" of the heavenly bodies connected to the latter's souls?

As far as I can see, Averroes did not spell out a theory describing how the "heats" of the stars contribute to generation. It can yet be assumed, I believe, that he did not invoke these "heats" lightheartedly, without having at least some preliminary ideas on how they work. In what follows I wish to submit a conjecture as to the possible contents of Averroes' ideas on this score. It cannot be entirely sustained by explicit statements of Averroes, but seems to me to be consistent with what he does say.

---

[42] See Gad Freudenthal, "Providence, astrology, and celestial influences on the sublunar world in Shem-Tov ibn Falaquera's *De'ot ha-Filosofim*", in Steven Harvey (ed.), *Medieval Hebrew Encyclopedias of Science and Philosophy* (Dordrecht, 2000), pp. 335-70. See also Genequand's brief note in *Ibn Rushd's Metaphysics*, p. 34, n. 3.

[43] *Supra*, p. 121.

In the Long Commentary on the *Metaphysics* Averroes affirms that all forms exist in actuality in the Prime Mover.[44] They exist there "in a manner similar to that of the existence of the artifacts in actuality in the soul of the maker", a stance expressed also in the Commentary on the *Book of Animals*.[45] But in Averroes' late philosophy, the intelligences and souls associated with the spheres of the planets depend on the Prime Mover as their formal cause "insofar as each intelligence enjoys a conception of the first being proportionate to the intelligence's level of existence".[46] These intelligences and souls thus harbor sets of forms which are subsets of the forms stored "in" the Prime Mover. This construal is particularly clear if we assume with Genequand that in the heavenly bodies, the intellect, qua higher faculty of the soul, is identical with the soul.[47] Perhaps this is the reason why Averroes thinks of the heavenly bodies as capable of supplying the forms of the animate beings. The *ultimate* agent satisfying the methodological condition that "the potential becomes actual only through something already actual in the same genus or species"[48] seems to be the souls or intelligences of the heavenly bodies, which depend on the Prime Mover. Averroes' scheme thus seems to be that the heavenly bodies (planets plus their souls), which are alive by virtue of their souls, transmit the forms that they "contain" in actuality to the sublunar matter. I thus submit that when Averroes refers to the contribution of the celestial bodies to the biological soul-heat he has in mind something crucially important, namely *the forms "in" the (souls of the) celestial bodies, which are carried and transmitted to the sublunar matter by the stars' "heats"*.

The underlying theory thus seems to be this: (i) *Just as the soul-heat of seed or semen carries the form existing in actuality in the male parent and transmits it to the offspring, so also the "heats"*

---

[44] Davidson, *Intellect*, p. 249, quoting Averroes, Long Commentary on the *Metaphysics*, 1505.

[45] Averroes, LC on the *Metaphysics*, 1505; Freudenthal, "Averroes' changing mind", sentences 31-37.

[46] Davidson, *Intellect*, p. 227.

[47] See Genequand, *Ibn Rushd's Metaphysics*, p. 40.

[48] Averroes, LC on the *Metaphysics*, p. 881; quoted after Davidson, *Intellect*, p. 245. Similarly: "Everything that [causes] something to pass from potentiality to actuality must somehow have in itself that which it brings forth" (LC on the *Metaphysics*, 1500:13-14).

*of the stars carry forms existing in actuality "in" the stars' souls
(or intelligences) to the nascent plant or embryo.* These forms
ultimately derive from the forms "in" the Prime Mover.[49] (ii)
Just as the soul-heat in the seed or the semen is different for
every plant or animal species, so also the "heats" emitted by
different stars are different and, indeed, the heat transmitted
by any individual star is different according to its motions and
different positions relative to other stars at the moment of emis-
sion of the rays.[50] In short: the stars' "heats" carry the formal
cause, in analogy with the soul-heat of semen and seed.

Averroes repeatedly takes care to emphasize that the active
factor in generation is not the soul-heat in seed or semen alone,
but rather the semen's heat "blended" with the heat of the sun
and the other heavenly bodies.[51] But how, we should now ask,
do the "heats" of the stars relate to the soul-heat deriving from
the parent? Averroes nowhere addresses this question, and any
tentative to answer it must remain strictly conjectural. Perhaps
he thought that the heat in the seed or semen is, as it were, the
sum-total of the soul-heat deriving from the progenitor plus the
heat originating in the celestial bodies, specifically the sun.[52]
Averroes indeed says that the "heats" of biological and of celestial origin somehow blend[53] in the seed and semen and there-
after inform matter conjointly. In the *Tahāfut al-Tahāfut*, he
at one point speaks of *"animal heat* which emanates from the
heavenly bodies" in which "there are [potentially] the souls
which create the sublunary bodies", thereby confirming that the

---

[49] Averroes, we already noted, substantiates the thesis that the "heats" of the stars
carry life-endowing forms, capable of moving the animate (plant and animal) forms in
matter from potentiality to actuality, by invoking the phenomenon of the "spontaneous generation", in which these "heats" educe such forms in inanimate matter
devoid of all prior soul-heat (e.g. mixtures of earth and water). See Averroes, LC on
the *Metaphysics*, 1501:18-1502:1; see also 1502:3-5 and Aristotle, *De gen. animal.*
3.11, 762a19 ff.

[50] Averroes, LC on the *Metaphysics*, 1502:8-12.

[51] Averroes, LC on the *Metaphysics*, 1464:17 ff.; 1501:16-1502:6 and Davidson,
Intellect, p. 248. See also Davidson, *ibid.*, p. 250. Davidson downplays the role of the
celestial bodies in sexual generation; see Davidson, *Intellect*, pp. 245-52. According to
Davidson, Averroes' view in the Long Commentary on the *Metaphysics* is that "[s]oul
heat or its surrogate suffices to bring living beings into existence" (Davidson,
Intellect, p. 248). Davidson repeatedly notes that the soul-heat is generated by the
parent plant or animal "in conjunction with the sun and the stars" (*ibid.*, p. 250) but
he does not spell out the role of the latter factor.

[52] Averroes, LC on the *Metaphysics*, 1501:16-17.

[53] Averroes, LC on the *Metaphysics*, 1501:1.

soul-heat carried by the seed or the semen derives from both the progenitor and the celestial bodies.[54] Again, in the Commentary on the *Book of Animals*, Averroes writes that the vital heat is "related" to the sun and that it "constitutes" and "generates" the animate bodies.[55] Perhaps Averroes considered the form carried by the heat of biological origin as the material or individual form of the offspring, and of the heat of celestial origin as carrying the (intelligible) form of its species.

IV. THE PHYSICS OF THE CELESTIAL BODIES' ACTIONS ON THE SUBLUNAR WORLD

Averroes' contention that the celestial bodies are involved in the generation of animals through their "heats" is startling, for of course in the Aristotelian philosophy of nature the celestial matter is a "stranger" to all sublunar generation and corruption[56] and in particular it is devoid of all qualities. Not even the sun, therefore, can have the quality "warm" and we must thus ask whence derive its heat and, eventually, the "heats" of the other stars? Aristotle, as is well known, was aware of the problem and sought to account for the sun's heat with two *ad hoc* "mechanical" explanations,[57] which one scholar has justly qualified as "both almost equally lame".[58] But these explanations evidently do not account for the sun's vivifying effects, which Aristotle nonetheless posits in his accounts of "spontaneous generation".[59] Averroes received this problématique in heritage, and we will now try to see how he came to terms with it. We will thus have to pose the following questions: How did Averroes construe the heat of the sun and, generally, the "heats" of the celestial bodies? What is his view on how the "heats" of the celes-

---

[54] Averroes, *Tahāfut al-Tahāfut*, p. 577; cf. Davidson, *Intellect*, pp. 252-3. A similar stance is expressed in a late addition to the Epitome of the *Metaphysics*; see Davidson, *Intellect*, p. 239.

[55] Averroes, Commentary on the *Book of Animals*, fol. 229a:22-23.

[56] *Supra*, p. 111, referring to Solmsen, *Aristotle's System of the Physical World*, p. 289.

[57] Aristotle, *De caelo* 2.7, 289a20 ff.; *Meteorologica* 1.3, 341a12-36.

[58] James Longrigg, "Elementary physics in the Lyceum and Stoa", *Isis*, 66 (1975): 211-29, on p. 214; see also Paul Moraux, "Quinta essentia'", in Pauly-Wissowa, *Realencyclopädie*, XXIV(1) (Stuttgart, 1963), pp. 1171-263, on pp. 1204-5; and John Thorp, "The luminousness of the quintessence", *Phoenix*, 36 (1982): 104-23.

[59] Freudenthal, *Aristotle's Theory* (*supra*, n. 2), pp. 25-6.

tial bodies reach the sublunar world and act on it? Specifically, is the transportation of these "heats" related to the rays of light reaching us from the stars, as Averroes at times insinuates?

Averroes addresses this issue in at least three of his writings, namely (in chronological order): (1) the Epitome of the *De caelo*; (2) the Middle Commentary on the *Meteorologica*; (3) the Middle Commentary on the *De caelo*. I will consider them in turn.

(1) *The Epitome of* De caelo. Averroes sets out from the postulate that "the substance [of the stars] is necessarily of the nature of the fifth body",[60] and then immediately anticipates an objection: "one may ask why the sun, insofar as it is not fire, nonetheless warms and shines? The same applies to all other stars".[61] Averroes then claims that the sun warms on two accounts: (i) because of its motion, and (ii) because of its light.

(i) Motion *qua* motion generates heat, Averroes says,[62] and to confirm this theory he reports Aristotle's alleged observation to the effect that the leaden head of an arrow melts through the heat produced by the friction with the air. But Averroes has a doubt on this theory: the analogy with the moving arrow is pertinent if and only if the moving body and the medium both warm together; but do the stars and the adjacent air heat? It is only when we suppose a positive answer to this question that the analogy holds. The assumption that motion is "one of the causes" bringing about "the heating of the stars and particularly of the sun" leads to other questions: why is the sun's heat greater than that of the other stars? And further: "what singles out the sun and the stars so that they warm by virtue of their motion, unlike the other parts of the spheres, seeing that they are all in motion?" Averroes briefly suggests that the body of the sun warms more than the other stars and more than the sun's sphere by virtue of its size, conjoined with its hardness. He then moves on to a discussion of the second cause of the stars' heat.

---

[60] Averroes, Epitome of *De caelo*, Ar., p. 48:1; Heb., fol. 62bb:7-8; Falaqera, *DF*, 227b:7-8. Similarly: Averroes, MC on *De caelo*, Ar., p. 229:5; Heb., fol. 48a:4-5; Falaqera, *DF*, 227b:7-8.

[61] Averroes, Epitome of *De caelo*, Ar., p. 48:7-8; Heb., fol. 62bb:16-18; Falaqera, *DF*, 227b:14-15.

[62] For what follows in this paragraph see Averroes, Epitome of *De caelo*, Ar., pp. 48:9-49:3; Heb., fol. 62bb:20-36.

(ii) "The specific thing about the stars' and the sun's heating",
Averroes writes "is their light".[63] How does this connect to the
heating? Averroes puts forward the following answer:

The light *qua* light warms the bodies down here by virtue of a divine force
(*quwwa ilāhiyya*; *koah 'elohī*) when it is reflected. This holds *a fortiori* when
the rays (Ar. *al-khuṭūṭ al-shu'ā'iyya*; Heb. *ha-qawwim ha-niṣoṣiyīm; ha-
qawwim he-sheviviyim*) fall on the warmed body in right angles, for then the
reflection is greater. In fact, when the reflection is stronger, the heating[64] is
stronger, as we can see in the cases of the burning mirrors and the glass
flasks which burn wool. *A fortiori* [this holds] when the reflecting body is
polished.[65]

The crucial point in this account is the claim that the warming
is caused by the reflected light, and that the heat is then produced by virtue of a *divine force*. Averroes is visibly not content
with Aristotle's "lame" accounts which ascribe the heating of
the celestial bodies to their motion and seeks an alternative
explanation. Thus, since the heavenly bodies are not themselves
warm, and since the claim that their heating is produced
through their motion seems dubious, Averroes concludes that
there must be something particular about the heavenly bodies
that allows them to warm. This is their light. But in the
Aristotelian natural philosophy there is no connection whatsoever between heat ("fire") and light ("the activity of what is
transparent *qua* transparent"[66])! Averroes' way out of the
quandary was to posit a "divine force", owing to which the heat
is generated by reflected light. Much hinges on that notion
(which Averroes apparently borrowed from Alexander of
Aphrodisias): it provides the theoretical foundation for the

---

[63] Averroes, Epitome of *De caelo*, Ar., p. 49:3-4; Heb., fol. 62bb:36-7; Falaqera, *DF*,
228a:14-15.

[64] The Arabic text of Averroes carries "luminosity" (*iḍā'a*), a reading confirmed by
both Ibn Falaqera's and Moses ibn Tibbon's Hebrew translations. Still, the context
makes clear that the correct is "heating". If this is not a slip of the pen of Averroes
himself, then the error must have crept into at least one family of manuscripts of the
text at an early date. In the MC on *De caelo* (Ar., p. 232:6-7; Heb., fol. 48b:4-6), to be
quoted below, Averroes writes: "It is witnessed by the senses that light warms
through reflection, and that the stronger the reflection the stronger the warming.
This is shown by the burning mirrors".

[65] Averroes, Epitome of *De caelo*, Ar., p. 49:4-10; Heb., 62bb:38-63aa:4; Falaqera,
DF, 228a:15-20. The passage is translated in Henri Hugonnard-Roche, "L'Épitomé
du *De caelo* d'Aristote par Averroès. Questions de méthode et de doctrine", *Archives
d'histoire doctrinale et littéraire du Moyen Age*, 51 (1984): 7-39, on pp. 29-30.

[66] Aristotle, *On the Soul* 2.7, 418b10 (trans. J.A. Smith); see also *ibid.* 419a9.

claim that the celestial bodies can influence natural processes in the sublunar realm.[67]

(2) In the Middle Commentary on the *Meteorologica*[68] Averroes writes:

The [...] question whether light warms or not requires profound investigation. For when one examines the pertinent observable [phenomena], such as the burning mirrors and the flask filled with water that burn cotton, it can be thought that the ray [*ha-niṣos*] burns essentially, especially when the reflection is in right angles, i.e. when the ray is reflected along the line of incidence or when many rays issuing from many points are reflected to a single point. Now when we take into consideration the first premises of this [discipline], namely that something causing something [else] to pass from potentiality to actuality is of the same nature in actuality, then it follows necessarily that heat is produced only from a hot body and fire from fire. Now we affirm that since it has been made clear that light is not a body, therefore if it has essentially a chafing potentiality, then what warms is the luminous body insofar as it is luminous, not the light *per se*. Since, then, what warms is the luminous body, and since it is evident that the stronger the body shines the stronger it warms, it is thought that the light is the cause of the warming. It follows that the warm [body] is not produced by a warm [body] like itself, nor fire from fire.

This being so, we must necessarily assent to one of the following [conclusions]: either (i) we do not accept that premise [viz. that "something causing something (else) to pass from potentiality to actuality is of the same nature in actuality"], or (ii) this [warming] effect pertains to the luminous body accidentally. (i) But it is necessary to accept this premise, for its claim has been established elsewhere and the doubts bearing on it have been dissipated. (ii) This being so, we should inquire how this [effect] can appertain to the luminous body accidentally. We affirm that if the luminous body were at

---

[67] In *On Providence*, Alexander of Aphrodisias explicitly construes the capacity of the celestial bodies to maintain the existence and order of the sublunary beings as due to the "divine force" emanating from them. See *On Providence* (*supra*, n. 32), p. 18 and 30-32 Thillet, and 124-6 and 150 Fazzo-Zonta. Professor Y. Tzvi Langermann kindly called my attention to the fact that the notion of 'divine force' appears in the work *Fī Mabādī' al-Kull* ascribed to Alexander of Aphrodisias; see Shlomo Pines, "The spiritual force permeating the cosmos according to a passage in the treatise *On the Principles of the All* ascribed to Alexander of Aphrodisias", in *idem*, *Studies in Arabic Versions of Greek Texts and in Medieval Science* (= *The Collected Works of Shlomo Pines*, vol. 2) (Jerusalem and Leiden 1986), pp. 252-5, on p. 253. According to this text, the "divine force" is a certain "pneumatic force (*quwwa rūḥāniyya*) that permeates all the parts of the world" thereby "bind[ing] (all) the (parts) of it to one another" (*ibid.*). This subject clearly calls for more research.

[68] In his Middle Commentary on Aristotle's *De caelo* Averroes refers to the Middle Commentary of the *Meteorologica* (see quotation below, p. 134), thereby indicating that the latter preceded the former.

rest, it would not have warmed, because it is not warm. It warms accidentally, [namely] because of its motion, as Aristotle says. Now when, by virtue
of the motion, heat is propagated in the body which is being warmed, then
what occurs to [the heat] when it is reflected along its line of incidence or
along right angles is analogous to what occurs to the ray of light. Therefore,
because [heat] is always concomitant with the ray of light, it is thought that
the ray of light is the cause of the warming. But this is not the case: rather,
it is by accident that more light implies [or: is followed by] more heat, which
leads one to think that the shining is the cause of the heat.

This kind of reflection occurring to the fiery heat in the air comes to pass
owing to the motion produced in the parts of the air because of their being
moved by the heavenly bodies. This heat is attached to the ray of light
because of the association [lit. relation; hityaḥasut] existing between them.
When the ray is in a straight [line] the heat is [propagated] in a straight
[line], and when the ray is reflected the heat is reflected, and when the ray is
refracted the heat becomes weaker, and when the motion is from many
points to one point, as is the case in the burning mirrors, the same occurs [to
the ray of light and to the heat].

What warms through its motion is also warming accidentally, for it is not
customary for motion to generate heat, inasmuch as it is not hot. The action
of motion is to prepare the substrate to receive heat. For the fire which is
produced, for instance, by striking certain bodies is generated by the heat
that is in the air through the motion of striking. It is not rare that owing to
a better preparation strong heat is generated through weaker heat; indeed,
an access of preparation can lead to a thing [i.e. heat] being held to have been
generated spontaneously [through itself]. This is the case with the heavenly
bodies: through their motion they generate heat in things that are not warm,
[namely,] through the intermediate warm bodies [viz. the air]. For this reason, no fire is ignited by stones in cold air and in cold places as is done in
warm places and in warm air. If motion had generated heat essentially, then
it would have been the case that something comes to exist through something of a different species. By the same token, rest is cooling accidentally: if
motion had warmed essentially, then rest, which is a privation, would have
cooled essentially. But the falsity of this is self-evident.

Thus the heavenly bodies warm upon approaching [the earth] and they cool
when they recede. And through these two operations they preserve the forms
of the elements. Thus, the celestial body preserves the elements, but it does
not generate them, as some people have imagined, especially apropos of
fire.[69]

The notion of a "divine power", by virtue of which light warms,
has entirely disappeared from this detailed account, and
Averroes has in fact severed all essential connection between

---

[69] Averroes, MC on *Meteorologica*, pp. 14-17; see also Falaqera, *DF*, 227:27-228a12.

light and heat. The fact that the celestial bodies are luminous has nothing to do with the fact that they (also) produce heat. Indeed, the heat reaching us from the celestial bodies is "fiery heat", i.e. the element fire dispersed in the air, which is put in motion by the celestial bodies; its concomitance with light is construed as due only to the fact that the propagation of the rays of light and of heat follows similar (presumably optical) rules.

(3) We lastly turn to the Middle Commentary on Aristotle's *De caelo*.[70] Averroes (again) rejects the idea that the stars are fiery: not everything that shines is fire; the luminosity of the stars in no way indicates that they are fiery or warm. How then do the celestial bodies warm? According to Aristotle, Averroes says, "the sun and the stars" warm through their motion.[71] Two confirmatory facts are adduced: the purported melting of a metallic arrow in motion indicates that motion generates heat in the air; and the phenomena of shooting stars, comets and the like show that the motions of the "celestial bodies" generate heat.[72] The greater the star and the closer it is to the earth, the stronger is its warming effect on the air.[73] "This is why we find that the heating due to the sun and the moon [!] is the greatest", Averroes adds, "namely owing to their size and proximity to the air".[74] Averroes emphasizes that the heavenly bodies can warm the air through their motion although they are not contiguous with it: the medium (in this case: the air) need not be moved with the same motion as that transmitted through it (e.g. the magnet attracts the iron without attracting the intermediate air). The luminosity of the stars thus has nothing to do with their warming. There are indeed, as has been made clear in the *De anima*, two distinct kinds of luminous bodies: the celestial ones (which are luminous but do not warm) and the sublunar fire (which is both luminous and warm).[75]

---

[70] Averroes, MC on *De caelo*, Ar., pp. 229 -233; Heb., fol. 48a-48b.

[71] Averroes, MC on *De caelo*, Ar., p. 229:6-11; Heb., fol. 48a:5-9.

[72] Averroes, MC on *De caelo*, Ar., pp. 229:11-230:3; Heb., fol. 48a:10-14.

[73] Averroes, MC on *De caelo*, Ar., p. 230:3-4; Heb., fol. 48a:14-15; see also Falaqera, *DF*, fol. 227a:15-20.

[74] Averroes, MC on *De caelo*, Ar., p. 230:4-6; Heb., fol. 48a:15-16; see also Falaqera, *DF*, fol. 227b:16-20. Similarly: Averroes, MC on *Meteorologica*, p. 9.

[75] Averroes, MC on *De caelo*, Ar., pp. 230:6-232:5; Heb., fol. 48a:16-48b:4; Aristotle, *De anima* 2.7, 418b10-13.

134

Averroes adds the following complementary consideration:

It is witnessed by the senses that light warms through reflection, and that the stronger the reflection the stronger the warming. This is shown by the burning mirrors. The causes thereof have been explained in the mathematical sciences. It has also been made clear in the *De anima* that the light is not a body. If this is so, then warming is effectuated by something that is not a body, and *a fortiori* by something that is not fire. This is the second manner in which apparently the stars warm what is below them. But concerning this manner there is a doubt which is not slight and we have already investigated it in [the Middle Commentary on] the *Meteorologica*.[76]

This account essentially repeats what had been said in the Middle Commentary on the *Meteorologica*, to which it explicitly refers. The luminosity of the heavenly bodies is in no way connected to their heating effect; the latter is due solely to their motion. Still, Averroes seems to waver when he refers to the second "manner" [*jiha, ṣad*] in which stars warm, namely when their light is reflected: although Averroes is careful to avoid the concept of the "divine force" by virtue of which reflected light generates heat, he seems dubious with regard to his thesis that "it is by accident that more light implies more heat". Although the concept of the "divine force" appears only in the Epitome of the *De caelo*, it seems to have remained close to the surface in later works.

Averroes thus has not one, but two accounts to offer for the heating effect of the celestial bodies. But do these accounts provide an adequate basis for the thesis that the generation of animals – notably "spontaneous", but also sexual – depends on the "heats" of the stars? As far as I am aware, Averroes does not address the issue. The reason for this is perhaps the following: Averroes drew on the theory of the "divine force" in the Epitome of the *De caelo*, in an early period, in which he was still committed to the theory of the active intellect as the Giver of Forms; in this context, the stars' "heats" had the limited role of providing the necessary thermal conditions for the existence of life on earth,[77] but had no role to assume *qua* formal cause. The theory that light heats by virtue of a "divine force" was adequate for the explanatory task it had at that early period. After

---

[76] Averroes, MC on *De caelo*, Ar., pp. 232:5-233:4; Heb., fol. 48b:5-9.

[77] Cf. Averroes' description of how the heavenly bodies are instrumental in exercising providence over generation and corruption in the passage quoted *supra*, p. 121 f., n. 32.

the demise of the Giver of Forms, when Averroes elaborated the thesis which ascribes the origin of animate forms to the "heats" of the stars, he did not see fit to rectify the Epitome of the *De caelo*, with a view to showing how the doctrine of the "divine force" can account for the role of the "heats" within the new theoretical framework. Worse, Averroes at some time at least formally even abandoned that doctrine and leaned toward an account of celestial heat that was entirely "mechanical" and thus totally unfit to account for the forms which, according to the later Averroes, are yet carried by the "heats" of the celestial bodies. In sum, therefore, none of Averroes' two theories of the heating by light was developed by Averroes so as to constitute a suitable basis for his later view of the biological role of the "heats" of the stars. Averroes' doctrine of the celestial influences on the sublunar world, although very central to a number of major concerns, remained without foundations; it is a glaring "anomaly" (in the sense of the late T.S. Kuhn) within Averroes' physical world-picture.[78]

CONCLUSION

Averroes changed his mind over the issue of the role of a separate form in generating animate beings: from the belief that all sublunar forms are due to a separate active intellect which imprints them directly unto matter, he moved to the idea that these forms are carried by seed or semen, which derive them from something that is itself alive – the male parent and the ensouled heavenly bodies. As we saw, Averroes himself recognized that his two solutions were not fundamentally different. For although in his later thought Averroes rejected the position that an intellect endows matter with forms directly, he still posited the existence of a separate Entity (or entities) to which all forms go back, the First Cause and the immaterial forms associated with the heavenly bodies. This Entity or entities are the "eternal cause" without which the explanation of the genesis of living forms would have to postulate an infinite series of

---

[78] With the doctrine of celestial influences transmitted via the rays of light Arab thinkers moved away from Aristotle, clearly yielding to the influence of "astrologically-colored" doctrines. This subject, and in particular the role of al-Kindī's theory of rays in this development, calls for more research.

causes. The difference between Averroes' two positions thus seems to be related above all to the *modus* through which the forms are imprinted in matter – directly, by emanation, in one case, through the mediacy of the seed or sperm and the "heats" of the stars in the second. Both theories *in fine* share the idea that the forms ultimately derive from the transcendent First Cause.

One may then wonder why Averroes at all changed his view on the role of the active intellect and even went out of his way to rectify early redactions of his works in the light of his later views. A suggestion concerning a possible "ideological" context of the debate over the theory of the active intellect has been made by Charles Genequand.[79] Averroes himself in fact is quite explicit on this point: writing in the later period of his life, after having dismissed the notion of the active intellect, Averroes in the Long Commentary on the *Metaphysics* opines that the theory that forms ensue in matter as the result of the action of a separate form such as the active intellect (this theory is associated with the names of Themistius and Avicenna) boils down to asserting that something comes to be out of nothing. He writes:

If one assumes that the forms are created, one is led to accept the theory of Forms and of the Giver of Forms. Their extremism in this assumption led the theologians [*mutakallimūn*], among people of the three faiths existing today, to the view that something can proceed from nothing, because if the form can be created, the whole can be created.[80]

Averroes' concern would thus seem to be that the emanationist theories invoking the active intellect appear to lend support to creationist theories and thus provide "the enemies of philosophy with a powerful weapon to support their claim".[81]

Averroes' two theories of the generation of animals, let me finally suggest, are both at variance with Aristotle's own view. For the idea that the generation of animate beings necessarily involves an eternal "mover from without" is exactly the stance against which Aristotle had addressed his adage "man gener-

---

[79] Genequand, *Ibn Rushd's Metaphysics*, pp. 27-9.

[80] Averroes, LC on the *Metaphysics*, 1503. In the Commentary on the *Book of Animals*, Averroes similarly expresses the view that "had there existed a separate Form, which would have created a form in matter, then it would have been possible that something come to be out of nothing". Freudenthal, "Averroes' changing mind", phrase 20.

[81] Genequand, *Ibn Rushd's Metaphysics*, p. 29.

ates man": as Klaus Oehler has shown,[82] it was directed against Plato's theory of Forms and made the point that forms exist only enmattered, namely in the different individual living beings. The generation of any individual plant or animal depends – in Aristotle's view – solely on the form which the seed or the semen transmit to matter, to the exclusion of any transcendent form. For Aristotle, the sun's role is merely to guarantee the continued generation and corruption in the sublunar world.

Averroes thought that the scientific and philosophical truth was consigned in Aristotle's writings. The search for truth was thus the search for the true meaning of Aristotle's works. Concerning the issue of the generation of animate beings Averroes believed he had succeeded in retrieving the original Aristotle, but in fact he could not see beyond the horizon of the neoplatonically tainted Aristotelianism in which he and his contemporaries were steeped.

Acknowledgements: For his thorough reading of, and helpful remarks on a draft of this paper I am very grateful to Professor Herbert A. Davidson (University of California at Los Angeles). I also thank Dr. Ruth Glasner (The Hebrew University of Jerusalem) for her very useful comments and suggestions.

[82] See n. 2 above.

PROVIDENCE, ASTROLOGY, AND CELESTIAL
INFLUENCES ON THE SUBLUNAR WORLD IN SHEM-TOV IBN
FALAQUERA'S *DE'OT HA-FILOSOFIM* *

Introduction

Maimonides' *Guide of the Perplexed* was rendered into Hebrew in
Southern France in 1204, triggering off a fairly strong interest in
natural philosophy and in metaphysics among Hebrew-reading
Jews.[1] This is the context in which two Jewish scholars knowledgeable in Arabic and familiar with scientific and philosophical
writings in Arabic—Judah ben Solomon ha-Cohen and Shem-Tov
ibn Falaquera—thought that it would be a worthy enterprise to
sum up in Hebrew significant portions of the available knowledge
in those domains in order to bring them to the reach of an
interested readership not knowing Arabic. Of these two thirteenth-
century works, Falaquera's *De'ot ha-Filosofim* (The Opinions of the
Philosophers)[2] is by far the more voluminous, detailed and lucid
one, and it is to it that I will devote most of my attention in what

* For helpful suggestions and criticism of an early version of this article I am
very grateful to Resianne Fontaine, Ruth Glasner, and Y. Tzvi Langermann.

[1] Isadore Twersky, "Aspects of the Social and Cultural History of Provençal
Jewry," *Journal of World History* 11 (1968): 185-207; Joseph Shatzmiller, "Rationalisme et orthodoxie religieuse chez les Juifs provençaux au commencement du
XIVe siècle," *Provence historique* 22 (1972): 261-86; Gad Freudenthal, "Les
sciences dans les communautés juives médiévales de Provence: Leur appropriation, leur rôle," *Revue des études juives* 152 (1993): 29-136; *idem*, "Science in the
Medieval Jewish Culture of Southern France," *History of Science* 33 (1995): 23-58;
and Mauro Zonta, *La filosofia antica nel Medioevo ebraico* (Brescia, 1996).

[2] *De'ot ha-Filosofim* survives in two manuscripts: Leiden, Bibliotheek der Rijks-
universiteit MS Or. 4758 (Warn. 20) (Institute for Microfilmed Hebrew Manuscripts [=IMHM] 17368); Parma, MS De Rossi 164 (IMHM 13897). References
to *De'ot ha-Filosofim* below are usually to only one manuscript (*DF*, L, or *DF*, P),
but longer passages were checked in both (where the two manuscripts carry the
relevant passage). I am grateful to Professor Steven Harvey and to Dr. Ruth
Glasner for having made copies of different portions of these two manuscripts
available to me.

336

follows, although I will occasionally draw also on Judah ben Solomon's *Midrash ha-Ḥokhmah* (The Quest for Wisdom)[3] in order to make a comparative point.

I will approach Falaquera's work from a very specific point of view, trying to analyze his handling of one particular *problématique* in the Aristotelian world-picture that Falaquera undertook to transmit to his coreligionists. This *problématique*, which I take to be a "strategic research site" in studies on medieval philosophy of nature, is the following inherent difficulty in the Aristotelian theoretical system: on the one hand, the sublunar realm as described by Aristotle is not a closed physical system, inasmuch as the explanation of certain natural phenomena draws on the assumption that the sun and the moon influence sublunar processes of generation and corruption; on the other hand, the celestial bodies are held to be constituted of the "fifth element," which, by its very definition, is held to be a "stranger" to all processes in the sublunar realm. I will investigate how Falaquera integrated this difficulty within his work, some sixteen centuries after Aristotle's death. Reading through Falaquera's work with the question concerning the types of causal connections between the supra- and the sublunar realms as a red thread, we will see that the sun's and the moon's actions upon sublunaries, far from being played down as an inconvenient, unsolvable riddle, instead became *paradigmatic* for the influences which *all* heavenly bodies were held to exert on sublunar substances. Moreover, in Falaquera's system, the very existence of an order in the natural universe, i.e. the existence of God's general providence, is held to hinge upon the emanations of the heavenly bodies. The physical system outlined by Falaquera thus owes an answer to the crucial problem that already beset Aristotle: how can the heavenly bodies, supposedly made of the impassive fifth element, influence the sublunar realm? We will see that as answers to these queries, Peripatetics introduced two new, non-Aristotelian notions: the notion of a "divine force" of the stars'

---

[3] The title is not easily translated because the word "midrash" alludes both to the literary genre of Midrash (homiletic interpretation, notably of the Bible) and to the root דרש meaning "to seek," "demand," or "request." For *Midrash ha-Ḥokhmah* I used Oxford, Bodleian, MS Michael 551 (Neubauer 1321) (IMHM 22135). References to *Midrash ha-Ḥokhmah* (*MH*) are to this manuscript. I am very much indebted to Dr. Resianne Fontaine for having kindly sent me a copy of this manuscript along with her transcription of parts of it.

rays, and the Neoplatonic one of an "overflow" from the "separate bodies" associated with the planets. By virtue of these two notions the cosmos could again be construed as interconnected.

It is well known that Falaquera was not an original thinker and that his voluminous summa consists mainly of long and short passages gleaned from Arabic sources and (skillfully) arranged and translated into Hebrew.[4] Thus most of the passages on which I will draw are borrowed from works by Averroes—notably his epitomes and his middle commentaries on Aristotle's treatises in natural science. Nonetheless I will treat De'ot ha-Filosofim as a unitary work, and will speak of Falaquera as if he were its author (although the notes will often indicate the sources of Falaquera's text). In one instance, though (below, sec. 2), I will show that Falaquera deliberately constructed his account out of conflicting statements that he found in Averroes' commentaries. One can, I think, rightfully consider Falaquera as expressing an *opinion moyenne* of certain erudite arabophone Jewish circles and, more important, De'ot ha-Filosofim was certainly read as expressing such a consensual opinion. My account will thus be more phenomenological than historical—it is the *Weltanschauung* displayed in De'ot ha-Filosofim that I will try to uncover in this article.

1. *Aristotle's Problematic Division of the Universe Into the Sublunar and the Supralunar Realms*

Two related features distinguish Aristotle's world-picture from those of his Presocratic predecessors: the absence of an over-arching teleology, and the separation of the cosmos into two distinct realms in which different natural laws obtain. The latter feature, in particular, was a momentous innovation, which, however, was not smoothly and consistently integrated within Aristotle's system; it rather remained the source of persistent difficulties—what Thomas S. Kuhn called an "anomaly." It will be at the center of the following discussions.

---

[4] See Steven Harvey, "Shem-Tov ibn Falaquera's *De'ot ha-Filosofim*: Its Sources and Use of Sources," in the present volume; Raphael Jospe, *Torah and Sophia: The Life and Thought of Shem Tov Ibn Falaquera* (Cincinnati, 1988); and Mauro Zonta, "Mineralogy, Botany and Zoology in Hebrew Medieval Encyclopaedias," *Arabic Sciences and Philosophy* 6 (1996): 263-315.

One of the most striking and pervasive phenomena of nature is surely the dependence of natural processes on the earth upon the celestial bodies and their motions. Specifically, growth and decay of plants are manifestly dependent upon the annual motion of the sun, and other life processes are suspended on the sun's daily motion. A number of Presocratic philosophers made this basic observation into a point of departure of their philosophies of nature. They combined it with another idea, according to which the celestial bodies, whose motions are so strikingly regular and which are not subject to any sort of decay, rank higher on the scale of being than the ephemeral and erratic earthy ones. The Presocratics thus posited that the lower and the higher worlds were originally one, but that subsequent to a primeval separation, the viler matter stayed at the center, while its noble part came to occupy the periphery. As Friedrich Solmsen once observed, "the modus and the agents of this separation, its causes and circumstances, vary from system to system. However, these details, while highly interesting in themselves, are secondary in importance to the idea of separation as such."[5]

Underlying this notion of separation was the idea that the primeval matter at the center was constituted of pairs of opposite powers which, through the separation, came to occupy their present distinct spatial regions. In this scheme, what is bright, warm, dry, and in motion is assigned to the upper region; the dark, cold, moist, and inert to the lower. The gist of the view is concisely summarized in the following passage from Plutarch's *De primo frigido* (955 B-C):

> We must recognise that the wise men and intellectuals of old set a gulf between terrestrial and celestial bodies: [they distinguished the two kinds of element] by the difference in their [intrinsic] powers. Bodies that are hot and shining, and swift and light, they assign to the being [*phusei*] that is deathless and endless. Bodies that are murky and cold, and slow [and heavy], they declare to be the hapless lot of creatures who dwell in the shadow of death below.[6]

---

[5] Friedrich Solmsen, "Aristotle and Presocratic Cosmogony," *Harvard Studies in Classical Philology* 63 (1958): 267; quoted from Gad Freudenthal, *Aristotle's Theory of Material Substance. Form and Soul, Heat and Pneuma* (Oxford, 1995), 79.

[6] Translated in Denis O'Brien, *Theories of Weight in the Ancient World, I: Democritus, Weight and Size* (Paris and Leiden, 1981), 366. The additions in square

Now this scheme had, so to speak, a "built-in" felicitous capital consequence. Since it was assumed that bits and pieces of what became the pure celestial matter were left behind, intermingled with the earthy matter, phenomena observed down here could be construed as being base replicates of what was the case in the heavens, which more often than not were construed as divine. For instance, the fact that the sun's rays have vivifying effects followed as a simple and natural corollary to the idea of a primeval separation. Moreover, the Presocratic monistic philosophies—say those of Heraclitus or of Diogenes of Apollonia—very naturally accounted for what they construed as the teleology operative in the world: the primeval single element was construed as a divine and intelligent omnipresent entity, and this easily allowed one to conceive of the scale of being as stretching from what is topologically low to what is topologically high.[7]

These two aspects of the Presocratic systems brutally fell to the ground in Aristotle's mature system. While in an early phase of his development—the one during which he wrote his lost dialogue *On philosophy*—Aristotle apparently still believed in the essential material unity of the cosmos and in the teleology operative in it, in his mature system the innovative introduction of what later generations were to call "the fifth body" or "ether" invalidated these ideas. In fact, when Aristotle introduced the new notion of ether, he thereby created a decisive and consequential rupture of the universe.[8] The sublunar and the supralunar realms were now sharply cut off from one another, and the impassive celestial matter was held neither to act on, nor to suffer from, the other four elements: as a "stranger to generation and destruction,"[9] it followed natural laws different from those pertaining to the sublunar elements. The sublunar world was to be construed as a closed system, whose functioning depends solely on the laws of physics bearing on the bodies constituted of the four sublunar elements.

brackets are O'Brien's. Quoted from Freudenthal, *Aristotle's Theory*, 79-80.

[7] See Willy Theiler, *Zur Geschichte der teleologischen Naturbetrachtung bis auf Aristoteles* (Zürich, 1924).

[8] The following paragraph follows Freudenthal, *Aristotle's Theory*, 104.

[9] Friedrich Solmsen, *Aristotle's System of the Physical World* (Ithaca, New York, 1960), 289.

But Aristotle quickly had to recognize that the separation of the cosmos into two independent realms could not be pushed through consistently. For the situation now was nothing less than paradoxical: the primary and pervasive fact that the heat of the sun is indispensable for life has become a rather uncomfortable anomaly within the Aristotelian physical theory: since the ether was supposed to be quality-less and thus not to be hot, Aristotle found himself compelled to account for the sun's warming effect through an *ad hoc* "mechanical" hypothesis. In the *On the Heavens* he ascribes the warmth of the sun to the friction between it and the air below it: "the heat and light which they [the heavenly bodies] emit are engendered as the air is chafed by their movement."[10] (Aristotle adduces as an analogy the case of missiles, whose leaden heads purportedly melt as a result of their motion through the air.) On this theory, Aristotle suggests, although the stars, notably the sun, are not themselves fiery, yet "the air which lies beneath the sphere of the revolving element is necessarily heated by its revolution."[11] Aristotle adds a reassuring remark to the effect that this heating occurs "especially in that part where the sun is fixed [unto the sphere]," but this is hardly warranted, seeing that the sphere of the moon lies between that of the sun and that of the air. In the *Meteorology*, this account is presupposed, and Aristotle tries to account for the disturbing fact that only the sun of all the heavenly bodies warms.[12]

The two parts of Aristotle's account of the fact that the sun warms are, as one scholar commented, "both almost equally lame."[13] Still, the flagrant fact *that* the sun warms could hardly be ignored and Aristotle draws on that effect in the explanations of a number of major phenomena. Two will in particular interest us here. First, Aristotle explains the existence of seasons and, more generally, the very existence of processes of generation and corruption, as resulting from the sun's motion along the "inclined

[10] Aristotle, *On the Heavens* II, 7 289a20. Trans. W.K.C. Guthrie (Cambridge, Mass., 1939).
[11] Aristotle, *On the Heavens* II, 7 289a30 (Guthrie trans.).
[12] Aristotle, *Meteorology* I, 3 341a12-36.
[13] James Longrigg, "Elementary Physics in the Lyceum and Stoa," *Isis* 66 (1975): 214; see also Paul Moraux, "Quinta essentia"," in Pauly-Wissowa, *Realencyclopädie*, XXIV(1) (Stuttgart, 1963), 1204-5; and John Thorp, "The Luminousness of the Quintessence," *Phoenix* 36 (1982): 104-23.

circle," i.e. the ecliptic: on this account, the alternation of generation and corruption is due to the alternate approaching and
retreat of the sun.[14] The same motion also continually mixes up
the elements, preventing them from settling down in four concentric layers, as they otherwise would long ago have done.[15] Second,
Aristotle accounts for the so-called spontaneous generation—the
supposed coming-to-be of living beings without semen or seed—by
attributing it to the specifically vivifying effect of the sun's heat,
which produces vital heat in small enclosed quantities of water.[16]

Let me repeat: Aristotle conceded that the sun's heating plays a
pivotal role in generation and corruption, and, in agreement with
the Presocratic tradition, that it is endowed with vivifying effects,
which ordinary fire does not have. Yet, the very fact *that* the sun,
made of the fifth element, warms is most problematical within
Aristotle's system; *a fortiori* Aristotle has no explanation for the
vivifying effects of the sun's heat. The theoretical grounding of the
accounts in question is thus quite uncertain.

Armed with this insight into a major "anomaly" of Aristotle's
system of the physical world, we would expect that Aristotelians
over the coming centuries would play down as much as possible
the sun's role in nature. The precise opposite is the case, however,
as we shall now see.

2. *The Connectedness of the Cosmos Reestablished:*
The "Divine Force" of Rays of Light

With the above sketch of the inner tensions within Aristotle's
system as a background, we now turn to Falaquera's *De'ot ha-
Filosofim*. The questions I will address are: Was Falaquera aware of
the existence of anomalies within the Aristotelian framework? To
what extent could his reader perceive the theoretical anomalies we
defined above? Were the issues addressed explicitly, and were
solutions offered? As noted earlier, I will treat Falaquera's text as a
synthetic whole, which, assembled from many sources by a knowledgeable and philosophically informed scholar, offered its Jewish

---

[14] Aristotle, *On Generation and Corruption* II, 10.
[15] Aristotle, *On Generation and Corruption* II, 10 337a8-12.
[16] Cf. Freudenthal, *Aristotle's Theory*, 25-6.

readers a consistent *cum* philosophical world-view, constructed out of occasionally conflicting materials found in Averroes.[17]

Consider first Falaquera's treatment of the dependence of generation and corruption upon the sun's motion along the ecliptic as given in book I, part II-A, chapter 9 of his work. Falaquera asks how one can reconcile the eternity of the rotatory translational motion with the alternating character of two opposite movements, viz. generation and corruption. He rejects the astrological interpretation: contrary to what the latter maintains, it is impossible that a translational motion *per se* produce now generation, now corruption.[18] The correct explanation is rather Aristotle's: the motion of the sun along the ecliptic can produce generation and corruption because it is an eternal continuous motion that incorporates two motions—a motion of approach and a motion of retreat.

So far so good—we are exactly where Aristotle had left off (*On Generation and Corruption* II, 10). Except for the short remark

---

[17] Commentaries by Averroes are referred to in the following editions: (1) *On Generation and Corruption*, *Epitome* and *Middle Commentary* (=*MC*): Hebrew translations (Moses ibn Tibbon and Kalonymus ben Kalonymus, respectively), *Averrois Cordubensis commentarium medium & epitome in Aristotelis De generatione et corruptione libros*, ed. Samuel Kurland (Cambridge, Mass., 1958); Eng. trans., *Averroes on Aristotle's "De generatione et corruptione," Middle Commentary and Epitome*, trans. Samuel Kurland (Cambridge, Mass., 1958). (2) *On the Heavens*, *MC*: Arabic, *Talkhîṣ al-samâ' wa-al-'âlam*, ed. Jamâl al-Dîn al-'Alawî (Fez, 1984); Hebrew (Solomon ibn Ayyub), Berlin MS 212 (Or. Qu. 811) (IMHM, 1769); *Epitome*: Arabic, in *Rasâ'il Ibn Rushd* (Hyderabad 1947); Hebrew (Moses ibn Tibbon), Paris, Bibliothèque Nationale de France, MS héb. 918 (IMHM 31960). (3) *Parva naturalia*, *Epitome*: Arabic, *Averrois Cordubensis compendia librorum Aristotelis qui Parva naturalia vocantur*, ed. Henry Blumberg (Cambridge, Mass., 1972); Hebrew (Moses ibn Tibbon), *Averrois Cordubensis compendia librorum aristotelis qui Parva naturalia vocantur*, ed. Harry Blumberg (Cambridge, Mass., 1954); Eng. trans., *Averroes, Epitome of "Parva naturalia,"* trans. Harry Blumberg, (Cambridge, Mass., 1961). (4) *Books of Animals*, treatise 16 (=*On the Generation of Animals*), *Epitome*: Hebrew (Jacob ben Makhir), Oxford, Bodleian, MS Opp. 1641 Qu (Neubauer, 1381) (IMHM, 22405). This text is not preserved in Arabic. (5) *Metaphysics*, *Epitome*: Arabic, *Compendio de la metafísica*, ed. (with Spanish trans.) Carlos Quirós Rodríguez (Madrid, 1919; reprint, with an introduction by Josep Puig Montada, Cordoba, 1998); German trans., *Die Metaphysik des Averroes*, trans. Max Horten (Halle, 1912). (6) *Meteorology*, *MC*: Hebrew (Kalonymus ben Kalonymus), "The *Middle Commentary* of Averroes on Aristotle's *Meteorologica*," ed. Irving Maurice Levey, (Ph. D. diss., Harvard University, 1947).

[18] *DF*, P, fol. 87v, lines 5-7 (= Averroes, *MC on On Generation and Corruption*, Heb., 89, ll. 25-26; Eng., 101). Here and below, the sign "=" means that Falaquera roughly follows (translates or paraphrases) the source indicated.

rejecting the astrological theory, Falaquera's text shows no signs of an evolution of the thinking on the questions at hand during the preceding millennium and a half. Yet precisely at this point the text has the following observation to add:

> And what Aristotle has affirmed with respect to the sphere of the sun should be understood as applying to the other oblique spheres too. In fact, although we know not the impression [*rishum*] made by each and every star on the existents down here, still, through a generalization [*ba-ma'amar ha-kelali*] it becomes clear that they are involved in generation and corruption. It is clear that some of the stars have each a specific action [*segullah*]. This is why we find that the Ancients have divided the existents in accordance with those stars and have assigned an existent X to be of the nature of a star Y. It can be seen that these stars take after the motion of the sun [*mit-mashlim bi-tenu'at ha-shemesh*], for the differences in the impressions [they produce] are mostly dependent on their proximity or distance from the sun.[19]

Seen from the vantage point of Aristotle's text, this doctrine is really remarkable: the sun's influence on the sublunar existents, which for Aristotle had been an embarrassing anomaly, has now become the *paradigm* for the relationship between the celestial bodies and the sublunar existents. There can be little doubt that this shift is to be ascribed to the infiltration of astrological motifs into Arabic natural philosophy, and we can thus identify here the results of what might be called the *astrologization* of the Aristotelian doctrine in the Middle Ages. But be this as it may, Falaquera's reader, perhaps even Falaquera himself, must have asked himself how this purported influence of the celestial bodies is to be explained: since the heavenly bodies are of the homogeneous impassive celestial fifth substance, how can they possibly have specific actions (*segullot*) on sublunaries?

Falaquera addresses this question in book I, part VII-C of his work, where he discusses the nature of the stars. His account is woven together from passages or even single sentences literally borrowed from three distinct works by Averroes, namely (in chronological order of composition): the *Epitome of On the Heavens*; the *Middle Commentary* (= *MC*) *on the Meteorology*; and the *MC on On*

---

[19]   *DF*, P, fol. 88r, ll. 2-9 (*DF*, L, fol. 162rb). (= Averroes, *ibid.*, 90, ll. 47-48; Eng., 102; and *Epitome of On Generation and Corruption*, Heb., 121, l. 68-122, l. 78; Eng., 133).

the Heavens. In these works Averroes offers two different views on our issue.[20] In the early *Epitome of On the Heavens* he holds that the heating by the sun (and the other stars) is due to two distinct but complementary processes: the heat-generating motion of the heavenly bodies and the heat-generating reflection of the light of the stars. By the time of writing the *MC on the Meteorology* the Commentator, ever more devoted to Aristotle, had changed his mind and upheld the original Aristotelian view, according to which the heating of the heavenly bodies is due to their motion alone. This stance is repeated also in the later *MC on On the Heavens.* We will now see how Falaquera used passages and sentences drawn from these three sources as bricks out of which he constructed a world-picture that fitted his own views, but which was at variance with those of Averroes.

It probably never occurred to Falaquera that Averroes' two diverging explanations of the heating by the celestial bodies could be the result of an evolution of his thought. Therefore, had Falaquera viewed his task as that of a "mere" encyclopedist, whose task is simply to put information and opinions at the disposal of his reader, he would have presented the two alternative views on the issue side by side, as they are in Averroes' *Epitome of On the Heavens.* He could have done so easily, namely by introducing the subject through the following sentences with which Averroes opens his discussion in the *Epitome*. "One may ask: how is it that the sun, which is not fire, warms and shines, with the same applying to all the other stars? We affirm: it has been found that the sun warms on two accounts: namely, owing to [its] motion, and owing to [its] light."[21] But Falaquera took a different path. He begins his exposition by giving an account of the motion theory of heat; then, with the help of sentences carefully selected from Averroes, he calls that theory into question; finally, he presents the alternative theory, according to which light is the cause of heat, as the only sound explanation for the heating by the heavenly bodies. Falaquera thus deliberately chose a route unlike those taken by Averroes: the Cordovan in one work had presented the two

[20] I develop this subject in greater detail in my forthcoming article: "Averroes' Later Thoughts on the Role of the Celestial Bodies in the Generation of Animate Beings."

[21] *Epitome of On the Heavens*, Ar., 48, ll. 9-10; Heb., fol. 62vb, ll. 16-20.

theories of heat as two complementary accounts (*Epitome*), and in the other two works he considered the motion theory of heat as the only adequate theory (the two middle commentaries). Falaquera constructs out of the very same texts a single account, which presents the light theory of heat as the only valid theory. This theory, I will suggest, had his preference because it established a firm link between natural philosophy and metaphysics and theology. Let us consider the details.

Falaquera begins his discussion with a sentence taken from the *MC on On the Heavens*, asserting the incontrovertible theoretical postulate underlying the entire *problématique*: "the substance of the stars is necessarily of the nature of the fifth body, and this is a general statement agreed upon by the early philosophers."[22] He then immediately switches to the *Epitome of On the Heavens*, reproducing a sentence to the effect that the stars and the spheres must be of the same matter (otherwise the stars would have a constrained motion, which could not be circular).[23] To introduce the issue in which he is interested, Falaquera next shifts back to the *MC on On the Heavens*: "We should not have a doubt concerning this matter, seeing that the stars heat, which might lead us to think that they are fiery."[24] Continuing the quotation from the *MC*, Falaquera writes that although all fire warms, not everything that warms is fire, for there are things that warm through motion.[25] In the *MC*, motion is presented as the only explanation of the stars' heating, and Falaquera continues to translate it faithfully: the heat of the stars is due to their motions or, more precisely, to the effect of the spheres' rotatory motions on the contiguous air, which is warmed through the "hitting" of its parts one against another. This is visible in such phenomena as shooting stars and meteors. The greater the star and the closer it is to the earth, the stronger is its warming effect on the air. "This is why we find that the heating due to the sun and the moon [!] is the greatest," and Falaquera,

[22] *DF*, P, fol. 227v, ll. 7-8 (= *MC on On the Heavens*, Ar., 229, l. 5; Heb., fol. 48r, ll. 1-2).
[23] *DF*, P, fol. 227v, ll. 15-16 (= *MC on On the Heavens*, Ar., 229, ll. 8-9; Heb., fol. 48r, ll. 6-8).
[24] *DF*, P, fol. 227v, ll. 16-20 (= *MC on On the Heavens*, Ar., 230, ll. 2-4; Heb., fol. 48r, ll. 12-16).
[25] See above, 340; *DF*, P, fol. 227v, ll. 21-22 (= *Epitome of On the Heavens*, Ar., 48, ll. 9-10; Heb., fol. 62vb, ll. 18-19).

following Averroes, adds without any apparent qualms, "namely owing to their size and proximity to the air."[26]

In the sequel of the *MC on On the Heavens* Averroes turns to a discussion of the question how it is possible that the motions of the spheres generate heat although the spheres and the stars cannot themselves become warm as well. Falaquera will come to this a little later through a long quotation from the *MC on the Meteorology*. At present he introduces a sentence that breaks the continuity of the discussion and is in fact quite confusing for the reader: "The sun warms on two accounts: namely, owing to [its] motion, and owing to [its] light."[27] The mention of light as a cause of heating by the heavenly bodies remains isolated in the present context, because Falaquera for a while continues his discussion of the heat generated through the motion of the heavenly bodies. Why, then, has he at all invoked light as a cause of heat at this point? The reason is that to sustain the thesis that the stars' motions generate heat, Falaquera wished to weave into his text confirmatory evidence as formulated in the *Epitome of On the Heavens*: this is the would-be fact, alleged by Aristotle, that the leaden head of an arrow in motion melts through the heat produced by the friction with the air.[28] Falaquera in the same move inserts into his text also Averroes' following remark, namely that this analogy is not really useful and does not prove that the heavenly bodies chafe the air through their motion, for in the case of the arrow, the moving body and the air warm together, whereas this cannot be the case with the heavenly bodies.[29] The reason that the reference to light has been introduced although it is out of context here is therefore that Falaquera wished to switch from the *MC* to the *Epitome of On the Heavens* in order to quote the two sentences bearing on the heat generated by the arrow in motion. He inadvertently (I suppose) began his translation a little too early, thus including the reference to light as a cause of heat together with the sequel of the

---

[26] *DF*, P, fol. 227v, ll. 12-14 (=*Epitome of On the Heavens*, Ar., 48, ll. 2-3; Heb., fol. 62vb, ll. 9-12).

[27] *DF*, P, fol. 227v, ll. 14-15 (= *MC on On the Heavens*, Ar., 229, ll. 6-8; Heb., fol. 48r, ll. 5-6).

[28] Aristotle, *On the Heavens* II, 7 289a24-26. *DF*, P, fol. 227v, ll. 21-27 (=*Epitome of On the Heavens*, Ar., 48, ll. 11-13; Heb., fol. 62vb, ll. 22-24).

[29] *DF*, P, fol. 227v, ll. 25-7 (= *Epitome of On the Heavens*, Ar., 48, ll. 13-15; Heb., fol. 62vb, ll. 24-7).

text in which he was interested and which indeed was in continuation of his earlier discussion.

Averroes' *Epitome of On the Heavens* has pointed out that the analogy with the moving arrow is defective, inasmuch as the celestial bodies in motion do not themselves heat. The *Epitome* leaves this objection to the motion theory of heat unanswered. This is presumably the reason that Falaquera next inserts a fairly long quotation (some twelve lines) from Averroes' *MC on the Meteorology*.[30] This passage begins with a distinction between warming essentially, and warming accidentally. Motion cannot warm essentially, because only something that is itself warm in actuality can warm essentially, but motion cannot be warm. When motion appears to warm, therefore, it in fact merely prepares the substrate to be heated. Take the hitting of stones against one another: the fire thereby produced derives from the "heat in the air", i.e. (I take it) from the element fire dispersed throughout the element air; the hitting only prepares the air to receive this heat.[31] At times the preparation is so good that something may be thought to have warmed spontaneously. Falaquera next translates the following, somewhat enigmatic sentence from the *MC on the Meteorology*: "This is also the case with the spherical [= heavenly] bodies, for through their motion they generate heat in things that are not warm, by means of the warm bodies." The quotation from the *MC on the Meteorology* concludes with the observation that the heavenly bodies warm upon approaching the earth and cool when they recede. Thereby they *preserve* the elements although they do not *produce* them, fire being the only exception to this rule.

Falaquera now switches again to the *Epitome of On the Heavens* in order to introduce a difficulty in the motion theory of heat, which will allow him to mark his distance from that theory which was the

---

[30] *DF*, P, fol. 227v, l. 27-228r, l. 12 (= *MC on the Meteorology*, 16, ll. 5-14).

[31] This reading is borne out by what Falaquera writes elsewhere. In a sort of excursus he again adduces the example of the glass flask used to ignite cotton, and comments that "it is evident that the rays have no action here, except that they prepare the air to receive the heat" (*DF*, P, fol. 250v, ll. 11-13). He explains that the motion "separates the particles which are dispersed throughout the air in a potentiality that is close to actuality," and adds that this "is the reason that blowing increases the substance of fire" (*loc. cit.*, ll. 15-16). This passage invokes three factors as being involved in generating fire: the heat dispersed in the elements, the heat of the stars, and the heat of the air itself (*loc. cit.*, ll. 8-10).

one explored until now: "One has to ask what singles out the sun and the stars so that they warm by virtue of their motion, unlike the other parts of the sphere, seeing that they are all in motion?"[32] Averroes' answer, as quoted by Falaquera, is that "the specific thing about the heating of the stars [Averroes has in addition: and the sun] is that they are luminous", i.e. emit light.[33] How does this fact connect to the heating? Falaquera adduces the following answer, given by Averroes in the *Epitome of On the Heavens*:

> The light *qua* light, when its luminosity (*nogho*) is reflected [Averroes: when it is reflected], warms the bodies down here by virtue of a divine force (*koah elohi*; translating *quwwah ilâhiyyah*). This holds *a fortiori* when the rays fall on the warmed body in a right angle, for then the reflection is greater as is the case when the sun is over our heads [the last 11 words not in Averroes]. In fact, when the reflection is stronger, the heating[34] is stronger, as we can see in the cases of the burning mirrors and the glass flasks which burn wool; this applies *a fortiori* to the case when the reflecting body is polished. The stars do not shine because they are fire, as some people had thought, for the light does not belong to fire insofar as it is fire.[35]

The train of thought here is the following: Since the heavenly bodies are not themselves warm, and since motion per se cannot account for heating (seeing that the entire sphere moves but that only the heavenly body affixed to it warms), there must be something particular about the heavenly bodies that causes them to warm. This is their light. But in the Aristotelian natural philosophy there is no connection between heat ("fire") and light ("the

---

[32] *DF*, P, fol. 228r, ll. 12-13 (= *Epitome of On the Heavens*, Ar., 48, ll. 16-49:1; Heb., fol. 62vb, ll. 30-34).

[33] *DF*, P, fol. 228r, ll. 14-15 (= *Epitome of On the Heavens*, Ar., 49, ll. 3-4; Heb., fol. 62vb, ll. 36-37).

[34] The context certainly calls for *heating* instead of *luminous*. However, the Arabic text of Averroes (see next note) also carries *idâ'ah* and both Falaquera's and Moses ibn Tibbon's Hebrew translation have *me'ir* (luminous). If this is not a slip of the pen of Averroes himself, then the error must have crept into at least one family of manuscripts of the text at an early date.

[35] *DF*, P, fol. 228r, ll. 15-21 (= *Epitome of On the Heavens*, Ar., 49, ll. 4-12; Heb., fol. 62vb, l. 38-63ra, l. 7). The passage is translated in part in Henri Hugonnard-Roche, "L'Épitomé du *De caelo* d'Aristote par Averroès, Questions de méthode et de doctrine," *Archives d'histoire doctrinale et littéraire du Moyen Age* 51 (1984): 29-30. Averroes' next sentence: "For this reason the fire, whose existence in the concavity of the sphere of the moon has been demonstrated, does not shine," is inserted by Falaquera later on, after another passage taken from the *MC on On the Heavens* (*DF*, P, fol. 228r, ll. 25-26).

activity of what is transparent *qua* transparent"[36])! In the *Epitome of On the Heavens*, Averroes' (and Falaquera's) way out of the quandary is to claim that the warming effect of the sun (and the other stars) is produced by their *reflected* light, and that the heat is then produced by virtue of a *divine force.* This notion has completely disappeared from Averroes' later works, probably because it was too "metaphysical."[37] Falaquera at this point of his discussion quotes from the *Epitome of On the Heavens*, but in his text, Averroes' statement entirely changes its significance. Falaquera, we saw, began his investigation of the heating by the celestial bodies with the thesis, quoted from the *MC on On the Heavens*, that its cause is motion. This account was followed by a quotation from the *Epitome of On the Heavens* ("one has to ask what singles out the sun and the stars so that they warm by virtue of their motion, unlike the other parts of the sphere, seeing that they are all in motion?"), which in Averroes' text allowed the transition from the discussion of one cause (motion) to the discussion of the second, complementary one (reflected light). In Falaquera's text, however, the very same argument is used as a *refutation* of the account that had presented motion as the cause of heat. In this context, the second part of the argumentation appears as presenting the only veritable cause of heating by the celestial bodies: the reflected light heats by virtue of a "divine force." Thus, the notion of "divine force", which for Averroes was one explanatory device out of two and which he had moreover abandoned in his later writings, emerges in *De'ot ha-Filosofim* as the only notion on which to ground the account for the heat emitted by the celestial bodies.

Why did Falaquera upgrade the notion of "divine force," which Averroes preferred to abandon? I submit that the reason for this is that this notion cements an intrinsic unity of natural philosophy and metaphysics. The motion theory of heat allows natural philosophy to stand on its own feet, independently of metaphysics. This was at odds with Falaquera's global Maimonidean project, in which the natural philosophy was to be the "maids" of the only veritable

---

[36] Aristotle, *On the Soul* II, 7 418b10 (trans. J. A. Smith); see also *ibid.*, 419a9.
[37] Averroes' movement away from metaphysical, often Neoplatonically colored notions and toward Aristotle's original conceptions is documented apropos of the theory of the intellect in Herbert A. Davidson, *Alfarabi, Avicenna, and Averroes, on Intellect* (New York and Oxford, 1992).

science, viz. the divine science of metaphysics *cum* theology. The notion of "divine force," by virtue of which reflected rays of light quasi-miraculously generate heat, gave Falaquera a welcome opportunity to thwart the independence of natural philosophy and to drive home the point that the "hand of God" is involved in the most common natural phenomena, whose understanding cannot therefore be achieved through purely physical notions.

It should also be remembered that the notion of "divine force" does not concern an isolated phenomenon of secondary importance. Rather, that notion in fact provides the theoretical foundation for the general claim that the celestial bodies influence the natural processes in the sublunar realm and it thus bears the onus of (re-) establishing the physical connectedness of the celestial and the earthy realms. This connectedness, which had been a natural feature of the Presocratic systems, was lost through the introduction by Aristotle of the fifth element, but was reinstated by the notion of a "divine force." Falaquera may have been aware of this and this may be another reason for his upholding it.[38]

It is this notion of "divine force" which perhaps also allowed Falaquera to introduce into his physical world-picture the alchemical idea of the planets' endowing each metal with its specific properties. Aristotle, as is well known, accounted for the generation out of "water" of metals in general through a theory that did not account for their specific forms.[39] This void was filled by the alchemical theory, one part of which is summarized by Falaquera.

---

[38] Professor Y. Tzvi Langermann kindly called my attention to the fact that the notion of "divine force" appears in the work *Fî mabâdi' al-kull* ascribed to Alexander of Aphrodisias; see Shlomo Pines, "The Spiritual Force Permeating the Cosmos According to a Passage in the Treatise On the Principles of the All Ascribed to Alexander of Aphrodisias," in *idem, Studies in Arabic Versions of Greek Texts and in Medieval Science* (= *The Collected Works of Shlomo Pines*, Vol. 2) (Jerusalem, 1986), 252-5, on 253. According to this text, the "divine force" is a certain "pneumatic force (*quwwah rûhâniyyah*) that permeates all the parts of the world" thereby "bind[ing] (all) the (parts) of it to one another" (ibid.). *Fî mabâdi' al-kull* thus allots the "divine force" explicitly the role it has implicitly in Falaquera's system, namely to reestablish the interconnectedness of the cosmos. The notion of "divine force" may be a descendent of the Stoic theory of *pneuma* (Pines, *ibid.*, 254-5), which indeed set a great stake on the unity of the cosmos. This subject clearly calls for more research.

[39] Aristotle, *Meteorology* III, 6 378a16-378b4. On this theory cf. D.E. Eichholz, "Aristotle's Theory of the Formation of Metals and Minerals," *Classical Quarterly* 43 (1949): 141-6; reprint, *idem, Theophrastus, De lapidibus* (Oxford, 1965), 38-47.

"Some say," he writes, that the colors of metals "are according to the planets [lit., stars] acting on them."[40] Specifically, the yellow color of gold is related to the sun, the white color of silver to the moon, etc. Falaquera, who follows here the *Ikhwân al-Ṣâfâ*,[41] makes no explicit attempt to integrate this alchemical doctrine into the framework of the Aristotelian philosophy of nature, which excludes any "privileged relationship" between a planet and a given metal or color, but the notion of "divine force" of the stars' rays offered a possible basis for such an integration.

Let us note a further important congenial feature of the notion of "divine force." Aristotle took only the sun to warm and he accounted for its capacity to do so by assuming that the friction between it and the air generates heat. We have noticed that this is a "lame" account. Now the account adduced by Falaquera, unlike Aristotle's, assumes, as we have seen, that not only the sun but all the celestial bodies influence sublunar substances.[42] Therefore the introduction of the notion of a "divine force" proves useful: On this basis, indeed, the generalized statement is warranted, inasmuch as the sun has no specific status with respect to the divine force underlying the stars' possibility to act on the sublunar world. While from a post-seventeenth-century perspective the concept of divine force appears as a typically scholastic question-begging obscuring notion, at the moment of its introduction it had the considerable merit of allowing for a systematization and a generalization of the theoretical framework. Introducing the notion of divine force was thus a consequential move that allowed Aristotelians to have their cake and eat it too: it allowed them to reconcile their adherence to the Aristotelian bipartition of the universe into two radically separated realms with an acceptance of the principles of natural astrology, which initially had been utterly incompatible with the foundations of Aristotelianism. Thanks to a study by Y. Tzvi Langermann we know that this move was followed in the fourteenth century by Gersonides.[43]

---

[40] *DF*, P, fol. 122r, ll. 1-4.

[41] Mauro Zonta, "Mineralogy, Botany and Zoology in Hebrew Medieval Encyclopaedias" (above, n. 4), 282.

[42] It must be remarked, though, that Falaquera does not explicitly ascribe this "divine force" to the stars other than the sun.

[43] Y. Tzvi Langermann, "Gersonides on the Magnet and the Heat of the Sun," *Studies on Gersonides—A Fourteenth-Century Jewish Philosopher-Scientist*, ed. Gad

In Falaquera's *De'ot*, to the celestial bodies are not only ascribed the capability to exert specific actions on the sublunar existents, but through the notion of a divine force this capacity has received a theoretical grounding. We may now look into the role that is attributed to the actions of the heavenly bodies on the sublunar ones.

3. *A Providential Universe*

Falaquera maintains that the double motion of the sun along the ecliptic is brought about not only by a material and a moving cause, but also by a *final cause*. He ascribes this view to Aristotle himself:

> Aristotle said: what we have said concerning the generation and corruption being always continuous and never ceasing to exist is not necessitated by the material and the moving causes, which we have just explained. Rather, it is evident that this is made necessary also by the final cause. The explanation of this is as follows: we observe that nature perpetually moves toward and strives after the most perfect state according to the possibility of each and every existent, [i.e. it desires] to bring each existent to the ultimate perfect[44] state that its nature is capable of attaining.[45]

But what is the degree of perfection sublunaries can attain? The most perfect things are those that exist necessarily and hence eternally, and those that exist necessarily as individuals are superior to those that exist eternally only *qua* species.[46] Now the highest degree of perfection cannot be attained by individual sublunar substances, the reason for this being that they are too far away from the First Principle (*ha-hathalah ha-rishonah*); they therefore inevitably perish.[47] Still, there is a degree of perfection attainable in the sublunar realm. Falaquera writes:

Freudenthal (Leiden, 1992), 267-84. Gersonides in all likelihood borrowed the notion of "divine force" from Averroes, with whose views he became familiar either through *De'ot ha-Filosofim* or, more probably, through Moses ibn Tibbon's Hebrew trans. of the *Epitome of the On the Heavens* (see above, n. 17).

[44] Add *ha-me'uleh* with *DF*, L.

[45] *DF*, P, fol. 88r, ll. 22-27 (*DF*, L, fol. 162va) (= Averroes, *MC on On Generation and Corruption*, Heb., 91, ll. 66-70; Eng., 103).

[46] *DF*, P, fol. 88r, l. 27-88v, l. 2 (= Averroes, *ibid.*, 91, ll. 70-72; Eng., 104).

[47] *DF*, P, fol. 88v, ll. 3-4 (= Averroes, *ibid.*, 91, ll. 73-74; Eng., 104).

The Deity, blessed be He, the source of Grace and Goodness, has made up for the deficiency which befalls these [sublunar] things, as far as it is possible to make up for the deficiency occurring to them. In fact, He has made generation perpetual. Indeed, it is only this way that the existence of these things can be continuous and perpetual, for the continuity of generation approximates the kind of continuity proper to the eternal bodies. And the cause of this continuity, through which the Deity (blessed be He) has made up for this deficiency, is the circular motion [*ha-ha' taqah be-haqqafah*]. It is for this reason that the continuity of these things is cyclic, as we see that air is generated from water and water from air, and so on perpetually.[48]

This passage marks a principled departure from Aristotle. The Stagirite had identified teleology in the workings of nature in individual living substances, as, for example, in the forms of the organs of living beings. But, in the view of the great majority of contemporary interpreters of Aristotle, he did not think of the workings of nature in different parts of the cosmos as teleological; much less did he ascribe any kind of teleology to intentions of the Deity.[49] In the medieval mind, however, Aristotle's matter-of-fact account which ascribes the periodicity of generation and corruption to the motion of the sun along the ecliptic, was transformed into a teleological account, according to which the sun's particular motion follows the *goal* of perfecting the sublunar world (although, as we shall see, Falaquera insists that the heavenly bodies do not exist *for the sake of* the sublunar ones). Falaquera's Neoplatonically-enriched Aristotelianism was obviously more congenial to thinkers who sought to introduce Greek philosophy into the playground of a monotheistic religion.

---

[48]   *DF,* P, fol. 88v, ll. 4-11 (*DF,* L, fol. 162va f.) (= Averroes, *ibid.,* 91, ll. 74-80; Eng., 104).
[49]   David M. Balme, "Teleology and Necessity," in *Philosophical Issues in Aristotle's Biology,* ed. Allan Gotthelf and James G. Lennox (Cambridge, 1987), 275-85; John M. Cooper, "Aristotle on Natural Teleology," in *Language and Logos: Studies in Ancient Greek Philosophy Presented to G.E.L. Owen,* ed. Malcolm Schofield and Martha Craven Nussbaum (Cambridge, 1982), 197-222; *idem,* "Hypothetical Necessity and Natural Teleology," in Gotthelf and Lennox, *Philosophical Issues,* 243-74; Allan Gotthelf, "The Place of Good in Aristotle's Natural Teleology," in *Proceedings of the Boston Area Colloquium in Ancient Philosophy,* Vol. 4, ed. John J. Cleary and Daniel C. Shartin (Lanham, Md., and New York, 1989), 113-39. For a different, more "medieval" view, cf. Charles H. Kahn, "The Place of the Prime Mover in Aristotle's Teleology," in *Aristotle On Nature And Living Things,* ed. Allan Gotthelf (Pittsburgh, 1985), 183-205.

The material, moving and final causes of the sun's double motion explain why generation and corruption are continuous.[50] Hence, all processes that hinge on that double motion are also explained by reference to the same causes. Thus, because the spheres and the elements are eternal, necessarily generation and corruption have been and will remain unceasing.[51] Also other facets of the world are attributable to the action of the spheres: were it not for the spheres, the four elements would long ago have settled in four concentric layers, with the heavy earth at the center and the light fire at the periphery.[52] The spheres can thus be said "to preserve these [elevated] places [i.e. land] *qua* species":[53] what for Aristotle had been a brute factual matter, has over the centuries become a providential arrangement through which the Deity, acting via the celestial bodies, watches over the lower world. This notion of providence is indeed explicitly introduced in Falaquera's text. It appears in a lengthy passage, explicitly ascribed to Averroes:

> Averroes said: The existence of things down here, i.e. the fact that their species persist, cannot be due to chance, as the Ancients had claimed. This can be seen if one considers how the motions of the spherical bodies are in agreement [*haskamah*] with the existence of

[50] *DF*, P, fol. 88v, ll. 19-21 (= Averroes, *MC on On Generation and Corruption*, Heb., 92, ll. 88-89; Eng., 105).

[51] *DF*, P, fol. 88v, ll. 25-26.

[52] "On this basis one can answer a doubt raised by some who have argued: how is it that the simple bodies [i.e. the elements] have not separated off from one another throughout this long lapse of time, so as not to be intermingled or mixed? In fact, they [the elements] are contraries and it is in the nature of each of them to be in its proper place. The reason that they are moved one to the other's place and that they are mixed with one another is only the translational motion. Were it not for this circular motion it would indeed have been necessary that they separate off during that long time. However, they are moved to one another's place through this double motion." (*DF*, P, fol. 88v, ll. 13-18 [*DF*, L, fol. 162vb] [= Averroes, *MC on On Generation and Corruption*, Heb., 92, ll. 82-87; Eng., 104]). On the history of this entire issue, see Gad Freudenthal, "(Al-) Chemical Foundations for Cosmological Ideas: Ibn Sînâ on the Geology of an Eternal World," in *Physics, Cosmology and Astronomy, 1300-1700: Tension and Accommodation*, ed. Sabetai Unguru (Dordrecht, 1991), 47-73.

[53] *DF*, P, fol. 89r, l. 3 (= Averroes, *Epitome of On Generation and Corruption*, Heb., 123, l. 1; Eng., 135). Also: "The ultimate causes [for the emergence of stretches of land from the sea, or the submergence of land in the sea] are the motion of the sun along its inclined sphere and the motion of the other stars, just as these are the ultimate causes of the generation of and corruption of all other [sublunar] existents" (*DF*, P, fol. 100v, ll. 19-21).

each and every thing coming to be in this [lower] world, and how the former preserve the latter. This can be observed first and foremost with respect to the sun and next also with respect to the moon. Indeed, concerning the sun it has been established that if its body had been greater than it is, or if it had been closer [to us], then it would have destroyed the plants and the animals through an excess of heat. Again, had it been farther away or smaller than it is, then [the plants and the animals] would have perished through cold. This is attested by the places that are uninhabited because of excessive heat or cold. Providence can also be clearly observed in the [existence of] the sun's inclined sphere [i.e. the ecliptic]. For if the sun did not have an inclined sphere, there would be no summer, nor winter, nor [variations in] heat and cold. Yet it is evident that the four seasons are necessary for the existence of the species of plants and animals. Providence is also very clearly observable in the [sun's] daily motion, for without it there would be no day and night, but rather there would be half a year night and half a year day. If this were the case, the existents would have perished by day due to an excess of heat, and at night to an excess of cold. Similarly, the operation of the moon can be observed in the existence of rains and the ripening of fruits. And it is clear that were the moon bigger than it is or smaller or farther away, or if its light were not borrowed from the sun, it would not have that effect. Again, if it [did not have] an inclined sphere it would not accomplish different operations at different moments in time. ... And what we have said with respect to the sun and the moon must be believed also with respect to the other stars. Indeed, Aristotle said that the motion of the stars takes after the sun, and he said so because it can be seen that the stars seek to resemble the sun. Therefore a truthful general statement [*pesaq amitti*] must be adopted, according to which all [heavenly] motions [were so designed as to] exercise providence over what is below them in this world.[54]

However, Falaquera (still quoting Averroes) adds a caveat to the effect that although the heavenly bodies exert providence over the sublunar world, they do not exist *for the sake of* doing so: "considering the nobility of these spherical [= celestial] bodies, it does not seem that their primary intention is to exercise providence over what is below them. For were this the case, then the eternal bodies would exist for the sake of the perishable ones, the perfect for the sake of the vile."[55] Elsewhere, however, Falaquera opines

[54] *DF*, P, fols. 288v, ll. 2-19 and l. 26-289r, l. 3 (= Averroes, *Epitome of the Metaphysics*, Ar., 166-8, pars. 74-77; German, 201-3). I am grateful to Professor Mauro Zonta for his help in locating this passage. Prof. Zonta added that the passage is inspired by Alexander of Aphrodisias' *De providentia*.
[55] *DF*, P, fol. 289r, ll. 3-5 (= Averroes, *Epitome of the Metaphysics*, Ar., 168, sec. 77;

that "the existence of the final[56] cause is more visible in the spherical [= celestial] bodies than in the things subject to generation and
corruption. Therefore the ignorance [or: stupidity; *sikhlut*] of the
one who ascribes the spherical [= celestial] bodies to chance is
greater than that of the one who ascribes to chance the things
subject to generation and corruption."[57]

4. *Astrology and the Doctrine of the Eternal Return in a Providential Universe*

The doctrine according to which the sun and the other celestial
bodies govern the sublunar processes has interesting corollaries. In
his *Generation of Animals* Aristotle himself had established a
connection between the lifespans and gestation periods of animals
and the revolutions of the sun and the moon.[58] Falaquera,
following Aristotle, remarks that when the sun approaches a thing,
generation is brought about, whereas the sun's retreat causes
waning or "growing old."[59] He then interjects the following
remark: "we shall therefore say that necessarily the number of
revolutions [*haqqafot*] during which the existent comes to be and
completes its growth is equal to the number of revolutions during
which occur its aging and its passing-away."[60] Noting that the
periods of growth and decay of many natural beings are measured
by the periods of the annual revolutions of the sun, Falaquera
adds: "Thus, as Aristotle had asserted, every existent has [a period
of] time and a definite span of life [*qeṣ mugdar*] defined by the
body moving periodically."[61] Since the revolutions of the sun not
only measuure time, but also *produce* the alternating periods of
growth and decay, it follows that the lifespan of each existent as
expressed by the number of revolutions of the sun is not merely

German 203-4).

[56] The MS has *ha-sibbah ha-po'elet* (l. 28); but at l. 26 we read *ha-ṭeva' po'el
mipnei devar mah.* Therefore, *ha-sibbah ha-po'elet* here does not refer to the
efficient cause, but to the cause acting "for some purpose," i.e. to the final cause.
[57] *DF*, L, fol. 210ra, ll. 28-33.
[58] Aristotle, *On the Generation of Animals* IV, 10, 777b17-778a10.
[59] *DF*, P, fol. 87v, ll. 21-22 (= Averroes, *MC on On Generation and Corruption*,
Heb., 89, ll. 38-39; Eng., 102).
[60] *DF*, P, fol. 87v, ll. 23-25 (= Averroes, *ibid.*, 90, ll. 42-43; Eng., 102).
[61] *DF*, P, fols. 87v, l. 26-88r, l. 1 (= Averroes, *ibid.*, 90, ll. 45-46 [*qeṣ manui*];
Eng., 102).

conventional, but represents the real cause determining that life-span. Since, as we have already noted, in Falaquera's view, what has been said concerning the sun's impressions on sublunar bodies holds also generally with respect to the other heavenly bodies, some existents will have their lifespans defined by the number of revolutions of other celestial bodies.[62] And indeed Falaquera asserts that "[the lifespans of] some existents are measured by the [revolutions of] the sun and [those of] some by [the revolutions] of the moon, ... and so it seems likely that for any other planet there are existents [whose lifespans] are measured by [the revolutions of] that planet."[63]

Judah ben Solomon's discussion of this issue concludes with a brief remark that highlights what is at stake here. Judah summarizes the point concerning the measuring of the lifespans through the number of revolutions of the heavenly bodies in one or two sentences and then interjects the following significant comment: "He [Aristotle] has made this matter [i.e. the lifespans and the periods of gestation] depend upon the periods of the stars, *in accordance with the opinion of the astrologers.*"[64] Just like Falaquera, Judah felt that the (Averroean) natural philosophy he was ascribing to Aristotle came very near the astrological doctrine. As we shall shortly see, this was quite welcome to him.

Elsewhere Judah explains the matter somewhat more fully: the moving cause of all existents subject to generation and corruption, he says, is the movement of the sun along the ecliptic. "The same holds," he continues,

> of the inclined spheres of the [other] planets, i.e. they are the cause of the generation of the things individualized by each one of them. Aristotle's view is that every existent has a limited lifespan, for it comes to be through something moving in a circle [i.e. cyclically]. Indeed, there are some existents whose lifespans are long, others whose lifespans are short.[65]

---

[62] *DF*, P, fol. 88r, ll. 2-12 (= Averroes, *ibid.*, 90, ll. 47-48; Eng., 102; and *Epitome of On Generation and Corruption*, Heb., 122-3; Eng., 134).

[63] *DF*, P, fol. 88r, ll. 10-11 (= Averroes, *Epitome of On Generation and Corruption*, Heb., 122, l. 85-123, l. 87; Eng., 134).

[64] *MH*, fol. 83r, l. 17.

[65] *MH*, fol. 59v, ll. 26-29.

At this point Judah adds again a revealing comment, not found in Falaquera, and which is certainly his own: "From here they entered [or: became involved in] astral determinism [*mikan nikhnesu ligezerat ha-kokhavim*],"[66] which echoes an almost identical remark by Maimonides.[67] We thus observe the following double development: First, under the influence of astrological doctrines, the original Aristotelian teaching has been "generalized" so as to ascribe to all the heavenly bodies properties which Aristotle himself had attributed only to the sun and to the moon. Second, this development, which moved the Peripaptetic doctrine closer to astrology and which we dubbed above the "astrologization of the Aristotelian view," created a vicinity between these initially unrelated doctrines: supporters of astrology such as Judah used it to strengthen their position, whereas for opponents of astrology, such as Maimonides and Falaquera, it was a source of embarrassment.[68]

The proximity between Aristotelian natural philosophy and astrology and related doctrines comes to the fore also in Falaquera's discussion of the doctrine of the eternal return. In two different contexts Falaquera sees fit to reject this doctrine, taught by the *ba'alei ha-haqqafot*, "the proponents of the cyclical recurrence."[69] The rationale of this theory should be quite clear now. If

[66] *MH*, fol. 59v, l. 29.
[67] Maimonides, *The Guide of the Perplexed*, trans. Shlomo Pines (Chicago, 1963), II,12, 280.
[68] On Maimonides' attitude to astrology see Y. Tzvi Langermann, "Maimonides' Repudiation of Astrology," *Maimonidean Studies* 2 (1991): 123-58; Gad Freudenthal, "Maimonides' Stance on Astrology in Context: Cosmology, Physics, Medicine, and Providence," in *Moses Maimonides: Physician, Scientist, and Philosopher*, ed. Fred Rosner and Samuel S. Kottek (Northvale, N.J., 1993), 77-90.
[69] Falaquera's source for this discussion is again Averroes (*Epitome of On Generation and Corruption*, Heb., 126; Eng., 137; and *Middle Commentary on On Generation and Corruption*, Heb., 97, ll. 82 ff.; Eng., 109-10). In the *Epitome of the Parva naturalia*, too, Averroes refers to the "proponents of the cyclical recurrence" (Arabic, 79, line 3; Heb., 51, ll. 11-15; Eng., 46). Very interestingly, Falaquera more than once states that the *ba'alei ha-haqqafot* are called in their (own) tongue: *ha-azdawânât* (or: *ha-izdawânât*) (*DF*, P, fol. 91r, ll. 25-26 [= *DF*, L, fol. 164rb]) or *ha-azrognât* (or: *ha-izrognât*) (*DF*, P, fol. 202r, ll. 25-26). On the difficulty of determining the Arabic original of this term, as found in Averroes, see the *Epitome of the Parva naturalia*, Ar., 79, l. 3, and the corresponding note (157-8, n. 44). Professor Dimitri Gutas most kindly undertook a research on the subject and informed me of its results as follows (January 23, 1999): "It appears that the unintelligible Arabic word is a transliteration of the Greek *horoskopion* (horoscope), of which the Arabic would be, as in the manuscripts, *huruzqâbât*, changing only the *waw* after the *zayn* into a *qaf*, a relatively common mistake in

the entire natural order, notably generation and corruption in the sublunar world, hinges upon the periodical movements of the sun and the other heavenly bodies,[70] does it not follow that whenever all the celestial bodies will occupy the same positions at different points of time, necessarily the very same individuals will come to be over and over again? Falaquera reports approvingly Alexander's contention that it is impossible that the heavenly bodies occupy exactly the same positions at different moments.[71] He rejects the thesis of the eternal return also on the grounds that what is generated and corrupted periodically are not individuals *qua* individuals, but rather individuals *qua* instantiations of their respective species, say clouds or rain. Put differently, the heavenly bodies may repeatedly produce in matter the same form, but the matter will be different, and hence the individuals will not be numerically

---

Arabic manuscripts. This is suggested by Helmut Gätje, who has a critical edition of Averroes' text, *Die Epitome der Parva naturalia des Averroes* (Wiesbaden, 1961), 84, line 2, and apparatus. This fits perfectly the context of the discussion in Averroes' text. The 'eternal return' interpretation of the Hebrew translators is an attempt to explain this word, but it is actually derived from what Averroes says just after this word, namely *wâjibûn an ta'uda al-ashkhâṣu al-maḥsûsâtu bi-a'yaniha*. The expression *'âda bi-aynihi* ("he reverted to a previous state") was glossed as *raja'a 'alâ a'qâbihi*, a roughly synonymous expression; this is why Samuel ibn Tibbon and Moses ibn Tibbon translate the two Arabic expressions with the same Hebrew expression, as Blumberg shows in his n. 44. Obviously, Moses ibn Tibbon and Falaquera, not knowing the expression *huruzqâbât* (however it was actually spelled in the manuscripts they were using), interpreted it in terms of what Averroes says right after it. . . . The big question, and I have no answer for it, is why Averroes would have used that transliterated term in his commentary. Obviously he must have had it in his archetype, a translation of the *Parva naturalia*. Since he does not explain it, perhaps he did not understand the word either. Now as far as I can tell, there are no horoscopes mentioned in *On Divination in Sleep* (the treatise Averroes seems to be commenting upon in that passage). What sort of translation of the *Parva naturalia* then was it? In this regard, Pines's article, "The Arabic Recension of *Parva naturalia* and the Philosophical Doctrine Concerning Veridical Dreams According to *al-Risâla al-Manâmiyya* and Other Sources" [*Israel Oriental Studies* 4 [1974]: 104-53; reprinted in *Studies in Arabic Versions of Greek Texts*, 96-145), is valuable: he suggests there was another recension of the *Parva naturalia* available in Arabic, one that has not been preserved in Greek. The present piece of evidence would seem to corroborate yet again Pines' keen eye."

[70] "It has become clear that the ultimate cause of all ordered motions of natural things—whether the existence upon earth of plants and animals, or the existence in the air of the meteorological phenomena—is the motion of the spherical bodies" (*DF*, P, fol. 92r, ll. 14-16).

[71] *DF*, P, fol. 91r, l. 27-91v, l. 10 (= Averroes, *Epitome of On Generation and Corruption*, Heb., 126, ll. 52-62; Eng., 137-8).

360

one.[72] The same considerations imply the untenability of astrology. According to Falaquera, the astrologers claim that by getting to know the "intelligible general natural order" (*teva' kelali muskal*) of the world, an essential part of which are the influences of the heavenly bodies on sublunar bodies, one can predict events relating to individuals. Falaquera rejects this claim with the same argument that refuted the claims of the upholders of the eternal return.[73] We thus note again that the theory of the celestial influences on the sublunar world as developed within the Aristotelian school has brought its upholders into uncomfortable vicinity with doctrines far removed from Aristotle's original teaching.

The point is again highlighted through a comparison with Judah ben Solomon's reaction to the same theoretical *problématique*. Judah first introduces the underlying theory and then comments:

> Since the revolving motion is continuous and perpetual, as has been repeatedly explained, therefore any coming-to-be which hinges on it is perpetual and continuous. . . . He [Averroes] said: since the revolving mover moves perpetually, it necessarily follows that the motion of things undergoing change will be cyclical. For instance, since the primary revolving motion is the periodical motion of the sun along the ecliptic, therefore the seasons of the year recur periodically, and the things that depend on them also change cyclically. This raises the question why a given animal—be it a man or another animal—should not change cyclically, i.e. that the same animal that died will live [again] at another time, just as the other things recur periodically.[74]

How does Judah relate to this argument? First, he reproduces the next sentence of his source (Averroes), which rejects the argument of the eternal return: "We say that the things passing away return *qua* species, not *qua* individuals."[75] But immediately afterwards Judah seeks to mark his distance from the opinion he has just quoted and he imparts to his reader his own view of the issue: "This is according to the opinion of Aristotle. But the one who examined minutely the science of astrology [lit. the science of the

[72] *DF*, P, fol. 91r, ll. 13-27 (= Averroes, *MC on On Generation and Corruption*, Heb., 97, l. 82-98, l. 91; Eng., 109-10).

[73] *DF*, P, fols. 202r, l. 6-203r, l. 2.

[74] *MH*, fol. 60r, ll. 4-5 and 12-19 (= Averroes, *MC on On Generation and Corruption*, Heb., 97, ll. 67-75; Eng., 109).

[75] *MH*, fol. 60r, ll. 19-20 (= Averroes, *ibid.*, 98, ll. 86-7; Eng., 109).

sphere] holds that necessarily this very same Reuben who died will be alive again."[76] The next two sentences make clear how far-reaching were the religious and eschatological implications of this doctrine: "And this is a pillar of our holy religion. And since things return cyclically, therefore our kingdom will return [i.e. be restored]. Would that this will happen quickly in our own days."[77]

5. *The Celestial Bodies in the Economy of the Natural World: The Preservation of the Elements and the Generation of Animals*

The world-picture drawn by Falaquera ascribes to the heavenly bodies further roles, and views these too as manifestations of divine providence. For one thing, the rotating spheres are directly involved in the existence of the four sublunar elements. While the rotatory movements do not *produce* the elements, they safeguard their continued existence. Falaquera in different contexts adduces slightly varying arguments to this effect. One argument, which shows that fire necessarily exists in the concavity of the sphere of the moon, is the following: "since there exists a revolving body, and since necessarily what has the greatest distance from it must have the utmost weight and sink downward, it follows with necessity that what is closest to the revolving body must be of the utmost purity and lightness, for it is in the nature of motion to produce this. And what corresponds to this description is necessarily fire."[78]

---

[76] *MH*, fol. 60r, ll. 21-22.

[77] *MH*, fol. 60r, ll. 22-24. Judah's statement recalls a similar claim made by Joseph ibn Kaspi in the beginning of the fourteenth century and studied by the late Shlomo Pines (see his "Joseph ibn Kaspi's and Spinoza's Opinions on the Probability of a Restoration of the Jewish State" [Hebrew], in *idem, Bein Maḥashevet Yisra'el le-Maḥashevet he-'Ammim* [Jerusalem, 1977], 277-305). Ibn Kaspi bases his prediction that a Jewish state will be reestablished on the consideration that the number of all possible states of affairs (the number of all possible combinations of the world's basic building blocks) is finite, so that necessarily every combination that had existed in the past will come to exist again. Judah's consideration is similar in that it postulates a finite number of possible stellar positions and, on the basis of the astrological doctrine, deduces from it that a Jewish kingdom will come to be again. Pines opined that Ibn Kaspi's belief is unrelated to theories of a cyclical return of time and of history (*ibid.*, 297), but Judah's case shows that a certain version of such a cyclical theory could be based on considerations similar to those of Ibn Kaspi.

[78] *DF*, P, fol. 66r, ll. 21-25 (= Averroes, *Epitome of On the Heavens*, Ar., 14; Heb., fol. 57ra). This argument obviously makes the celestial spheres, or rather their motions, into the "remote cause" of the earth's being at the center of the world.

It should be noticed that when fire is in its natural place it is very pure and this is why it cannot be seen; the visible fire, by contrast, is that mixed up with other elements.[79] However, even in its natural place fire is not of the utmost purity and simplicity; had the fire in the concavity of the moon been utterly simple, then it would be more powerful than the other elements and would have overpowered them. The reason for this is that the other elements, because they are never absolutely pure, are always compounded and that, generally, the compounded is weaker than the simple and is overcome by it.[80] Falaquera concludes: "It should be believed that out of its [natural] place fire is not pure, on a par with the other elements. For the spherical [= celestial] bodies mix it up [until it becomes] a mean—i.e., balanced—mixture ['eruv memuṣa']."[81] Were it not for this action of the celestial spheres, Falaquera adds, there would be no generation and corruption, and hence no life on earth: witness the places under the poles, where the absence of motion entails that elements are nearly pure, and where consequently no generation and corruption take place.[82] Although Falaquera does not say so explicitly here, the corollary, formulated elsewhere, obviously is that the celestial bodies "preserve" the elements.[83]

A related argument is adduced in another context.[84] The point of departure now is the eternal rotatory motion of the heavens. That motion necessarily requires an immobile point at the center.

Cf. *DF*, P, fol. 70v, ll. 11-20.

[79] *DF*, P, fol. 66r, l. 25-66v, l. 1 (= Averroes, *ibid.*, Ar., 14; Heb., fol. 57ra; see also Ar., 49; Heb., fol. 63ra). A part of this passage is translated in Hugonnard-Roche, "L'Épitomé du *De caelo* d'Aristote par Averroès" (above, n. 35), 25. See also *DF*, P, fol. 191r, ll. 6-9: "The white color is generated when the pure fire combines with the element that is of the utmost purity, namely the air; the black color is generated out of turbid fire, when it combines with the [element] earth, which is of little purity." Falaquera here quotes Averroes, *Epitome of the Parva naturalia*, Ar., 13, ll. 6-8; Heb., 9, ll. 5-7; Eng., 10.

[80] *DF*, P, fol. 66v, ll.17-20.

[81] *DF*, P, fol. 66v, ll. 25-27.

[82] *DF*, P, fol. 67r, ll. 3-4. Falaquera notes that this conclusion is in accordance with Alexander's affirmation, but he remarks that Alexander made his statement for the wrong reason (*DF*, P, fol. 67r, ll. 1-2).

[83] *DF*, P, fol. 228r, ll. 10-11; cf. also immediately below.

[84] For what follows see *DF*, P, fol. 226v (= Averroes, *MC on On the Heavens*, Ar., 199ff.; Heb., fol. 43a); cf. Ruth Glasner, "The Early Stages in the Evolution of Gersonides' *The Wars of the Lord*," *Jewish Quarterly Review* 87 [1996]: 23, n. 89). See also Aristotle, *On the Heavens* II, 3.

This is the earth. And since the earth so defined is an absolutely heavy body, its existence implies that of an opposite, namely an absolutely light element, which is the fire. The existence of these opposites in turn implies the existence of intermediary bodies, namely the elements water and air. Thus the existence of all the elements (opposite and intermediary) follows with necessity from the existence of the revolving eternal celestial sphere. Now where there are opposites, necessarily there are also generation and corruption, which must be eternal, in fact coeval with the celestial sphere. This in turn implies that the celestial motions must be more than one, for (as we have already seen) a single rotatory motion would not result in periodically alternating generation and corruption. In fact, the motions in question must be different from one another, so as to be able to set in motion each of the sublunar natures, as the sun sets in motion the fire, and the moon the water. Falaquera's text gives no indication as to how the motions of the heavenly bodies, all consisting of the fifth body, can have such differentiated effects on the sublunar elements. In this context, Falaquera formulates explicitly the contention that the celestial spheres *preserve* the elements, although they do not *produce* them.[85]

The sun and the other heavenly bodies further bring about phenomena discussed in the *Meteorology*. Inasmuch as these phenomena depend on the sun's raising the two exhalations, the sun can be said to be the cause of generation and corruption of these phenomena too.[86] Also the moon—of which Falaquera says expressly that "its nature is cold and moist" (he does not seem to be embarrassed by the fact that the heavenly matter is considered to be qualityless)—has an incidence on these phenomena, specifically on the production of rain.[87]

Having made the motions of the celestial bodies into guarantors of the providential order down here, Falaquera introduces the final cause also through another slot. Aristotle himself already had graded the elements and had claimed that their topological order paralleled their rank, i.e. the upper ones are like forms to the lower ones.[88] Aristotle, in line with his strict division of the cosmos

---

[85] *DF*, P, fol. 228r, ll. 10-11; see above, 347.
[86] *DF*, P, fol. 95r, l. 27-95v, l. 26.
[87] *DF*, P, fol. 95v, l. 26-96r, l. 3.
[88] Aristotle, *On the Heavens* IV, 3 310b14-15; see also Mary Louise Gill, *Aristotle*

XVI

into a supra- and a sublunar region, of course left the fifth element
out of this scheme. But in the cosmos described by Falaquera the
heavenly bodies are closely related to the teleology operative
within nature and hence to the final cause. Aristotle's scheme is
therefore extended: all the "simple parts" of the world, viz. the
elements, are like forms, or perfections, to one another. Thus
"earth exists for the sake of the water, water for the sake of the air,
air for the sake of fire, and fire for the sake of the sphere."[89] The
four sublunar elements therefore relate to the fifth element as
matter relates to form, and thus their existence without the fifth
element is not conceivable. Conversely, "the existence of the
spherical body requires these [elements], just as [generally] forms
necessarily require matter."[90] This construal of the relationship of
the elements to one another confirms the notion of the cosmos as
an inteconnected whole, pace Aristotle's division of it into two
strictly unrelated realms.

Another context in which the celestial bodies, or rather the
celestial element, function as a final, or formal, cause is that of
generation of plants and animals. This instance is particularly
interesting because Aristotle himself invokes the celestial substance
in this context. In the *Generation of Animals*, Aristotle devotes a
lengthy investigation to the question of the origin of what he
identifies as the nutritive soul operative within semen. He refutes
the hypothesis that this is simple fire, and argues instead that
within semen there is a physical substance—"connate pneuma" or
"vital heat"—which carries its *formative* power. Of this substance
Aristotle says expressly although enigmatically that it is "more
divine" than the simple elements and also that it is "analogous to
the element which belongs to the stars."[91] The cases of so-called
"spontaneous generation"—i.e. generation not from seed or
semen, but rather from some other residue of the body, or from
matter enclosed in a frothy bubble and heated by the sun—are
essentially not different: they involve either vital heat accumulated

on Substance (Princeton, New Jersey, 1989), 239.
[89] *DF*, P, fol. 271v, ll. 14-17 (= Averroes, *Epitome of the Metaphysics*, Ar., 122 [par. 70]; German, 145). See also *DF*, P, fol. 85r, ll. 17-20, and 111v, ll. 21-22.
[90] *DF*, P, fol. 250v, ll. 18-22.
[91] Aristotle, *On the Generation of Animals* II, 3 736b30 ff. Cf. Freudenthal, *Aristotle's Theory* (above, n. 5), 37 ff., 106-119.

in a residue (owing to its prior heating by vital heat within the body), or specifically the heat of the sun.[92]

Falaquera's text here is obscured because (as he himself states) it incorporates two conflicting versions of Averroes' account, belonging to two different periods of his life.[93] What is important in the present context, however, is the following. Falaquera states that it is both wrong to say that there is no (nutritive) soul in semen and to say that there is (nutritive) soul in semen. The correct way to put things is to say that semen has a soul-power (koah nafshi), which is "the power through which the semen brings about the formation of the members and their shapes [yesirat ha-evarim ve-tavnitam]."[94] Falaquera notes that this power is precisely what Galen had called "formative power."[95] After a short digression,[96] Falaquera adds: "concerning the vital heat and [the heat] attributable to the sun, it seems that they generate the ensouled bodies. And the bodily [part] in which this vital heat can be seen to subsist is the semen of generating [animals], or some residue of the natural residues in the case of those animals which do not give birth [i.e. animals generated through "spontaneous generation"]."[97] All this is Aristotle—highly condensed Aristotle, but still Aristotle. In the following sentence things change, however. Falaquera writes: "this [soul-] power [in the semen] is not enough to endow [a living being] with soul without [the assistance of] the sun and the sphere."[98] This is explained by saying that "it is impossible that infinite causes will not depend upon an eternal principle."[99] This idea is not only un-Aristotelian, but straightforwardly anti-Aristotelian. Aristotle repeatedly had rehearsed the

---

[92] Freudenthal, Aristotle's Theory, 25-6.

[93] See Gad Freudenthal, "Averroes' Changing Mind on the Generation of Animate Beings," in the Proceedings of the Congreso internacional VIII centenario de Averroes (Cordoba, 9-11 December 1998), ed. Ahmed Hasnaoui (forthcoming).

[94] DF, L, fol. 246ra, ll. 12-14, and 246rb, ll. 3-4. Falaquera here follows Averroes, Epitome of the Book of Animals, Heb., 229r.

[95] DF, L, fol. 246rb, ll. 3-4. Cf. Galen, On the Natural Faculties 1.6 (trans. A. J. Brock [Cambridge, Mass., 1916], 25).

[96] DF, L, fol. 246va, ll. 16-22.

[97] DF, L, fol. 246va, ll. 22-27.

[98] DF, L, fol. 246va, ll. 30-31.

[99] DF, L, fol. 246va, ll. 31-33.

catch phrase "man is generated by man" in order to counter the Platonic theory of Ideas, which contended that the generation of material individuals, specifically living beings, presupposes the existence of eternal, transcendent "prototypes." In Aristotle's view, by contrast, generation was self-contained, an eternal series of sexual generation of individuals belonging to the same species whose existence depends only on the continuous generation through semen and seed.[100] Thus, in Falaquera's exposition of Aristotle's views, the Platonic transcendent component that had been evicted by Aristotle through the front door has crept back through the chimney.

A more theoretical statement of the theory involved here is the following:

> Considering all things that come to be, whether they derive from nature or from art, we observe that the agent [ha-po'el] ... must necessarily be of one essence or definition with the substrate [pa'ul], or at least they must resemble [domeh] [one another]. This can be seen in most composite natural substances, such as the generating animals and the plants. For the genitor gives rise to [an individual] the same as itself in species, as a man who is procreating a man, and a horse a horse; or else, they resemble [each other], as the donkey begetting a mule. The same [principle] holds of simple [bodies] too, for fire in actuality generates fire in actuality. In view of this, there is a doubt concerning the animals and plants born spontaneously, and concerning the fire generated through motion. The same [doubt] applies where there are [other] movers which are not of the species of the moved: for instance, semen which moves menses until a man is formed, or a bird hatching eggs and moving them until a bird develops from them.
> We say [concerning the relationship of the mover to the moved] that it can be observed in most of these generated [creatures] that they are brought to perfection through more than one mover. For instance, the progenitor moves the semen, and the semen moves the menses. It thus follows that the mover which must necessarily be of one [or] similar essence with the moved [body] is the ultimate mover, for it is it that provides the proximate mover with the power with which it moves. Thus, the ultimate mover in the semen is the progenitor, and in the case of the egg—the bird. However, these [ultimate movers, too,] are wanting without an

---

[100] For a forceful statement of this, see Klaus Oehler, "Ein Mensch zeugt einen Menschen: Über den Missbrauch der Sprachanalyse in der Aristoteles-forschung," in his *Antike Philosophie und byzantinisches Mittelalter* (München, 1969), 95-145. Cf. also Freudenthal, *Aristotle's Theory*, 38-9, 45-6, 188.

external principle [*hathalah*], namely either the spherical bodies as
Aristotle opined—and this is the truth—or the active intellect, as
many of the later philosophers think.
As for the cases of animals and plants coming to be spontaneously,
their ultimate mover is either the spherical bodies, acting through
the soul-powers [*kohot nafshiyyim*] which overflow [*shofe'im*] from
them according to Aristotle's view, or else it is the active intellect
according to the opinion of the later [philosophers].[101]

This passage clearly shows two things. First, the action of the
celestial bodies as ascribed to Aristotle is strictly equivalent to that
of the active intellect as construed by the "later philosophers," i.e.
Avicenna, Ibn Bâjjah, and those who embraced that theory. In
other words, Falaquera's Aristotle construes the celestial bodies as
donating the forms to the sublunar existents. They function as
their formal cause. Second, Falaquera's Aristotle maintains that
the heavenly bodies are the source of "soul-powers" overflowing
from them and through which they exercise their role as a formal
cause. Falaquera's reader must have asked himself how bodies
consisting of the fifth element can possibly be the source of such
emanations, but Falaquera offers no answer to this quandary.[102]
Clearly, these "soul-powers" just as the "divine force" of the rays of
the stars, bridge over the Aristotelian hiatus separating the supra-
from the sublunar realm, thus (re-) instituting the unity of the
entire cosmos.

6. *Conclusion*

In this article I have presented views put forward by Shem-Tov ibn
Falaquera in his work *De'ot ha-Filosofim* with respect to one, albeit
important, issue. The issue in question is that of the causal dependence of major natural processes in the sublunar world upon
influences emanating from the supralunar realm. Falaquera's opus

---

[101]   *DF*, P, fol. 250r, l. 14-250v, l. 5. This entire passage is taken from (the later
version of) Averroes, *Epitome of the Metaphysics*, Ar., 50 (cf. also apparatus, 293-4);
German, 57. On the existence of two versions of this passage, see Davidson,
Alfarabi, Avicenna, and Averroes, on Intellect (above, n. 37), 239. For Averroes'
changing views on the role of the active intellect, see also the article referred to
above, n. 93. See also *DF*, P, fol. 85r, l. 17.
[102]   Falaquera seems to write loosely here, for in fact these powers overflow not
from the heavenly bodies, made of the fifth substance, but from the intellects or
souls of the heavenly bodies and their spheres.

was compiled with a view to giving its reader a coherent view of the universe as it could be gained from the sources Falaquera himself selected as the most salient ones. These sources are predominantly in the Aristotelian tradition (Averroes' commentaries, notably), and Falaquera deliberately excluded from his *vade mecum* many other sources with which he was familiar, such as, most importantly, Avicenna's works. Similarly, in *De'ot ha-Filosofim* Falaquera recorded no doctrines related to magic or astrology, although he was familiar with them and indeed referred to them in other works.[103] This can be interpreted as showing that Falaquera did not consider these views as belonging to the opinions of the philosophers. It would indeed seem that *De'ot* was forged with a view to reflecting the largely Aristotelian worldview of a milieu of Muslim and Jewish intellectuals in Spain (or southern France) of the middle of the thirteenth century. Although Aristotelianism had undergone a certain "astrologization" (see below), still Falaquera regarded straightforward astrological and magical doctrines to be outside its confines.

In *De'ot ha-Filosofim* (and in kindred works) Falaquera thus adopted the humble posture of the "seeker of truth," who searches for possible fragments of knowledge on behalf of his reader and who contents himself with taking stock of the "philosophers' opinions" (in a certain tradition), without passing a personal judgment on them. We have seen, however, that this posture does not always reflect reality. Falaquera at times very shrewdly tinkered with his sources so as to produce what in fact was a personal account, albeit one presented as a consensual teaching. Apropos of the delicate problem of explaining how the sun (and the stars) heat although they cannot be warm, Falaquera uses isolated passages taken from three commentaries by Averroes in order to construct an account that is not upheld by any of them. Averroes in the *Epitome of On the Heavens* maintained that the sun heats by virtue of its motion and its light, whereas in the *MC on On the Heavens* and the *MC on the Meteorology* he argued that it heats

---

[103] See Jospe, *Torah and Sophia* (above, n. 4), 153, referring to *Sefer ha-Mevaqqesh* (Amsterdam, 1775), 82-3; Dov Schwartz, *Astrologyah u-Magyah* (Ramat-Gan, 1999), chap. 4; *idem*, "Is it Possible to Write a History of Jewish Thought?" (Hebrew), *Jewish Studies* 38 (1998): 136-7.

through its motion alone. Falaquera, by contrast, draws a picture according to which the heating by the heavenly bodies is due only to the reflection of their light. Although Falaquera never explicitly speaks in his personal name and never overtly declares a preference for one Averroean account over another, he in fact deliberately promotes a theory not at all found in Averroes. Falaquera presumably did so, as we have seen, for philosophical-theological reasons, namely because this theory allotted to God an important role in the workings of nature.

My approach to Falaquera's text has advisedly been synchronic rather than diachronic. In view of Falaquera's educational aim, it was meaningful, I think, to identify in *De'ot ha-Filosofim* significant passages relevant to our concerns and determine where they signal a departure from Aristotle's original doctrine: although this leaves open the historically interesting and important question concerning the ultimate origin of the various doctrinal elements separating Falaquera from Aristotle, it nonetheless enriches our insight into the world-picture of thirteenth-century Hebrew-reading intellectuals by pointing out how an issue that had been problematic already in Aristotle's original system is tackled within it.

During the sixteen centuries separating Aristotle and Falaquera, the Aristotelian doctrine integrated ideas that had initially been developed within a quite distinct tradition, viz. astrology. The Aristotelian doctrine thus underwent what I have called a process of *astrologization*. As a consequence, Peripatetism became considerably closer to astrology, specifically to natural astrology, than was the original Aristotelian doctrine. We have thus seen that at central junctures, Judah ben Solomon observed apropos of the doctrines he was discussing—and which are the very same ones described by Falaquera (Averroes being the common source for both)—that they lent credence to astrology. Aristotelians and proponents of astrology thus shared overlapping sets of beliefs concerning the dependence of sublunar processes upon the supralunar bodies (and their intellects and souls). (They parted company over the questions such as whether and how the operations of the upper bodies on the lower realm can be known, whether human actions are predictable, and whether the astrological constructs such as aspects, constellations, etc. have an objective, real, existence.)[104] This made the theoretical problem of defining a criterion of demarcation between natural philosophy and natural astrology into an uneasy one.

The "astrologized" Aristotelianism reestablished a feature of the world view that had characterized Presocratic cosmologies, as well as, later, the Stoic world-picture: the fundamental inner unity of the cosmos. Whereas this tenet had been forsaken by Aristotle by separating the cosmos into two almost unrelated parts, it was reenacted by Aristotle's medieval followers: the stars' various influences bound together the two parts of the cosmos and thus recreated its interconnectedness. This move hinges entirely on the *ad hoc* introduction of the notion of "divine force," through which reflected rays are held to exercise their influence (specifically: to warm), and that of "soul-powers" emanating from the heavenly bodies. Only these concepts vouchsafed general divine providence within a globally teleological world view. This allowed for a relatively smooth integration of the Aristotelian teaching with the traditional religious monotheistic beliefs.

Additional Note

My short remarks on "divine force" on p. 350 of this essay appear to me now as embarrassingly uninformed. I devote some attention to the history of this notion in: "The Medieval Astrologization of the Aristotelian Cosmos: From Alexander of Aphrodisias to Averroes," scheduled for publication in a forthcoming volume edited by Alan C. Bowen and Christian Wildberg, to be entitled *A Companion to Aristotle's Cosmology: Collected Papers on the De caelo.*

---

[104] See e.g. Thérèse-Anne Druart, "Astronomie et astrologie selon Farabi," *Bulletin de philosophie médiévale* 20 (1978): 43-7; *idem*, "Le second traité de Farabi sur la validité des affirmations basées sur la position des étoiles," *Bulletin de philosophie médiévale* 21 (1979): 47-51.

Index of Names

The names of Aristotle and Ibn Rushd (Averroes) have not been included–explicitly or implicitly they are referred to on almost every page. The indication of a page number is understood to include the immediately following page(s).